Satellite Systems for Personal and Broadband Communications

Springer

Berlin
Heidelberg
New York
Barcelona
Hong Kong
London
Milan
Paris
Singapore
Tokyo

E. Lutz · M. Werner · A. Jahn

Satellite Systems for Personal and Broadband Communications

With 211 Figures and 64 Tables

 Springer

Dr. Erich Lutz
Markus Werner
Dr. Axel Jahn

Institute for Communications Technology
German Space Center DLR
82230 Weßling, Germany

TK
5104
L84
2000

ISBN 3-540-66840-3 Springer-Verlag Berlin Heidelberg New York

Library of Congress Cataloging-in-Publication Data
Lutz, E. (Erich), 1950- Satellite systems for personal and broadband communications /
E. Lutz, M. Werner, A. Jahn, p. cm.
Includes bibliographical references.
    ISBN 3540668403
1. Artificial satellites in telecommunication. I. Werner, M. (Markus), 1965-
II. Jahn, A. (Axel), 1965- III. Title.
TK5104.L84 2000   621.382'5--dc21   00-030751

Springer-Verlag is a company in the BertelsmannSpringer publishing group
© Springer-Verlag Berlin Heidelberg 2000
Printed in Germany

Typesetting: Camera-ready by authors
Cover layout: Medio, Berlin
SPIN: 10752502   Printed on acid free paper   62 / 3020 hu - 5 4 3 2 1 0

To Dorothee, Brigitte, and Annette

Preface

The important role that satellites play in the field of communications will be manifested by a large number of new satellite systems to be implemented within the next few years. These new systems will basically belong to two categories: (i) satellite networks for mobile/personal communications, mainly with handheld terminals, and (ii) satellite networks for broadband multimedia communications, mainly for fixed but also for portable and mobile terminals.

This book gives an overview of both families of satellite systems. In Part I, the basics of geostationary and non-geostationary satellite constellations are dealt with, as well as the principles of satellite communications. Part II deals with satellite systems for mobile/personal communications and addresses various aspects of networking (multiple access, cell structure, routing, etc.); it also deals with technology, regulation, and financing. Part III is dedicated to future satellite systems for broadband communications (Internet, multimedia) and discusses satellite-specific aspects of broadband communications, in particular on the basis of ATM and TCP/IP. A survey of existing and planned satellite systems completes the book.

The authors of this book are scientists at the German Aerospace Center, DLR (Deutsches Zentrum für Luft- und Raumfahrt) and work in the Digital Networks group of the Institute for Communications and Navigation.

In recent years, many research projects have been undertaken here, dealing with various aspects of satellite communication systems for mobile telephony and broadband services, and a large number of journal papers have resulted from these activities, covering a large part of this area. Thus, the idea arose to concentrate the experience gained into a book on satellite communications. Some activities have especially encouraged the realization of this idea:

Since 1993, the authors have been giving lectures on mobile satellite communications for the Carl-Cranz-Gesellschaft in Oberpfaffenhofen, Germany, developing their know-how and presentation material in this field. Since 1996, Erich Lutz has been lecturing on mobile satellite communication networks at the Technical University Munich. Parts I and II of this book are widely based on the notes of these lectures. In 1998, the authors were involved in writing an application document for a research project dealing with ATM-based satellite multimedia communications. Through this activity, insight had been gained which proved to be valuable for Part III of this book.

Axel Jahn has contributed interesting material in the area of satellite channel characterization, resulting from his project activities. The section on intersatellite link routing is an excerpt of the research work pursued by Markus Werner over recent years. Also, the chapter on network dimensioning is widely based on his research activities.

As the authors extensively cooperate with their colleagues in the Digital Networks group of the Institute for Communications and Navigation, in this sense, the whole group has contributed to the book.

The book is intended to cover a wide area of modern satellite communications. It should be easily understood by graduate students in the communications field. Also, it is a valuable source of information for professionals working in the areas of communications and/or satellites.

The authors would like to thank Prof. Joachim Hagenauer and Springer-Verlag, who endorsed the realization of the book. Also, we thank Ursula Hiermeyer for producing a large number of illustrations and for translating parts of the text into LATEX. Finally, we thank our families, who supported our work and tolerated our additional absence.

Please feel free to address any feedback regarding errors, suggestions, and updated information on the satellite systems mentioned in this book to Erich.Lutz@dlr.de.

Oberpfaffenhofen, March 2000

Erich Lutz

Markus Werner

Axel Jahn

Table of Contents

Part I. Basics

1. **Introduction** 3
   1.1 Mobile and Personal Satellite Communications 5
       1.1.1 Applications of Mobile Satellite Communications 7
       1.1.2 Personal Satellite Communications 7
       1.1.3 UMTS, IMT-2000 8
   1.2 Broadband Multimedia Satellite Communications 10
   1.3 Frequency Bands 11
   1.4 Key Aspects of Satellite Communication Systems 12

2. **Satellite Orbits, Constellations, and System Concepts** 15
   2.1 Satellite Orbits 15
       2.1.1 Elliptical and Circular Orbits 15
       2.1.2 Satellite Velocity and Orbit Period 17
       2.1.3 Orientation of the Orbit Plane 18
       2.1.4 Typical Circular Orbits 19
       2.1.5 Orbit Perturbations 21
       2.1.6 Ground Tracks 21
   2.2 Satellite – Earth Geometry 23
       2.2.1 Geometric Relations between Satellite and Earth Ter-
             minal 24
       2.2.2 Coverage Area 26
   2.3 Satellite Constellations 27
       2.3.1 Inclined Walker Constellations 30
       2.3.2 Polar Constellations 31
       2.3.3 Asynchronous Polar Constellations 34
   2.4 GEO System Concept 34
       2.4.1 Inmarsat-3 35
       2.4.2 EAST (Euro African Satellite Telecommunications) ... 36
   2.5 LEO System Concept 37
       2.5.1 Globalstar 38
       2.5.2 Intersatellite Links and On-Board Processing 38
       2.5.3 Iridium 41

2.6   MEO System Concept 41
    2.6.1   ICO 42
2.7   Satellite Launches 42

3. Signal Propagation and Link Budget 47
3.1   Satellite Link Budget 48
    3.1.1   Antenna Characteristics 48
    3.1.2   Free Space Loss and Received Power 51
    3.1.3   Link Budget 53
    3.1.4   Spot Beam Concept 56
3.2   Peculiarities of Satellite Links 58
    3.2.1   Dependence on Elevation 58
    3.2.2   Time Dependence of Satellite Links 58
    3.2.3   Faraday Rotation 61
3.3   Signal Shadowing and Multipath Fading 61
    3.3.1   Narrowband Model for the Land Mobile Satellite Chan-
        nel 62
    3.3.2   Satellite Channels at Higher Frequencies 67
    3.3.3   Wideband Model for the Land Mobile Satellite Channel 68
3.4   Link Availability and Satellite Diversity 74
    3.4.1   Concept of Satellite Diversity 74
    3.4.2   Correlation of Channels 76
    3.4.3   Link Availability and Satellite Diversity Service Area . 79
3.5   System Implications 79

4. Signal Transmission 83
4.1   Speech Coding 83
    4.1.1   Quality of Coded Speech 84
    4.1.2   Overview of Speech Coding Schemes 84
4.2   Modulation 87
    4.2.1   Modulation Schemes for Mobile Satellite Communica-
        tions 87
    4.2.2   Bandwidth Requirement of Modulated Signals 92
    4.2.3   Bit Error Rate in the Gaussian Channel 94
    4.2.4   Bit Error Rate in the Ricean and Rayleigh Fading
        Channel 95
4.3   Channel Coding (Forward Error Correction, FEC) 98
    4.3.1   Convolutional Coding 100
    4.3.2   Block Coding 102
    4.3.3   Error Protection with Cyclic Redundancy Check (CRC) 105
    4.3.4   RS Codes 105
    4.3.5   Performance of Block Codes 106
    4.3.6   Performance of Block Codes in Fading Channels 108
4.4   Automatic Repeat Request (ARQ) 110
    4.4.1   Stop-and-Wait ARQ 110

4.4.2   Go-Back-N ARQ 111
4.4.3   Selective-Repeat ARQ 112
4.5   Typical Error Control Schemes in Mobile Satellite Commu-
       nications 112

Part II. Satellite Systems for Mobile/Personal Communications

5.   **Multiple Access** 117
5.1   Duplexing 117
       5.1.1   Frequency-Division Duplexing (FDD) 117
       5.1.2   Time-Division Duplexing (TDD) 117
5.2   Multiplexing 118
5.3   Multiple Access 120
5.4   Slotted Aloha Multiple Access 122
       5.4.1   The Principle of Slotted Aloha 122
       5.4.2   Throughput of Slotted Aloha 122
       5.4.3   Mean Transmission Delay for Slotted Aloha 124
       5.4.4   Pure Aloha Multiple Access 125
5.5   Frequency-Division Multiple Access, FDMA 126
       5.5.1   Adjacent Channel Interference 127
       5.5.2   Required Bandwidth for FDMA 127
       5.5.3   Intermodulation 128
       5.5.4   Pros and Cons of FDMA 129
5.6   Time-Division Multiple Access, TDMA 129
       5.6.1   Bandwidth Demand and Efficiency of TDMA 131
       5.6.2   Burst Synchronization in the Receiving Satellite 133
       5.6.3   Slot Synchronization in the Transmitting TDMA Ter-
               minals 134
       5.6.4   Pros and Cons of TDMA 136
5.7   Code-Division Multiple Access, CDMA 136
5.8   Direct-Sequence CDMA (DS-CDMA) 137
       5.8.1   Generation and Characteristics of Signature Sequences 138
       5.8.2   Investigation of Asynchronous DS-CDMA in the Time
               Domain 142
       5.8.3   Investigation of Asynchronous DS-CDMA in the Fre-
               quency Domain 145
       5.8.4   Multi-Frequency CDMA, MF-CDMA 150
       5.8.5   Qualcomm Return Link CDMA (Globalstar) 150
       5.8.6   Synchronous Orthogonal DS-CDMA with Coherent
               Detection 151
5.9   CDMA Receivers 153
       5.9.1   PN Code Synchronization in the CDMA Receiver 153
       5.9.2   Rake Receiver 155
       5.9.3   CDMA Multiuser Detection 158

5.10 Characteristics of CDMA 163
5.11 CDMA for the Satellite UMTS Air Interface 165
    5.11.1 The ESA Wideband CDMA Scheme 165
    5.11.2 The ESA Wideband Hybrid CDMA/TDMA Scheme .. 169

6. Cellular Satellite Systems 171
  6.1 Introduction 171
    6.1.1 Concept of the Hexagonal Radio Cell Pattern 173
    6.1.2 Cell Cluster and Frequency Reuse 173
  6.2 Co-Channel Interference in the Uplink 176
    6.2.1 Co-Channel Interference for FDMA and TDMA Uplinks 178
    6.2.2 Co-Channel Interference for an Asynchronous DS-CDMA
        Uplink 181
  6.3 Co-Channel Interference in the Downlink 190
    6.3.1 Co-Channel Interference for FDMA and TDMA Down-
        links 192
    6.3.2 Co-Channel Interference for CDMA Downlinks 194
  6.4 Bandwidth Demand and Traffic Capacity of Cellular Satellite
    Networks 194
    6.4.1 Total System Bandwidth 194
    6.4.2 Traffic Capacity per Radio Cell 195
    6.4.3 Traffic Capacity of the System 196
    6.4.4 Required User Link Capacity of a Satellite 197
    6.4.5 Overall Network Capacity Considerations 198

7. Network Aspects 201
  7.1 Architecture of Satellite Systems for Mobile/Personal Com-
    munications 201
  7.2 Network Control 203
    7.2.1 Tasks of Network Control 203
    7.2.2 Signaling Channels of the Air Interface 204
  7.3 Mobility Management 206
    7.3.1 Service Area of a Gateway Station 207
    7.3.2 Location Area 208
    7.3.3 Registration and Location Update 209
  7.4 Paging 211
  7.5 Call Control 212
    7.5.1 Setup of a Mobile Originating Call 212
    7.5.2 Setup of a Mobile Terminating Call 212
  7.6 Dynamic Channel Allocation 214
    7.6.1 C/I-Based DCA 217
    7.6.2 DCA Using a Cost Function 218
  7.7 Handover 219
    7.7.1 Handover Decision 221
    7.7.2 Handover Procedure 221

      7.7.3   Channel Allocation at Handover 222
  7.8 Call Completion Probability 224
  7.9 Routing 225
      7.9.1   Routing in LEO/MEO Satellite Networks 225
      7.9.2   Off-Line Dynamic ISL Routing Concept 228
      7.9.3   On-line Adaptive ISL Routing 236
  7.10 Integration of Terrestrial and Satellite Mobile Networks 240

8. Satellite Technology 243
  8.1 Satellite Subsystems 243
  8.2 Antenna Technology 246
      8.2.1   GEO Antennas for Mobile Links with Spot Beams 246
      8.2.2   LEO/MEO Antennas 248
  8.3 Payload Architecture 250

9. Regulatory, Organizational, and Financial Aspects 257
  9.1 Allocation of Frequency Bands 257
  9.2 Licensing/Regulation 262
      9.2.1   Granting a System License 262
      9.2.2   Licensing in the USA 263
      9.2.3   Licensing in Europe 264
      9.2.4   Common Use of Frequency Bands by Several Systems . 264
      9.2.5   Global Licensing and Political Aspects 265
  9.3 Financing and Marketing of S-PCN Systems 266
  9.4 Operation of S-PCN Systems 268

Part III. Satellite Systems for Broadband Multimedia Communications

10. Multimedia Communications in Satellite Systems 273
  10.1 Types of Broadband Communication Networks 273
      10.1.1 Traditional Circuit-Switched Networks and the Packet-
             Switched Internet 273
      10.1.2 New Multimedia Satellite Systems Using New Satellite
             Orbits 274
  10.2 Multimedia Services and Traffic Characterization 274
      10.2.1 Video Traffic and MPEG Coding 275
      10.2.2 Self-Similar Traffic 276
  10.3 ATM-Based Communication in Satellite Systems 277
      10.3.1 Principles of ATM 277
      10.3.2 Implications for ATM-Based Satellite Networks 285
  10.4 Internet Services via Satellite Systems 287
      10.4.1 Principles of TCP/IP 287
      10.4.2 Internet Protocol (IP) 288

10.4.3 Transport Control Protocol (TCP) 290
10.4.4 TCP/IP in the Satellite Environment 294
10.4.5 IP over ATM in the Satellite Environment 297

11. ATM-Based Satellite Networks 299
11.1 System Architecture 299
11.2 Services 300
11.3 Protocol Architecture 303
11.4 ATM Resource Management 304
    11.4.1 Connection Admission Control and Usage Parameter
          Control 304
    11.4.2 Congestion Control, Traffic Shaping, and Flow Control 305
11.5 Multiple Access for ATM Satellite Systems 306
    11.5.1 TDMA-Based Multiple Access 308
    11.5.2 CDMA-Based Multiple Access 309
11.6 Radio Resource Management 311
11.7 Error Control 312

12. Network Dimensioning 315
12.1 Spot Beam Capacity Dimensioning for GEO Systems 315
    12.1.1 Motivation and Approach 315
    12.1.2 Market Prediction 316
    12.1.3 Generic Multiservice Source Traffic Model 318
    12.1.4 Calculation of the Spot Beam Capacity Requirements . 320
    12.1.5 System Bandwidth Demand Calculation 321
    12.1.6 Applied Spot Beam Capacity Dimensioning: A Case
          Study 322
12.2 ISL Capacity Dimensioning for LEO Systems 328
    12.2.1 Topological Design of the ISL Network 328
    12.2.2 ISL Routing Concept 332
    12.2.3 Network Dimensioning 332
    12.2.4 Numerical Example 336
    12.2.5 Extensions of the Dimensioning Approach 341

Appendix

A. Satellite Spot Beams and Map Transformations 345
A.1 Map Projections and Satellite Views 345
A.2 Generation of Satellite Spot Beams 348

B. Parameters of the Land Mobile Satellite Channel 353
B.1 Narrowband Two-State Model at L Band 353
B.2 Narrowband Two-State Model at EHF Band 356
B.3 Wideband Model at L Band 357

C. Existing and Planned Satellite Systems 361

    C.1 Survey of Satellite Systems 361

    C.2 ACeS (Asia Cellular Satellite) 367

    C.3 Astrolink 369

    C.4 EuroSkyWay 372

    C.5 Globalstar 375

    C.6 ICO (Intermediate Circular Orbits) 379

    C.7 Inmarsat-3/Inmarsat mini-M 386

    C.8 Iridium 389

    C.9 Orbcomm 396

    C.10 SkyBridge 399

    C.11 Sky Station 403

    C.12 Spaceway 406

    C.13 Teledesic 407

References 411

Index 422

Part I

Basics

1. Introduction

The idea of satellite communications was born a long time ago, in 1945, when Arthur C. Clarke wrote his famous paper "Extra-Terrestrial Relays" [Cla45]. In this paper, Clarke anticipated the concept of a geostationary satellite constellation consisting of three approximately equidistantly spaced artificial satellites in a specific orbit, such that the satellites revolve synchronously with the earth. Clarke also foresaw manned spaceflight, and indeed thought that his satellites would house a crew which would be provisioned and relieved by a regular rocket service. Moreover, he anticipated the use of directive satellite antennas and linking the satellites by radio or optical beams. He intended to use these satellites for bidirectional communications and for broadcast services.

Why should Mr. Clarke have thought of satellite communications at all? The reason is that satellite communications has some distinct advantages:

- Satellite communications is independent of the bridged distance within the coverage area of a satellite. Therefore, satellite communications is economic for serving large areas and for worldwide communications.
- Satellites have collecting and broadcasting characteristics. Inherently, they set up networks with star topology. Therefore, satellites are useful for direct communication with a large number of fixed and mobile users.
- Satellite networks can be established in a relatively short time and have a flexible architecture.

For today's satellite communication systems, different kinds of satellite constellations are used. *Geostationary (GEO) satellites* are placed in an equatorial orbit approximately 36 000 km above the earth, such that the satellites orbit synchronously with the rotating earth and seem to be fixed in the sky. Because of their high altitude, global coverage (excluding the polar regions) can be provided by just three geostationary satellites. However, the large distance to the satellites causes a very high signal attenuation and a large propagation delay (0.5 s round-trip delay from one earth station to another and back), which may be annoying in a telephone conversation.

Low earth orbiting (LEO) satellites at 700–1500 km avoid the large signal attenuation and delay of the geostationary orbit. However, a large number of satellites is needed to cover the earth's surface. On the other hand, LEO satellites have lower weight and are less complex than GEO satellites. LEO satellites have orbit periods of roughly 2 hours, and thus are non-geostationary. As a consequence of this, non-real-time messages can be transported by a satellite in a store-and-forward manner. On the other hand, during real-time connections, it may become necessary to switch to another satellite antenna beam (spot beam) or to another satellite.

Satellites in *medium earth orbits (MEO)* around 10 000 km avoid the large signal attenuation and delay of the geostationary orbit and still allow a global coverage with a small number (e.g. 10) of satellites.

Due to their advantages mentioned above, satellites are attractive for a variety of applications. The main applications of communications satellites are listed below, roughly in the order of their date of introduction:

- For intercontinental telephone and TV trunking, satellites have been used in a point-to-point topology. Rather than competing with an increasing number of submarine cables, satellites can provide backup links or carry peak traffic load.
- Satellites for radio and television broadcasting exploit the inherent broadcasting feature of satellites for providing direct-to-home television throughout a wide geographic area. A suitable compression and transmission scheme is the DVB/MPEG standard.
- With increasing importance, satellites are used for mobile communications with maritime, land mobile, aeronautical, and personal terminals which can be laptop-sized or notebook-sized terminals, or handheld mobile phones. Here, satellites can provide mobile communication services with wide-area or global coverage, which are most attractive for remote users and for travelers. The underlying standards are the GSM standard or the future UMTS standard which will adopt satellites as an integral part of the Universal Mobile Telecommunication System.
- Currently, satellites are going to be used for Internet services based on the TCP/IP protocol. For end users, they provide high-speed download of rich contents; Internet service providers use satellites for direct access to the Internet backbone.
- Satellites have also been used in fixed networking applications. VSAT (very small aperture terminal) satellite networks are e.g. used to connect sites of multinational companies. In the near future, broadband wide area networks (WANs) for multimedia communications (including Internet services) will be introduced, which are most probably based on the ATM transfer mode. Here, satellites are especially suitable for the distribution or multicast of multimedia content and for broadband communications to mobile or remote users. In the light of this, satellites will be an indispensable component of the future Global Information Infrastructure (GII).

The last four application areas of satellite communications are illustrated in Fig. 1.1. Of course, this simple picture cannot comprise the complete field of satellite communications; rather it highlights the most important features of satellites with regard to the different applications.

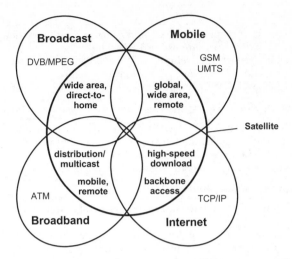

Fig. 1.1. Application areas of satellite communications

In order to give a brief overview of the history and the future of mobile and broadband satellite communications, a selection of historic and expected future milestones is presented in Table 1.1.

This book will concentrate on the application of satellites to mobile/personal communications and to broadband multimedia communications (including the Internet). Both of these areas will become extraordinarily important during the next few years.

1.1 Mobile and Personal Satellite Communications

As mentioned above, major applications of satellites are mobile communications (communications with people in vehicles) and personal communications (communications with people using personalized portable or handheld terminals).

In order to assess the market opportunities of such satellite communication systems competing with terrestrial communication systems, the strengths of satellite systems must be identified. Table 1.2 lists the advantages and drawbacks of mobile and personal satellite systems compared to terrestrial mobile radio systems. It can be concluded that mobile satellite systems are a very suitable complement to terrestrial mobile radio systems. Thus, there is no direct competition between these two types of systems, rather they address different market segments.

Table 1.1. Evolution of mobile and broadband satellite communications

The beginnings

1945:	Paper by Arthur C. Clarke, proposing "extra-terrestrial relays"
1957:	Sputnik: the first satellite (a LEO satellite)
1960:	Echo: a reflecting satellite
1964:	SYNCOM III: the first geostationary satellite
1965:	INTELSAT I, "Early Bird": the first commercial geostationary satellite

First generation of mobile satellite communication systems: analog technology

1976:	Three geostationary MARISAT satellites for maritime communications: the first mobile communications satellites (ship earth station: 40 W transmit power, 1.2 m antenna diameter)
1982:	Inmarsat-A: the first mobile satellite telephone system (maritime)

Second generation: digital transmission technology

1988:	Inmarsat-C: the first land mobile satellite system for data communications
1993:	Inmarsat-M and mobilesat (Australia): the first digital land mobile satellite telephone systems (briefcase terminals)
1996:	Inmarsat-3: satellite telephony with laptop terminals

Third generation: handheld terminals

1998:	Iridium: the first global LEO satellite system for telephony using handheld terminals
2003:	Universal mobile telecommunication system (UMTS, IMT-2000) with integrated satellite component

Broadband satellite systems: Internet and multimedia communications

2000:	ASTRA satellite system for high-speed Internet download. Requests are placed via a satellite return channel.
>2001:	Satellite systems for fixed, portable and mobile multimedia communications (Spaceway, EuroSkyWay, SkyBridge, Teledesic, etc.)

Table 1.2. Comparison of terrestrial and satellite mobile radio systems

Terrestrial mobile radio systems	Mobile satellite systems
– covered areas grow successively with the deployment of the infrastructure	– large areas can be covered quickly and completely
– no worldwide usage because of incompatible standards	– worldwide usage
– high bandwidth efficiency because of small radio cells	– lower bandwidth efficiency
– terrestrial radio links offer ample link margin to compensate for signal shadowing → suitable for urban environments	– signal shadowing deteriorates satellite links → suitable for rural environments
cost effective for *limited areas* with *high user density* and *high traffic density*	cost effective for *large areas* with *low user density* and *limited traffic density*

1.1.1 Applications of Mobile Satellite Communications

The following main market segments or applications can be deduced from the characteristics of mobile satellite systems :

- Personal communications for professional travelers (reporters, etc.) and business travelers. These people will be able to overcome the problem of incompatible terrestrial standards for cellular systems (GSM, AMPS, IS-95, etc.).
- Mobile radio services in areas which are not covered by terrestrial mobile radio systems (geographic extension). By the end of 1999, more than 80% of the world's land surface and about 40% of its population are likely to be without terrestrial mobile radio coverage [Bai99].
- Basic communications in less developed countries for areas which have no access to the telephone network. The ITU estimates that in the year 2003 60% of the world's population will never have placed a phone call or used a fax.
- Fleet management for trucks (tour planning, data exchange, just-in-time transport, cargo control, anti-theft, emergency calls).
- Support for emergency and security forces and authorities.

Of course, it is difficult to predict the size of the future mobile satellite communications market. Until 1999, market studies were rather optimistic and for the year 2004 expected 15 ... 24 million users of mobile satellite systems for voice and data communications, generating a revenue of US-$ 18 ... 25 billion [Ass99, MSN99]. After the difficulties experienced by the first LEO and MEO mobile satellite systems the market expectations decreased substantially: For the year 2005, KPMG now expects a revenue of US-$ 1 ... 2 billion from traditional mobile satellite services to handhelds, plus a revenue of US-$ 3 billion from satellite Internet services to laptop and palmtop terminals (Space and Satellite Finance 2000 conference, Jan. 2000).

1.1.2 Personal Satellite Communications

The central idea of personal communications is the ability of a mobile subscriber to set up and to receive a call at any place and time, using his or her own (personalized and typically handheld) terminal. In this context, satellite personal communications networks should provide a range of services with acceptable quality and at affordable costs:

- mobile telephony, typically with net bit rates of 4.8 kb/s
- fax, typically group 3
- mobile real-time data communications, typically 2.4 or 4.8 kb/s
- store-and-forward data communications (email, voice-mail, still pictures, control and measurement data, etc.)
- paging/messaging (with or without acknowledgement)

- position determination and reporting (e.g. to a dispatch center, for fleet management)
- supplementary services (call forwarding, etc.)
- value-added services (online services, data retrieval, etc.).

Global personal communications is implemented by a number of terrestrial personal communications networks (PCNs) being supplemented by satellite personal communications networks (S-PCNs).

Global S-PCNs are based on constellations of non-geostationary satellites in low earth orbits or medium earth orbits. Typical examples are the Iridium system with 66 LEO satellites in 6 polar orbits at an altitude of 780 km, Globalstar with 48 LEO satellites in 8 inclined orbits at 1414 km, and ICO with 10 MEO satellites at an altitude of 10 390 km. Orbcomm with 36 satellites at 825 km is an example of a LEO satellite system dedicated to data communications.

Regional S-PCNs use geostationary satellites because the coverage area of a GEO satellite is fixed and can be efficiently tailored to the intended service area of the system. The high signal attenuation must be compensated by a very high-gain satellite antenna, which results in large antenna dimensions. Due to the narrow beam of such a satellite antenna, a large number of spot beams is necessary to fill the coverage area of the satellite, further increasing antenna complexity. An example of a regional GEO S-PCN is the Asian Cellular Satellite (ACeS) system with a satellite antenna diameter of 12 m and 140 satellite spot beams.

Handheld terminals for satellite personal communications must fulfill stringent technical requirements and thus represent a challenge to the manufacturers:

- The terminal must be suitable for terrestrial cellular networks and satellite networks (dual-mode).
- The terminal must provide a direct communication link to the satellite.
- The terminal must be of small size and low weight (less than 500 g).
- The terminal must be equipped with an omnidirectional antenna exhibiting virtually no antenna gain.
- The terminal must work with a small battery and must have long standby and talk times. Also, it must exhibit low radiation. Therefore, the average transmit power will be low (less than 0.5 W).
- The terminal must be available at an affordable price.

In addition to handheld terminals, mobile (vehicle-mounted) terminals with approximately 10 W transmit power and portable (laptop, notebook) terminals with approximately 2 W transmit power are used.

1.1.3 UMTS, IMT-2000

Within the next few years, a paradigm shift in the way people communicate will occur:

i) The number of mobile phones and other mobile or portable terminals will increase dramatically. It is expected that the growth rate of mobile phones will exceed the growth rate of fixed phones.

ii) Voice conversation will no longer be the dominating form of communications. Data traffic and multimedia traffic will become of increasing importance and will benefit from new packet-oriented or Internet-based transmission schemes.

In order to prepare for this evolution, a new (third) generation of mobile communication networks is being developed, whose central idea is to combine mobility with multimedia services. These networks will form a global family of networks, called Universal Mobile Telecommunication System (UMTS) or International Mobile Telecommunication - 2000 (IMT-2000). UMTS/IMT-2000 will also include an integrated satellite component, commonly designated as Satellite-UMTS (S-UMTS). The main features of UMTS/IMT-2000 are:

a) With regard to services and scenarios:

- higher data rates (up to 2 Mb/s indoor, 384 kb/s outdoor, and 144 kb/s for mobile applications); according to ITU and ETSI, the satellite component of UMTS/IMT-2000 should support data rates up to 144 kb/s
- time-varying data rates and asymmetric data rates
- circuit-switched services, packet data, real-time and non-real-time services
- mobile multimedia services, mobile Internet access
- improved quality of service (e.g. ISDN speech quality)
- communications in all types of environments (indoor, urban, wide area)
- worldwide roaming, using a unique subscriber number
- personal communications, virtual home environment (VHE), service portability
- user-controlled service profile.

b) With regard to user terminals:

- a great variety of terminal types (basic terminals, PDAs, audio-visual (multimedia) terminals)
- terminals which are reconfigurable by the user or via radio
- download of applications.

c) With regard to the network:

- convergence of fixed networks, mobile networks, and the Internet
- flexible network architecture (generic radio access network concept, generic core network concept)
- interoperability and integration with second-generation mobile networks (e.g. GSM)
- a large network capacity to satisfy the increasing demand.

In 1998 five proposals for the satellite component of UMTS/IMT-2000 have been submitted to the ITU, including a proposal for the S-UMTS air interface which was developed by the European Space Agency, ESA (see Sect. 5.11).

1.2 Broadband Multimedia Satellite Communications

Within the next few years, a large number of new satellite systems for broadband multimedia communications will be developed, mainly for fixed terminals, but also for portable and mobile terminals. These broadband satellite systems will provide a much wider range of services. They will allow video telephony, Internet access, and other (interactive) multimedia applications with asymmetric and time-varying data rates up to several tens of Mb/s. The user terminals can be fixed terminals (USATs), laptop terminals, or mobile terminals.

Typical applications of satellite multimedia communications are:

- video telephony and video conferencing
- high-speed Internet access and high-speed on-line services
- e-commerce
- email and fax
- direct-to-home video
- distance learning and training
- telemedicine
- shared applications
- distributed games, etc.

Compared to fixed terrestrial broadband networks, multimedia satellite systems have some distinct advantages:

- Satellite systems can provide global multimedia services to end users, long before they will be available through terrestrial networks.
- In many regions the provision of terrestrial multimedia services will not be commercially viable.
- Satellite systems are especially efficient for multimedia broadcasting services and for distributing multimedia contents within a wide geographic area.

However, satellite networks can never compete with the tremendous capacity of fixed terrestrial trunk networks (backbone networks). Thus, similar to satellite networks for mobile communications, satellite multimedia networks will be dedicated to wide-area or global traffic, while terrestrial networks will concentrate on areas with high traffic density.

Spaceway is an example of a global broadband GEO satellite system aimed at fixed users. However, also broadband GEO systems for mobile users (mainly aboard ships or aircraft) are being planned, such as the regional Eu-roSkyWay system.

Teledesic is a pioneering LEO concept, planning to use 288 satellites (this number may be reduced to approx. 100) in 12 polar orbits at an altitude of 1375 km. Based on the ATM transfer mode, it should provide high-speed Internet access, interactive multimedia communications, video conferencing, and other broadband services, mainly to fixed users.

Because of the high data rates required on the satellite link, multimedia satellite terminals must use directive antennas, and there must be a line of sight to the satellite. For LEO systems, the terminal antennas must be steerable, and two antenna beams are necessary for continuous communications. These constraints will certainly have an influence on the possible user application scenarios for multimedia satellite communications.

Anyway, a recent market research report by Merrill Lynch estimates that the revenue from broadband satellite services is likely to rise to more than US-$ 30 billion in the year 2008 [Wat99].

1.3 Frequency Bands

As any kind of radiocommunications, satellite communications is dependent on the availability of radio spectrum, a limited and precious resource. At a series of ITU World Radio Conferences frequency bands were allocated for the communication between satellites and users (user link or service link):

Non-geostationary satellite systems for data transmission ("little-LEOs") are allowed to use VHF and UHF frequencies around 150 and 400 MHz. Non-geostationary voice systems ("big-LEOs", such as Iridium and Globalstar) can use L and S band frequencies in the ranges 1.610–1.6265 GHz and 2.4835–2.500 GHz for the mobile up- and downlink. The band 1.6138–1.6265 GHz is allocated for both directions and is used by Iridium in time-division duplex (TDD). L band frequencies from 1.525 to 1.559 GHz (downlink) and from 1.6265 to 1.6605 GHz (uplink) have been allocated to geostationary mobile satellite systems.

For S-UMTS systems, frequency bands at 1.980–2.010 GHz (uplink) and 2.170–2.200 GHz (downlink) have been allocated, which can be globally used after the year 2000 (as intended by the ICO system).

For the provision of video services and high-speed Internet access, multimedia satellite systems must provide up- and downlinks with high data rates in the order of several Mb/s. This requires broad frequency bands which are not available below 10 GHz. Therefore, multimedia satellite systems will use frequencies around 11/14 GHz (Ku band), 20/30 GHz (K/Ka band), or even 40/50 GHz (V band, also denoted EHF band).

Frequencies at 150/400 MHz, 4–7 GHz, 15 GHz, and 20/30 GHz are foreseen for the communication between satellites and fixed earth stations (feeder links).

Some S-PCN or multimedia systems (Iridium and Teledesic, e.g.) use intersatellite links for direct communications between satellites. These intersatellite links may work at 23 GHz, 60 GHz, or at optical frequencies.

Figure 1.2 gives an overview of frequency bands available for satellite communications.

1.4 Key Aspects of Satellite Communication Systems

There is a large number of aspects to be taken into account when developing a satellite communication system:

- The space segment consists of one or more satellites, which are placed in suitable orbital planes. This satellite constellation must be designed such that the satellites cover the intended service area of the system with sufficient availability and quality. Actually, the design of the whole system concept depends on the choice of the satellite constellation. Thus, we distinguish between LEO, MEO and GEO system concepts. The aspects of satellite constellations and related system concepts will be discussed in Chap. 2.
- From the communications point of view, the radio link between a satellite and a mobile or handheld terminal is the most important aspect of satellite communications, and many questions arise in this respect. Which frequency bands can be used? What effects will appear for the signal transmission in such links? Which transmission schemes are appropriate? How can we manage the multitude of communication links between the user community and a satellite (multiple access)? What interferences will arise between these communication links, between different satellites, and between different satellite systems? Such questions will be investigated in Chaps. 3–6.
- A large number of aspects are related to networking. Mobility management must be used to stay in contact with mobile users. Connections must be set up, maintained, and closed down. The mobility of users and of non-geostationary satellites require the handing-over of active connections. In systems with intersatellite links, the signals must be efficiently routed through the ISL network. Last but not least, the satellite networks must be connected to or integrated with terrestrial fixed and mobile networks. Chapter 7 will deal with such aspects.
- The technology of communications satellites for new systems must meet enormous challenges with regard to multibeam antennas, on-board processing, and intersatellite link technology. The first two aspects will be addressed in Chap. 8.
- Special requirements arise when satellite systems have to be designed for broadband multimedia communications. Chapters 10 and 11 give an introduction to this field and specify the corresponding issues to satellite

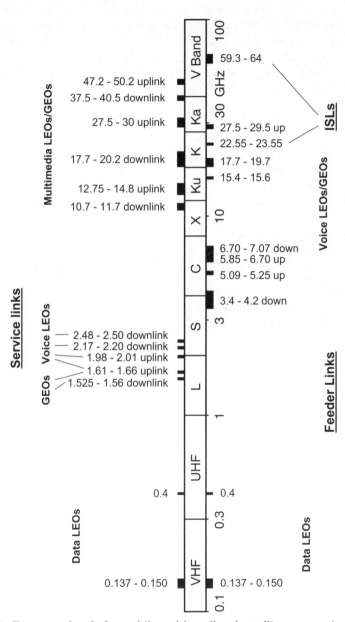

Fig. 1.2. Frequency bands for mobile and broadband satellite communications

systems. In particular, Chap. 12 deals with network dimensioning of multiservice satellite systems.

– Finally, there are a number of aspects which are not directly related to technical questions but none the less are of upmost importance: the allocation of frequency bands for satellite systems, the granting of licenses, cooperation agreements between network operators (e.g. roaming agreements), and financing and marketing of the systems. Chapter 9 is devoted to such aspects.

Overall, the book is organized into three parts:

– Part I provides the fundamentals of satellite communications: satellite constellations, signal propagation, and transmission schemes.
– Part II deals with satellite systems for mobile and personal communications with limited data rates.
– Part III introduces satellite systems for broadband multimedia communications.
– In the Appendices detailed aspects of satellite - terminal geometry are discussed and parameters of the land mobile satellite channel are listed. Also, a survey of operational and planned satellite systems is given.

Of course, most of the aspects mentioned above are related to mobile systems as well as to broadband systems. Therefore, several aspects will be addressed more than once, within different contexts.

2. Satellite Orbits, Constellations, and System Concepts

At this point of the book we present the subject of satellite orbit mechanics to discuss some geometric relations between satellites and ground terminals. We also introduce different satellite constellations which can be used to build up regional or global satellite systems.

2.1 Satellite Orbits

In the early 17th century Johannes Kepler discovered some important properties of planetary motion that have come to be called Kepler's laws:

- *First law (1602):* the planets move in a plane; the orbits around the sun are ellipses with the sun at one focal point.
- *Second law (1605):* the line between the sun and a planet sweeps out equal areas in equal intervals of time.
- *Third law (1618):* the ratio between the square of the orbit period T and the cube of the semi-major axis a of the orbit ellipse, T^2/a^3, is the same for all planets.

These laws can be applied to any two-body system subject to gravitation, and thus also describe the motion of a satellite around the earth. Extensive treatments of orbit mechanics can be found in the textbooks [BMW71, MB93, Dav85].

2.1.1 Elliptical and Circular Orbits

Figure 2.1 shows the geometry of an elliptical satellite orbit according to Kepler's first law. The satellite orbit has an elliptical shape with the earth at one focal point. The ellipse is defined by two parameters: the semi-major and semi-minor axes a and b. The shape of the ellipse can also be described by the numerical eccentricity

$$e = \sqrt{1 - \frac{b^2}{a^2}} \qquad \text{with} \qquad 0 \le e < 1 \ . \tag{2.1}$$

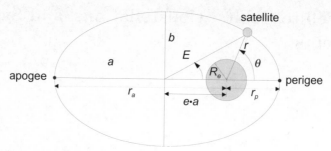

Fig. 2.1. Parameters of elliptical orbits

With this parameter the distance of the focal points from the ellipse center can be expressed as $e \cdot a$. The distance of the satellite from the earth's center is the radius r. The point of the orbit where r is smallest is called perigee with $r = r_p$. The point with largest r is denoted apogee with $r = r_a$. From Kepler's second law we can deduce that a satellite moves quickly near perigee and slowly near apogee. According to Fig. 2.1 and using Eq. (2.1) we can set up the following relations:

$$
\begin{aligned}
a &= \frac{r_a + r_p}{2} \\
e &= \frac{r_a - r_p}{r_a + r_p} \\
r_a &= a(1 + e) \\
r_p &= a(1 - e) \,.
\end{aligned}
\tag{2.2}
$$

The angle θ between the perigee and the satellite as seen from the earth's center is commonly called the *true anomaly*. It can be used to determine the satellite radius r along the elliptical orbit:

$$
r = \frac{a(1 - e^2)}{1 + e \cos \theta} \,.
\tag{2.3}
$$

The angle between perigee and the satellite with respect to the ellipse center is denoted the *eccentric anomaly* E, which is related to θ through

$$
\cos \theta = \frac{a}{r}(\cos E - e) = \frac{\cos E - e}{1 - e \cos E} \,.
\tag{2.4}
$$

The time t after perigee passing t_p can be related to the eccentric anomaly E through

$$
\frac{2\pi}{T}(t - t_p) = E - e \sin E \,,
\tag{2.5}
$$

where T is the orbit period of the satellite and the term $2\pi(t - t_p)/T$ is called the *mean anomaly*. Using Eq. (2.5) and Eq. (2.4) the time can be derived as a function $t(\theta)$. However, since the inverse function of Eq. (2.5) cannot be solved, the time behavior of $\theta(t)$ must be determined numerically.

The satellite altitude h above the earth's surface is

$$h = r - R_e \qquad (2.6)$$

with R_e being the radius of the earth. Accordingly, the orbit altitude at apogee is $h_a = r_a - R_e$ and the altitude at perigee is $h_p = r_p - R_e$. Actually the earth is not an ideal sphere but exhibits some flattening at the poles. In the following, we will use $R_e = 6\,378$ km representing the mean equatorial radius[1].

Circular Satellite Orbits. A circular satellite orbit is a special case of an elliptical orbit with zero eccentricity, $e = 0$. Thus, $a = b = r = r_a = r_p$. The earth is at the center of the circular orbit, and the satellite altitude $h = r - R_e$ is constant. Furthermore, it follows for the time behavior of the true anomaly that

$$\theta(t) = \frac{2\pi t}{T} \ . \qquad (2.7)$$

2.1.2 Satellite Velocity and Orbit Period

Isaac Newton extended the work of Kepler and in the year 1667 discovered the law of gravity. This law states that two bodies with masses m and M at a distance r attract each other with the gravitational force

$$F_g = G\,\frac{m\,M}{r^2} \ . \qquad (2.8)$$

Here, $G = 6.6732 \cdot 10^{-11}$ N m^2/kg^2 is the universal gravitation constant.

For a satellite orbiting around the earth the mass m represents the satellite mass and $M = M_e = 5.9733 \cdot 10^{24}$ kg is the mass of earth. The total mechanical energy consisting of the potential energy and the kinetic energy is constant:

$$\frac{mv^2}{2} - \frac{\mu m}{r} = -\frac{\mu m}{2a} \ , \qquad (2.9)$$

where $\mu = GM_e = 398\,600.5$ km^3/s^2. Thus, the velocity v of a satellite in an elliptic orbit may be obtained by

$$v = \sqrt{\mu\left(\frac{2}{r} - \frac{1}{a}\right)} \qquad (2.10)$$

which can be simplified for circular orbits ($r \equiv a$) to

$$v = \sqrt{\frac{\mu}{r}} \ . \qquad (2.11)$$

---

[1] The polar earth radius amounts to $6\,357$ km whereas the mean radius averaged over the earth's surface is $6\,371$ km.

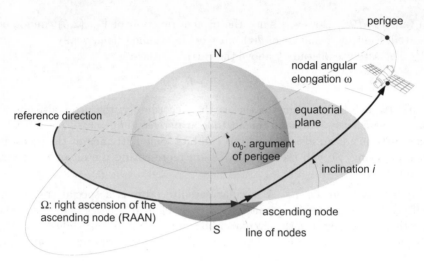

Fig. 2.2. Orientation of the orbit plane in space. The reference direction points from the earth to the sun at vernal equinox

Equation (2.11) states that the velocity of satellites in circular orbits is constant, coinciding with Kepler's second law. The orbit period can now be derived as

$$T = \frac{2\pi r}{v} = 2\pi \sqrt{\frac{r^3}{\mu}} \tag{2.12}$$

which for elliptical orbits generalizes to

$$T = 2\pi \sqrt{\frac{a^3}{\mu}} \tag{2.13}$$

according to Kepler's third law.

The orbit mechanics discussed so far are idealized in the sense that they assume a spherical and homogeneous earth, empty space, and the absence of any gravitational forces from sources other than the satellite and the earth. For this ideal scenario the satellite orbit will remain constant for all times.

2.1.3 Orientation of the Orbit Plane

In this section we deal with the orientation of the orbit plane in space. For the ideal scenario mentioned above, this orientation is sidereally fixed (i.e. fixed with respect to the stars) and is independent of the earth's rotation. Figure 2.2 shows the parameters that characterize the orbit orientation:

– The **inclination** i defines the angle between the orbit plane and the equatorial plane. It is counted positively with respect to the ascending satellite orbit track. The line of intersection between the two planes is called the *line of nodes*. The *ascending node* is passed when the satellite enters the northern hemisphere.

– The **right ascension of the ascending node (RAAN)** Ω determines the angle between a reference direction and the line of nodes. The reference direction is given by the direction from the earth's center to the sun at vernal equinox. Equivalently, this direction corresponds to the intersection between the equatorial plane and the plane of the ecliptic. The reference direction remains fixed in space.[2]

– The **argument of perigee** ω_0 is the angle between the line of nodes and the semi-major axis of the ellipse. This parameter is relevant only for elliptical orbits.

The position of the satellite is thus completely determined by a set of six orbital parameters: the semi-major axis a of the ellipse, the eccentricity e, the inclination i, the right ascension of the ascending node Ω, the argument of perigee ω_0, and the true anomaly θ. These parameters are often referred to as the *Kepler elements*. The Kepler elements of satellites will change during a satellite's lifetime due to orbital perturbations. Official databases (for instance of North American Aerospace Defense (NORAD)) are updated and regularly distributed for all existing satellites.

2.1.4 Typical Circular Orbits

Besides the differentiation between elliptical and circular orbits, the altitude h and inclination i are the most important orbit characteristics. For circular orbits, the relation between period T and altitude h can be derived from Eq. (2.12) as

$$h = \sqrt[3]{\mu \left(\frac{T}{2\pi} \right)^2} - R_e \ . \tag{2.14}$$

It is advantageous to choose orbit periods that are integer divisors of a day ($T = 1, 2, 4, 6, 12, 24$ h) since in this case the satellite positions reiterate periodically day by day. However, some periods may not be used since the associated altitude falls into the *Van Allen belts*, which are regions of the ionosphere with high ion concentration, thus reducing the satellite's lifetime. Figure 2.3 shows the relation between satellite period and altitude using Eq. (2.14). Three regions of operation can be identified:

– **low earth orbit** (LEO), with altitudes from 500–1 500 km above the earth and periods of approximately 2 h,
– **medium earth orbit** (MEO[3]), with altitudes from 5 000–10 000 km above the earth and periods of approximately 4–6 h, and

---

[2] The intersection of the planes varies somewhat due to the perturbations of the terrestrial rotation. For details see [MB93].

[3] A synonym for MEO is ICO: intermediate circular orbit.

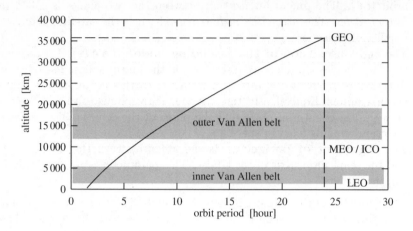

Fig. 2.3. Orbit classification according to altitude

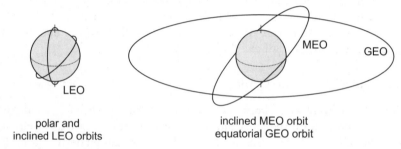

Fig. 2.4. LEO, MEO, and GEO orbit types

– **geostationary** orbit (GEO) at 35 786 km altitude and 24 h period[4].

Since a GEO satellite has the same period of eastward revolution as the earth, the satellite appears to be stationary above the equator as seen from an observer on earth. It should be noted that the GEO has an inclination angle $i = 0°$. For $h = 35786$ km but an inclination $i \neq 0$ the satellite is no longer stationary, however, it has still the same revolution period as the earth. The orbit is then called geosynchronous. In contrast, LEOs and MEOs with periods $T < 24$ h are non-geosynchronous orbits.

Another orbit classification follows the inclination characteristics:

– *equatorial orbits* are not inclined ($i = 0°$), whereas
– *inclined orbits* usually have inclination angles $i = 40°$–$80°$.
– *Polar orbits* use inclination angles around $90°$. Sometimes the inclination exceeds $90°$, cf. Sect. 2.1.5.

---

[4] The precise GEO period is $T = T_e = 23$ h 56 min 4.1 s.

2.1.5 Orbit Perturbations

The deviation of the earth from a homogeneous sphere in mass distribution and shape produces additional higher-order force terms of gravity affecting the satellite. The forces cause the ellipse to rotate slowly in the orbit plane and the orbit plane to rotate about the earth's north–south axis.

Drift of the RAAN. It can be shown [MB93] that the drift of the RAAN Ω amounts to

$$\dot{\Omega} = -\frac{9.964°}{(1 - e^2)^2} \left(\frac{R_e}{a} \right)^{3.5} \cos i \qquad (2.15)$$

in degrees per solar day. For $i < 90°$ the ascending node is drifting to the west $(\dot{\Omega} < 0)$, for $i > 90°$ to the east $(\dot{\Omega} > 0)$. For polar orbits $(i = 90°)$ the orientation of the ascending node remains fixed with respect to the stars.

Sun-Synchronous Orbits. The earth moves around the sun with an angular speed of $0.9856°$ per day. For inclination angles $i > 90°$ one can find values of i such that the eastbound drift of the ascending node according to Eq. (2.15) compensates for this change. Thus, the orientation of the orbit plane with respect to the line from the earth to the sun remains fixed and has some advantages for the illumination of the satellite.

Drift of the Argument of Perigee. The flattening of the earth causes perigee to move in the ellipse plane with rate

$$\dot{\omega}_0 = \frac{4.982°}{(1 - e^2)^2} \left(\frac{R_e}{a} \right)^{3.5} (5 \cos^2 i - 1) \qquad (2.16)$$

in degree per solar day [PSN93]. For $i = 63.4°$ and $i = 116.6°$ the orientation of the semi-major axis remains constant. These inclination angles are thus of special interest for satellite orbits. For instance, the *Molnija* $(T = 12 \text{ h})$ and *Tundra* $(T = 24 \text{ h})$ orbits have an inclination of $i = 63.4°$. The Ellipso system uses sun-synchronous elliptical orbits with an inclination of $i = 116.6°$.

Other causes of orbit perturbations are the gravitation of the sun and the moon, solar radiation pressure, atmospheric drag (effective up to 750 km altitude [Ric99]), and distortions due to the satellite propulsion.

2.1.6 Ground Tracks

This section derives the time-variant coordinates of a satellite position. First, we want to introduce coordinate systems that allow us to relate the satellite's position to spherical or Cartesian geocentric coordinates, cf. Fig. 2.5. In an inertial spherical coordinate system with earth-centered origin any point on the earth's surface is specified by two angular coordinates, the latitude φ and the longitude λ, and a distance r from the origin.[5] The longitude λ $(0° \leq$

---

[5] For simplicity, the distance r on the earth's surface can be assumed constant, equal to the mean equatorial earth radius $R_e = 6\,378$ km.

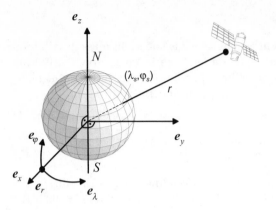

Fig. 2.5. Earth coordinate systems

$\lambda < 360°$) is counted positively from the prime "Greenwich" meridian to the east. The latitude φ $(-90° \leq \varphi \leq 90°)$ determines the position on great circles with constant longitude that run through the north and south poles. The latitude is counted positively from the equator $(\varphi = 0°)$ to the north.

Alternatively, an inertial system can be described by Cartesian coordinates. Let e_x, e_y, and e_z be the orthogonal normalized base vectors of the Cartesian coordinate system with earth-centered origin. The z-axis points to the geographic north pole and the x-axis to the prime meridian. The coordinate systems are inertial systems with respect to the earth's rotation, i.e. a fixed point P on earth does not change its geocentric coordinates (x_p, y_p, z_p) or (λ, φ, r) during an earth revolution.

The Cartesian and spherical coordinate systems can be transformed into each other by

$$
\begin{pmatrix} x \\ y \\ z \end{pmatrix} = \begin{pmatrix} r \cos\varphi \cos\lambda \\ r \cos\varphi \sin\lambda \\ r \sin\varphi \end{pmatrix} \quad \text{or} \quad \begin{pmatrix} \lambda \\ \varphi \\ r \end{pmatrix} = \begin{pmatrix} \arctan 4\, y/x \\ \arctan 4\, z/\sqrt{x^2 + y^2} \\ \sqrt{x^2 + y^2 + z^2} \end{pmatrix} ,
$$

$$(2.17)$$

with the definition of the arc tangent for four quadrants

$$
\arctan 4\, y/x \doteq \begin{cases} \arctan y/x & \text{for} & x \geq 0 \\ \arctan y/x + \pi/2 & \text{for} & x < 0, y \geq 0 \\ \arctan y/x - \pi/2 & \text{for} & x < 0, y < 0 . \end{cases}
$$

$$(2.18)$$

The goal is now to calculate the satellite track $(\lambda_s(t), \varphi_s(t))$, i.e. the coordinates of the satellite with respect to the rotating earth. The procedure is depicted in Fig. 2.6. With spherical trigonometry one can express the satellite latitude $\varphi_s(t)$ at time t as a function of the elongation $\omega(t)$ denoting the angle from the ascending node to the satellite,

$$\sin\varphi_s(t) = \sin i \sin\omega(t) . \tag{2.19}$$

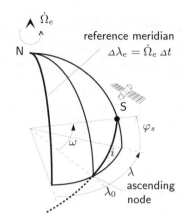

Fig. 2.6. Ground track of the satellite on the rotating earth

The elongation ω is related to the true anomaly θ by

$$\theta(t) = \omega(t) - \omega_0 \tag{2.20}$$

(cf. Sect. 2.1.1). Furthermore, by spherical trigonometry one can find

$$\cos\varphi_s(t)\cos(\lambda(t) - \lambda_0) = \cos\omega(t) . \tag{2.21}$$

Here, $\lambda(t)$ is the longitude of the satellite with respect to the non-rotating earth and λ_0 denotes the longitude of the ascending node. Let t_0 denote the time when the satellite passes the ascending node. To include the rotation $\dot{\Omega}_e$ of the earth we have to consider that at time t the reference meridian has changed its sidereal orientation by $\Delta\lambda_e = \dot{\Omega}_e(t - t_0)$ with $\dot{\Omega}_e = 2\pi/T_e$ being the angular frequency of the earth. Therefore, we have for the geographic satellite longitude λ_s

$$
\begin{aligned}
\lambda_s(t) &= \lambda(t) - \dot{\Omega}_e(t - t_0) \\
&= \lambda_0 + \arccos\frac{\cos\omega(t)}{\cos\varphi_s(t)} - \frac{2\pi}{T_e}(t - t_0) .
\end{aligned}
\tag{2.22}
$$

In general, satellite tracks are not closed after a revolution due to the earth's rotation. Closed tracks can be achieved for orbit periods $T = T_e \cdot \frac{m}{n}$, with integers m, n. If m an n are not divisors of each other the tracks will close after n revolutions. Figure 2.7 shows typical examples of satellite ground tracks. The point of the track with highest latitude is called the vertex and corresponds to the inclination i.

2.2 Satellite – Earth Geometry

This section deals with basic geometric relations between satellites and user terminals. Other geometric aspects such as the generation of satellite spot beams and the transformation of perspective plots are given in App. A.

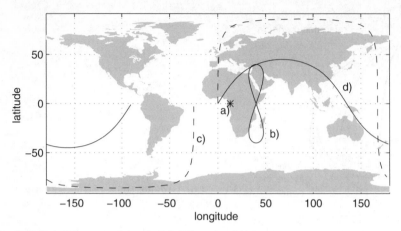

Fig. 2.7. Satellite ground tracks of different orbit types:
a) geostationary, $\lambda_0 = 13°$, $T = T_e$, inclination $i = 0°$
b) geosynchronous, $\lambda_0 = 40°$, $T = T_e$, inclination $i = 40°$
c) polar LEO, $\lambda_0 = 0°$, $T = 6\,000$ s, inclination $i = 86°$
d) inclined MEO, $\lambda_0 = 0°$, $T = T_e/4$, inclination $i = 45°$

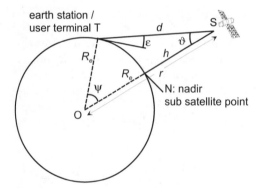

Fig. 2.8. Geometric relations in the plane STO (satellite S, earth station T, earth center O)

2.2.1 Geometric Relations between Satellite and Earth Terminal

The geometry of a satellite and a user terminal on earth is illustrated in Fig. 2.8. The projection of the satellite to the earth is called the sub-satellite point (SSP) or *nadir*. Important parameters are:

– the **elevation** angle ε at which a user can see the satellite above the horizon,
– the **nadir** angle ϑ that gives the deflection of the user from nadir as seen from the satellite,
– the earth central angle ψ between the sub-satellite point SSP and the user, and
– the **slant range** d denoting the distance between the user terminal and the satellite.

The sine and cosine laws for the triangles STN and OTS yield the relations between the earth central angle ψ, the elevation ε, and the nadir angle ϑ:

$$\psi = \frac{\pi}{2} - \vartheta - \varepsilon = \arccos\left(\frac{R_e}{r}\cos\varepsilon\right) - \varepsilon$$

$$= \arcsin\left(\frac{r}{R_e}\sin\vartheta\right) - \vartheta , \tag{2.23}$$

$$\vartheta = \arcsin\left(\frac{R_e}{r}\cos\varepsilon\right)$$

$$= \arctan\left(\frac{\sin\psi}{\frac{r}{R_e} - \cos\psi}\right) , \tag{2.24}$$

$$\varepsilon = \arctan\left(\frac{\cos\psi - \frac{R_e}{r}}{\sin\psi}\right)$$

$$= \arccos\left(\frac{r}{R_e}\sin\vartheta\right) . \tag{2.25}$$

The slant range d can be calculated from

$$d = \sqrt{R_e^2 + r^2 - 2R_e r \cos\psi} . \tag{2.26}$$

Dependency of Nadir, Elevation, and Earth Central Angle from Geographic Coordinates. Let λ_t, φ_t denote the longitude and latitude of the subscriber terminal. Then from spherical considerations it follows that for the earth central angle between the user and the nadir [MB93]

$$\cos\psi = \sin\varphi_t \sin\varphi_s + \cos\varphi_t \cos\varphi_s \cos(\lambda_t - \lambda_s) . \tag{2.27}$$

For a GEO satellite the latitude is equatorial ($\varphi_s = 0$) and Eq. (2.27) simplifies to

$$\cos\psi = \cos\varphi_t \cos(\lambda_t - \lambda_s) . \tag{2.28}$$

The values of the elevation ε, the satellite nadir angle ϑ, and the distance d can be calculated as a function of the terminal and satellite position using Eq. (2.27) and Eqs. (2.23) – (2.26).

Dependency on Time. Since non-geostationary satellites move relative to an earth-fixed station, the geometric relations vary with time. Using the equations for the satellite track (Eqs. (2.19) and (2.22)) the time-dependent distance and elevation can be calculated. Figure 2.9 shows the time-varying elevation for a MEO and LEO satellite overhead pass and for 70° and 40° passages.

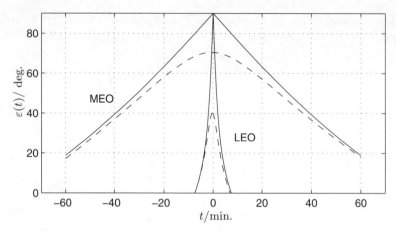

Fig. 2.9. Time dependency of the elevation angle ε for an overhead pass of a MEO and LEO satellite and a $70°$ MEO and $40°$ LEO passage

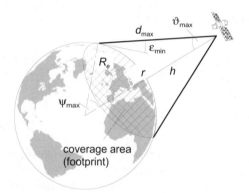

Fig. 2.10. Coverage area (footprint) of a satellite

2.2.2 Coverage Area

The *coverage area* or *footprint* of a satellite is defined as the area on the earth's surface where a satellite is seen with an elevation angle ε greater than a given *minimum elevation* $\varepsilon_{\min}$, see Fig. 2.10. The threshold $\varepsilon_{\min}$ defines the border of the coverage area. For a given orbit altitude a *coverage angle* $\psi_{\max}$ corresponds to a certain minimum elevation $\varepsilon_{\min}$. The minimum elevation $\varepsilon_{\min}$ is an important system parameter since it impacts the required number of satellites and orbits for global coverage of a system. The reader should note that the minimum elevation is driven by the satellite antenna and the link budget (cf. Sect. 3.1.1), since the coverage can be also regarded – from a communications point of view – as the area in which the satellite can provide sufficient signal strength for information transmission.

The coverage area of a satellite as defined by geometry is a spherical cap on the earth's surface. Its contour is determined by one of the following

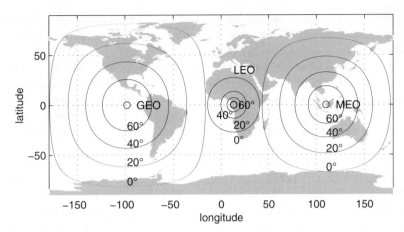

Fig. 2.11. Coverage areas of satellites in different orbit types for several minimal elevation angles

parameters which are related to each other through the equations in Sect. 2.2.1: the minimum elevation $\varepsilon_{\min}$, the maximum nadir angle $\vartheta_{\max}$, the maximum slant range $d_{\max}$, or the coverage angle $\psi_{\max}$. From Fig. 2.10 it can be seen that the coverage area increases, the higher the satellite altitude h and the lower the minimum elevation $\varepsilon_{\min}$. The extent of the coverage area (arc length) is given by

$$D_A = 2\psi_{\max}R_e \tag{2.29}$$

with $\psi_{\max}$ expressed in radians. Examples for the coverage area of satellites in different orbit types are shown in Fig. 2.11 for several minimum elevation angles. The maximum duration of the satellite visibility $T_{\text{vis.},\max}$ that can be experienced during an overhead flight is

$$T_{\text{vis.},\max} = \frac{\psi_{\max}}{\pi}T \tag{2.30}$$

neglecting the earth's rotation.

The area of the spherical cap can be calculated as

$$A = 2\pi R_e^2(1 - \cos\psi_{\max}) . \tag{2.31}$$

The proportion of the coverage area with respect to the total earth surface $A_e = 4\pi R_e^2$ is obtained by

$$\frac{A}{A_e} = \frac{1}{2}(1 - \cos\psi_{\max}) . \tag{2.32}$$

2.3 Satellite Constellations

Single satellites can only provide service in limited areas. To extend the coverage, a satellite system may use a number of satellites. The composite of all satellites in the system is called a *constellation*. The satellites in a constellation usually have equal orbit types, but some systems (e.g. Ellipso, Orbcomm) use a mixture of different orbit types.

When a constellation with several satellites is considered, the total coverage area consists of the union set of the coverage areas of all the satellites. Due to overlapping the constellation coverage area is in general smaller than the sum of all satellite coverage areas. Moreover, the constellation coverage area may vary with time when the satellites are non-geostationary. Then, the coverage can be described by the *instantaneous coverage areas* which are given by the current position of the satellites. The *guaranteed coverage area* of a constellation is defined as the regions on earth in which at least one satellite is visible for 100% of the time. The guaranteed coverage area is a function of latitude and longitude and depends on the orbit and constellation type. Typically, satellites in geostationary and highly elliptical orbits (HEOs) provide a regional coverage that can be extended to a multiregional service by using several satellites. With GEO satellites full global coverage cannot be achieved since the polar regions cannot be reached from a GEO position. Furthermore, the elevation angle to a GEO satellite decreases at higher latitudes. Here, inclined or polar MEOs, LEOs, and HEOs can provide service with reasonably high elevation angles.

Multiple coverage is given if satellite footprints overlap, i.e. if a user in the considered area sees more than one satellite simultaneously. This multiple visibility can be used to improve the availability and quality of the service through the concept of satellite diversity, cf. Sect. 3.4.

Number of Satellites and Orbit Planes for Global Coverage. To provide global coverage several satellites have to be placed in different orbits. Theoretically, the minimum required number of satellites for single coverage can be estimated using spherical hexagons without overlapping. The area of such a hexagon is [WJLB95]

$$A_h = 6R_e^2 \left(2\alpha - \frac{2\pi}{3} \right) , \qquad (2.33)$$

where

$$\alpha = \arctan \left(\frac{\sqrt{3}}{\cos \psi_{\max}} \right) \qquad (2.34)$$

is the edge angle of one of the six spherical triangles in the hexagon. With the earth surface $A_e = 4\pi R_e^2$ and A_h the minimum required number[6] of satellites N is now determined by

$$N = \left\lceil \frac{A_e}{A_h} \right\rceil = \left\lceil \frac{\pi}{3\alpha - \pi} \right\rceil , \qquad (2.35)$$

---

[6] It should be noted that Eq. (2.35) assumes a "stationary" satellite constellation without taking into account the time-variant orbit characteristics.

where the operator $\lceil x \rceil$ denotes the smallest integer greater than or equal to x. Another approximation can be derived using the surface of a sphere cap $2\pi R_e^2 (1 - \cos \psi_{max})$ and considering an overlapping of 21%. Thus,

$$N \approx \frac{2.42}{1 - \cos \psi_{max}} . \qquad (2.36)$$

Using Eq. (2.35) and Eq. (2.23) the number of satellites can be derived as a function of orbit altitude and minimum elevation angle. Figure 2.12 shows this approximation together with the number of satellites of actual commercial satellite constellations. In the figure the range between single and double coverage is indicated by shading. It can be seen that the constellations Iridium

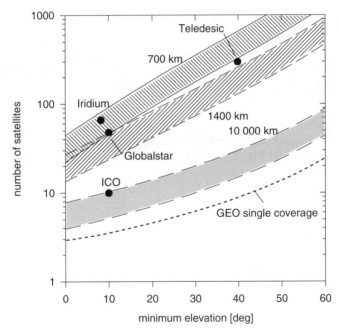

Fig. 2.12. Number of satellites required for single to double global coverage, versus orbit altitude and minimum elevation angle

and Globalstar both lie at the upper limit of the shaded range. However, we will see (cf. Fig. 2.24) that Iridium is only providing single coverage due to its polar orbits whereas the inclined Globalstar constellation provides double coverage in wide areas (cf. Fig. 2.21). This shows that polar constellations are less efficient than inclined ones with respect to the required number of satellites.

Another simple consideration yields the minimum required number of orbit planes for global constellation coverage. The situation at the equator

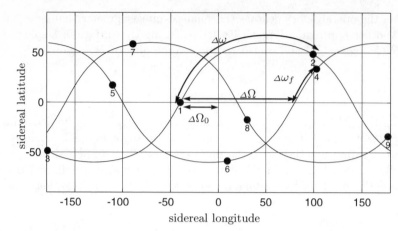

Fig. 2.13. Walker constellation $9/3/1$: $N = 9$, $P = 3$, $F = 1$, $\Delta\omega_f = 2\pi F/N = 40°$, $\Delta\omega = 2\pi P/N = 120°$, $\Delta\Omega = 2\pi/P = 120°$, $i = 60°$. The offset $\Delta\Omega_0 = -40°$ of the first orbit plane from the reference meridian defines the orientation of the Walker constellation with respect to earth for a given time.[7] (Note: the satellite track is displayed without earth rotation in sidereal coordinates.)

is investigated. Ideally the hexagons form a continuous string around the equator. Each orbit plane contributes two satellites to the equatorial coverage and covers an arc of $3R_e\psi_{\max}$. Thus, the minimum required number of orbit planes P in a constellation is

$$P = \left\lceil \frac{2\pi}{3\psi_{\max}} \right\rceil . \tag{2.37}$$

2.3.1 Inclined Walker Constellations

J. G. Walker [Wal77] developed constellations for global coverage using N satellites in inclined circular LEOs with equal period. A Walker constellation consists of P equally inclined orbit planes with their ascending nodes being equally spaced along the equator. The constant longitude offset $\Delta\Omega$ between the planes is $2\pi/P$. On each plane a number of satellites $S = N/P$ are equally distributed with an angular spacing of $\Delta\omega = 2\pi/S$. Besides the inclination angle i and the period T three other parameters are needed to describe the constellation, typically denoted by a triplet, called the

Walker Notation: $N/P/F$

where N is the number of satellites in the constellation, P is the number of orbit planes, and F is the phasing factor ($F = 0, 1, \ldots, P - 1$).

---

[7] In effect, the definition of $\Delta\Omega_0$ is related to the RAAN Ω of the first satellite in the Walker constellation.

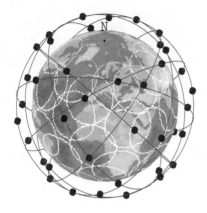

Fig. 2.14. The Globalstar system, Walker constellation 48/8/1, altitude 1414 km

The phasing factor F determines the angular offset $\Delta\omega_f$ between the satellites in adjacent orbit planes:

$$\Delta\omega_f = 2\pi \cdot \frac{F}{N} \ . \tag{2.38}$$

The parameters of an exemplary Walker 9/3/1 constellation are illustrated in Fig. 2.13. Table 2.1 provides the Walker notation for two commercial satellite constellations. One of them, the Globalstar constellation, is shown in Fig. 2.14.

Table 2.1. Satellite constellations with Walker notation

System	Period	Walker notation	Inclination	Figure
ICO	6 h	10/2/0	45°	Fig. 2.25
Globalstar	113.5 min	48/8/1	52°	Fig. 2.14

2.3.2 Polar Constellations

The Iridium system (see Fig. 2.15, or Sect. 2.5.3) is an example of a polar constellation. The orbit planes in polar constellations are arranged in such a way that satellites in adjacent orbits revolve in the same direction. Consequently, there must be an orbit that has an adjacent orbit with opposite rotation when the last orbit plane meets the first orbit plane. The region between the two counter-rotating orbits is called a *seam*.

The following parameters characterize the constellation:

- Let P be the number of orbit planes ($\approx 90°$ inclination), and N the number of satellites in the constellation.

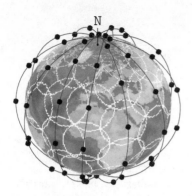

Fig. 2.15. Polar LEO constellation Iridium (six orbits, each with 11 satellites, 780 km altitude)

- $S = N/P$ denotes the number of satellites per orbit plane. The satellites are equally distributed in the orbit plane. The angular difference between neighbors in one plane is $2\pi/S$.
- The satellites in adjacent orbits must be shifted relative to each other to provide coverage without gaps. The phase shift in co-rotating orbit planes is π/S.
- The spacing of the orbit planes is explained in Fig. 2.16. Let ψ denote the one-sided earth central angle of the satellite footprint (coverage angle, see also Fig. 2.8). Then, for continuous coverage of co-rotating orbits the planes must not be spaced by an angle separation larger than $\psi + \Delta$ where Δ can be calculated using spherical geometry:

$$\cos \Delta = \frac{\cos \psi}{\cos(\pi/S)} \ . \tag{2.39}$$

- For continuous coverage the spacing between the two counter-rotating orbits must be smaller. From Fig. 2.16 it can be seen that the maximum spacing is 2Δ.

Hence, a condition for global coverage can be formulated:

$$\pi = (P - 1) \cdot (\psi + \Delta) + 2\Delta \ . \tag{2.40}$$

Table 2.2 shows some solutions of Eq. (2.39) and (2.40) for different values of P and S. In order to achieve a given minimum elevation $\varepsilon_{\min}$, the orbit altitude h is adjusted according to the earth central angle ψ which follows from the number of planes P:

$$h = R_e \left(\frac{\cos \varepsilon_{\min}}{\cos (\psi + \varepsilon_{\min})} - 1 \right) \ . \tag{2.41}$$

The required number of satellites in a polar constellation can be approximated [Bes78] by

$$N \approx \frac{4}{1 - \cos \psi_{\max}} \tag{2.42}$$

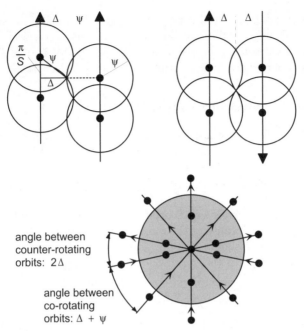

Fig. 2.16. Co-rotating orbits, counter-rotating orbits with seam, and angle between the orbit planes

which is almost double the number of an ideal "stationary" constellation, cf. Eq. (2.36).

Table 2.2. Optimum polar constellations for continuous coverage [Bes78, Nel95, RM96a]

N	P	S	ψ	Δ	$h(\varepsilon_{\min} = 10°)$
15	3	5	42.3°	23.9°	3 889 km
24	4	6	33.6°	15.85°	2 293 km
36	4	9	27.6°	19.4°	1 551 km
66	6	11	19.9°	11.5°	868 km
77	7	11	18.46°	8.66°	767 km

An example for a 66-satellite polar constellation is the Iridium system (Fig. 2.15) with $P = 6$ orbit planes inclined with $i = 86°$, each with $S = 11$ satellites. The orbit spacing is 31.6° ($2\Delta = 22.0°$), similar to the 66-satellite constellation shown in Table 2.2. The orbit altitude is $h = 780$ km with a minimum elevation of $\varepsilon_{\min} = 8.2°$.

A collision risk is involved with polar constellations since the orbits cross at the poles. To avoid satellite collision two provisions can be taken: (i) the inclination is chosen to be nearly polar (e.g. $i = 86°$ for Iridium), or (ii) the different orbit planes are staggered in altitude (e.g. Teledesic, cf. App. C.13).

2.3.3 Asynchronous Polar Constellations

In LEO systems with large numbers of satellites it may be difficult to control the orbital phasing of each satellite. Asynchronous polar LEO constellations can be used in this case. Here the orbital phase is controlled only within an orbit plane whereas the phase shifts between different orbit planes are not controlled.

The constellation consists of P orbit planes which are equally distributed over 180° longitude. Because of the random orbit phasing, the angle between adjacent orbit planes should not be larger than 2Δ, as in the case of the counter-rotating rings in the synchronous polar constellation, cf. Fig. 2.16. Therefore,

$$2\Delta = \frac{\pi}{P} \ . \tag{2.43}$$

The S satellites in each orbit are equally distributed; the phase difference between them is $2\pi/S$.

Example. The former Teledesic concept with 840 satellites was an example of an asynchronous polar LEO constellation. Table 2.3 shows that Teledesic approximates a theoretical asynchronous polar constellation.

Table 2.3. Comparison between the former Teledesic constellation and a similar asynchronous polar constellation

	Teledesic	Asynchronous constellation
No. of orbit planes	$P = 21$	$P = 21$
No. of satellites per plane	$S = 40$	$S = 40$
No. of satellites	$N = 840$	$N = 840$
Orbit height	$h \approx 700$ km	$h = 700$ km
Inclination	$i = 98.2°$	$i = 90°$
Orbit spacing	$2\Delta = 9.5°$	$2\Delta = 8.57°$
Coverage angle	$\psi = 6.35°$	$\psi = 6.21°$
Min. elevation	$\varepsilon_{min} = 40°$	$\varepsilon_{min} = 40.7°$

2.4 GEO System Concept

GEO constellations (Tab. 2.4) use an orbit period $T = T_e = 23$ h 56 min 4 s $= 86\,164$ s corresponding to an altitude of $h = 35\,786$ km. The orbits are not

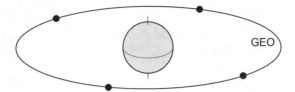

Fig. 2.17. GEO satellite constellation

inclined; thus the satellite revolves eastward around the earth with the same period, and appears to be stationary for an observer on earth, a prerequisite property allowing fixed earth terminals with high-gain antennas for VSAT or DirectTV. Few GEO satellites (three or four) can achieve global coverage with the exception of the polar regions, cf. Fig. 2.17.

Table 2.4. Pros and cons of GEO constellations

Pros	Cons
+ mature satellite technology	− GEO is densely occupied
+ little or no terminal antenna steering	− no coverage at polar regions
+ constant propagation delay	− long propagation delay
+ small Doppler shifts	− low elevation for high latitudes
+ no satellite handover	− high free space propagation loss
+ few satellites for global coverage	↪ large satellite antennas
↪ few launches	
↪ few earth stations	

GEO satellites have been in use for more than 35 years. Today, more than 180 GEOs are in commercial use, most of them for fixed high data rate services and broadcasting. GEO systems for mobile services suffer from high free space propagation losses, the compensation of which requiring large satellite antennas. Thus, they were used for less demanding applications such as low-rate data communications (Inmarsat-C), mobile telephony with vehicles for limited regions (MSAT in North America), telephony with briefcase/laptop terminals (Inmarsat-M/Inmarsat-3), or fleet management (Omnitracs, Euteltracs).

2.4.1 Inmarsat-3

The Inmarsat-3 system is an example of a global GEO satellite system for portable mobile telephony. Most of the earth land masses are covered by four GEO satellites with few large spot beams, cf. Fig. 2.18. User terminals are laptop sized and are equipped with medium-gain antennas that have to be steered manually. The main characteristics of Inmarsat-3 are summarized in Tab. 2.5 and App. C.7.

Table 2.5. Inmarsat-3 system at a glance

Services	voice (4.8 kb/s), real-time data (2.4 kb/s) G3-fax, short message service (SMS)
Constellation	four GEO satellites
Coverage	global, except for polar regions one global beam and five spot beams per satellite
Ground segment	approx. 20 gateway stations

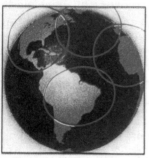

Atlantic Ocean Region West

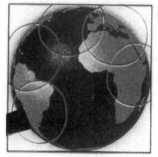

Atlantic Ocean Region East

Indian Ocean Region

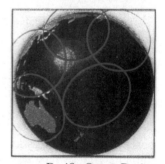

Pacific Ocean Region

Fig. 2.18. Coverage of the Inmarsat-3 GEO constellation

2.4.2 EAST (Euro African Satellite Telecommunications)

The EAST system has been proposed by Matra Marconi Space and the Cyprus PTT authority [Tro97]. It will provide regional coverage to Africa and the Near and Middle East for fixed and mobile telephony. The offered services will include voice and low-rate data from handheld and portable devices. The coverage area is depicted in Fig. 2.19, and the EAST features are summarized in Tab. 2.6.

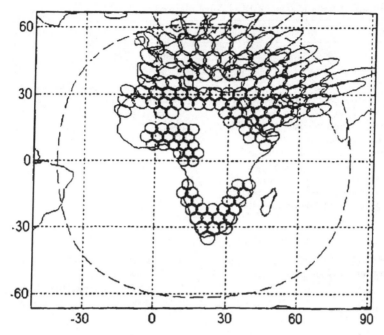

Fig. 2.19. EAST coverage: an example of market-oriented coverage for mobile services [Tro97], © 1997 IEEE.

Table 2.6. EAST system at a glance

Services	speech, data , fax, short message service
Constellation	two co-located GEO satellites
Coverage	approx. 150 spot beams per satellite

2.5 LEO System Concept

The main drawback of using GEO satellites for mobile telephony lies in the long signal propagation delay. To overcome this problem low earth orbits may be used. LEO orbits utilize altitudes between 700 and 1 500 km in inclined or polar planes reducing the round-trip delay to a few milliseconds. Furthermore, GEO satellites may become very complex when new technologies like on-board switching or processing are employed. Since LEO satellites cover smaller areas with fewer subscribers, satellite requirements with respect to the communications payload are more relaxed; however, the number of satellites has to be increased substantially. Table 2.7 presents the advantages and drawbacks of LEO constellations.

Table 2.7. Pros and cons of LEO constellations

Pros	Cons

Consequences of the low altitude

Pros	Cons
+ small coverage areas	− small coverage areas of satellites
↪ lower capacity per satellite	↪ many satellites
↪ smaller satellites	↪ many earth stations
↪ low launch cost per satellite	↪ or intersatellite links required
+ short propagation delay	
+ low free space propagation loss	
↪ low transmit power	
↪ small satellite antennas	
↪ omnidirectional terminal ant.	
↪ handheld terminal easily achievable	

Consequences of the non-geostationary orbit

Pros	Cons
+ high elevation at higher latitudes	− full constellation needed for operation
+ better subscriber coverage	− complex satellite control
	− short satellite lifetime (fuel, battery)
	− earth stations need fast steerable ant.
	− propagation delay variations
	− high Doppler shifts and variations
	− short satellite visibility
	↪ frequent satellite handovers

2.5.1 Globalstar

The LEO Globalstar system represents a relatively simple LEO system concept to extend terrestrial cellular networks. The system has been operating since November 1999. Dual-mode handsets support terrestrial and satellite services for voice, paging, and real-time data. Globalstar utilizes 48 satellites in inclined orbits, see Fig. 2.14. The constellation provides seamless global coverage with the exception of polar regions as shown in Fig. 2.20. The constellation was designed to provide good visibility (cf. Fig. 2.21) for the temperate climate zones with main business relevance. Multiple satellite coverage is exploited to improve signal transmission quality by the use of satellite diversity. Table 2.8 and App. C.5 summarize the main features of Globalstar.

2.5.2 Intersatellite Links and On-Board Processing

Because of the small footprints of LEO satellites a large number of gateway stations would be required for global coverage since for real-time communications each satellite needs a permanent connection to a ground earth station (e.g. Globalstar will use approx. 50 gateways all over the continents). An

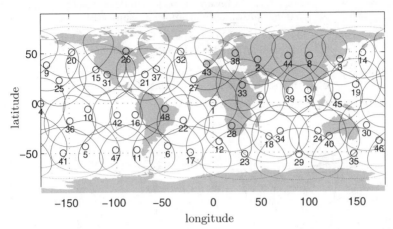

Fig. 2.20. Coverage of Globalstar inclined LEO constellation (minimum elevation 10°)

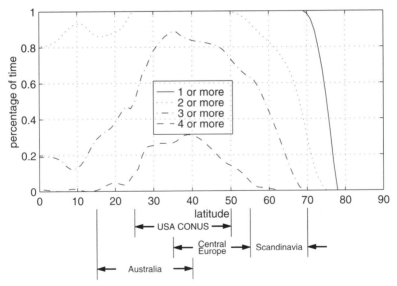

Fig. 2.21. Multiple satellite coverage of Globalstar inclined LEO constellation (minimum elevation 10°)

Table 2.8. Globalstar system at a glance

Services	voice (1.2 ... 4.8 kb/s), real-time data (9.6 kb/s), fax paging, position determination using GPS
Constellation	48 LEOs in eight circular orbits, 1414 km altitude 52° inclination, 48/8/1 Walker constellation
Coverage	global, up to ±70° latitude 16 spot beams per satellite, 10° minimum elevation
Ground segment	50 gateway stations

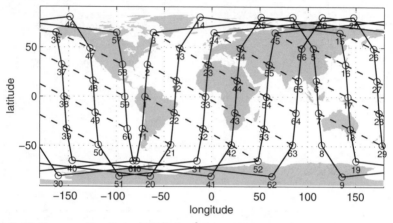

Fig. 2.22. Intersatellite links (ISLs) for a polar LEO constellation (Iridium). Solid lines denote the intra-orbit links and dashed lines denote the inter-orbit links

alternative is to interconnect the satellites by microwave or optical links, the intersatellite links (ISLs). In LEO systems employing ISLs a gateway station can be reached from any satellite in the constellation. Thus, the number of gateway stations can be reduced and their position can be chosen arbitrarily. Moreover, traffic can be routed via ISLs until reaching the destination area, thus avoiding expensive terrestrial long-distance lines. Another advantage of ISLs is the possibility of bypassing terrestrial networks, which is especially interesting for governmental and military users.

In general one can distinguish between (i) intra-orbit ISLs between satellites in the same orbit plane and (ii) inter-orbit ISLs between satellites in different orbit planes (Fig. 2.22). Intersatellite links represent a technological challenge, especially because of the required antenna steering. The pointing, acquisition, and tracking (PAT) must compensate for satellite attitude changes and vibrations. Moreover, in LEO orbits the pointing angle between satellites in different orbit planes will vary due to the changing geometric relations. Figure 2.23 shows the pointing angles of an Iridium inter-orbit ISL as a function of time. It is also important to mention that in polar constellations ISLs cannot be established over the constellation seam (cf. Sect. 2.3.2) due to the high relative speed of the satellites in the counter-rotating orbit planes.

Without ISLs, usually transparent satellite payloads are used, just performing frequency conversion and signal amplification. The use of ISLs requires the application of signal processing and switching on board the satellites (on-board processing, OBP) in order to route the calls within the ISL network and into the appropriate spot beams. If the signal processing is done in the baseband, modulation methods, channel coding, and link protocols can be adapted to the user link or gateway link, respectively. The switching technology on board the satellite represents another technological challenge.

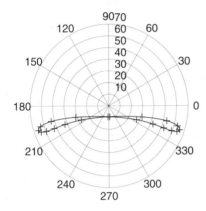

Fig. 2.23. Pointing angle of an Iridium inter-orbit ISL

2.5.3 Iridium

A LEO system utilizing ISLs and OBP is Iridium, which has been operating since November 1998. As for Globalstar, the offered services are global telephony with handheld dual-mode terminals. The 66 satellite constellation is shown in Fig. 2.15. Eleven polar orbits cover the whole earth including the polar caps as depicted in Fig. 2.24. The main features of Iridium are listed in Tab. 2.9 and App. C.8.

2.6 MEO System Concept

A compromise between LEO and GEO are the medium earth orbits (MEO) with altitudes ranging from 5 000 to 10 000 km. Global coverage can be achieved with 10–15 satellites. Figure 2.25 shows an example of a MEO constellation. MEOs combine the advantages of both GEOs and LEOs:

+ moderate number of satellites,

Table 2.9. Iridium system at a glance

Services	voice (2.4 kb/s), real-time data (2.4 kb/s), fax short messaging, paging
Constellation	66 LEO satellites in six circular polar orbits altitude 780 km, 86.4° inclination
Coverage	global, 48 spot beams per satellite 8.2° minimum elevation
Ground segment	12 gateway stations

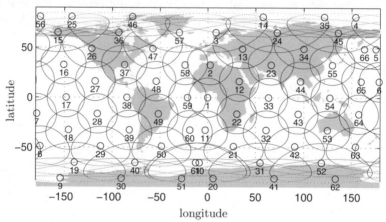

Fig. 2.24. Coverage of Iridium polar LEO constellation (minimum elevation 8.2°)

+ satellite technology is state of the art,
+ acceptable signal delay (1-hop round-trip delay approx. 0.1 s),
+ good satellite elevation at higher latitudes, and
+ few ground earth stations without a need for ISLs.

2.6.1 ICO

The ICO constellation uses 10 satellites in two orbit planes yielding global coverage (Fig. 2.26) for elevation angles higher than 10°. The details of ICO are listed in Tab. 2.10 and App. C.6.

2.7 Satellite Launches

Although satellite launching is a key issue for the deployment of satellite services, only some fundamental aspects will be addressed here (cf. [MB93] for details). Especially in non-GEO satellite systems, launching has a great impact on the system cost and the time to market since the installation of a

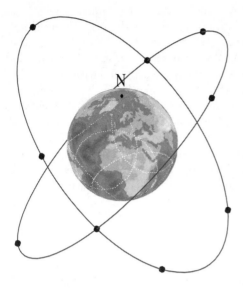

Fig. 2.25. MEO constellation ICO with inclined orbits

Table 2.10. ICO system at a glance

Services	voice (3.6 kb/s), real-time data (2.4 kb/s), fax, paging
Constellation	10 MEO satellites in two circular orbits, altitude 10 390 km 45° inclination
Coverage	global, 163 spot beams per satellite 10° minimum elevation
Ground segment	12 satellite access nodes

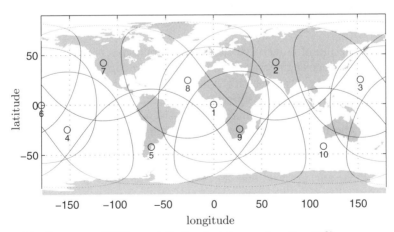

Fig. 2.26. Coverage of ICO constellation (minimum elevation 10°)

complete constellation requires a long time schedule. Big launchers might be required to inject several satellites into one orbit. On the other hand small launchers are also important to replace single defective satellites over the short term. Thus, different requirements for payload mass and dimensions must be considered. Furthermore, choice of the geographic latitude φ of the launch site is affected by the orbit inclination i. The launch site should be on the satellite track; thus, $\varphi \leq i$. It would be ideal to have a launch site with $\varphi = i$ since the earth's rotation contributes with a velocity vector of $v = 2\pi R_e/T_e \cos\varphi$ in an eastern direction to the launcher at lift-off. If lower inclination angles $i < \varphi$ are wanted, launch maneuvers consuming high energy will be required. The marine-based launch platform Sea Launch or airborne rockets such as Pegasus can easily accommodate various inclinations.

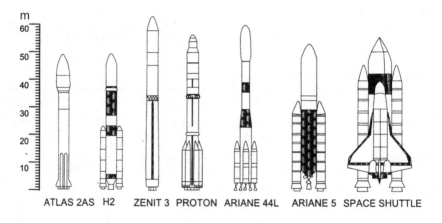

Fig. 2.27. Comparison of several launchers

The structure of launch vehicles typically employs several stages, cf. Fig. 2.27. A main stage engine with liquid oxygen and hydrogen propellant is used to escape from the strong earth's gravitation to low earth orbits. Additional external solid boosters can support the thrust of the main stage. Some of the launchers (e.g. Proton) may also consist of several stages. The upper section of the launch vehicle contains an equipment bay for storage of the payload. A fairing protects the payload from damage during passage through the atmosphere. Once the launcher has left the atmosphere the fairing no longer serves any purpose and its two shells are jettisoned by means of a pyrotechnical device. The transfer of the payload to its final orbit is done using a perigee and/or apogee engine. Figure 2.28 ([MB93]) illustrates the launch procedure for a transfer to GEO via a geosynchronous transfer orbit (GTO). The vehicle's main engine and boosters are ignited shortly before

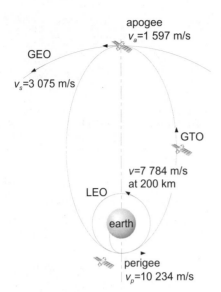

apogee
v_a=1 597 m/s

GEO

v_s=3 075 m/s

GTO

v=7 784 m/s
at 200 km

LEO

earth

perigee
v_p=10 234 m/s

Fig. 2.28. Injection via a geostationary transfer orbit (GTO) into GEO [MB93]
© John Wiley & Sons Limited, reproduced with permission

lift-off. After a couple of minutes of flight the boosters and the main stage are cut off. This will carry the launcher into a LEO orbit. From LEO, a burn of the upper stage engine at perigee transfers the payload into the elliptical GTO. The apogee of the GTO has the same altitude as GEO, and thus a second burn of the upper stage engine will place the satellite in GEO. Some launchers may inject directly into GTO whereas the Proton launcher injects directly into GEO. Table 2.11 shows the main characteristics of some launch systems.

To calculate the required energy boosts for the acceleration phases during a launch the energy equilibrium of potential and kinetic energy in an elliptical orbit is used, cf. Eq. (2.9). The injection into an equatorial LEO orbit with $h = 200$ km requires a satellite velocity of $v = 7\,784$ m/s. At perigee of the GTO the satellite is accelerated up to a speed of $v_p = 10\,234$ m/s. At GTO apogee the satellite speed is again increased to $v_a = 3\,075$ m/s. The reader should note that the engine at perigee and apogee can burn from several minutes (apogee motor) to several hours (perigee motor). In this way perigee and apogee are raised successively to geosynchronous altitude.

Table 2.11. Survey of launchers

Launcher	Provider	Launch site	Price[a] US $
Ariane 44-L	Arianespace	Kourou	70 Mio.
Ariane 5	Arianespace	Kourou	100 Mic
Space Shuttle	NASA	Cape Canaveral	
Delta II	Boeing	Cape Canaveral Vandenberg AFB	50 Mio.
Delta III	Boeing	Cape Canaveral Vandenberg AFB	
Atlas III (B)	ILS	Cape Canaveral	
Atlas V 5xx	ILS	Cape Canaveral	
Proton M	ILS	Baikonur	40–50 M
Zenith 2 SL-16	Sea Launch (Boeing)	maritime platform	
Long March	China	Xichang	1 000/kg
H-2A	Japan	Tanegashina	70 Mio.
Pegasus XL	Orbital Sciences	aircraft	50 000/l

[a] Prices should be considered as rough estimates.

3. Signal Propagation and Link Budget

This chapter deals with various phenomena occurring when a signal propagates between a satellite and a user terminal. The basic effect is signal attenuation due to free space loss. Since the attenuation increases with distance it is especially severe in the satellite scenario. Together with the antenna characteristics of the transmitter and receiver, the attenuation determines the required transmission power to achieve a reliable signal transmission. This topic is covered by the *link budget*.

In the second part of the chapter, time-dependent signal shadowing and multipath fading will be discussed, which add to free space loss when mobile/personal users communicate via satellite.

The radio links in satellite communications are known under specific terms which are compiled in Fig. 3.1. The most critical link with regard to communication quality is the link between the user and the satellite; common names for this link are *user link* (*mobile user link* in the case of mobile communication), *subscriber link*, or *service link*. The link between the satellite and a fixed earth station is called the *feeder link* (the satellite is "fed" with, for example, a broadcast signal) or *gateway link* (in particular if the fixed earth station represents an interface to a terrestrial network).

Both links can be separated into an *uplink* and a *downlink*. Finally, the series of unidirectional links from the fixed earth station to the user is called the *forward link*, and the links in the opposite direction constitute the *return link*.

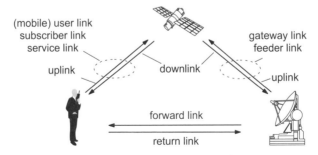

Fig. 3.1. Link designations for mobile satellite networks

3.1 Satellite Link Budget

In this section, we introduce the basics of link budgets. For simplicity, we restrict our considerations to

– a single link between a satellite and an earth station or terminal, and
– a single continuous digital transmission stream.

3.1.1 Antenna Characteristics

Antennas are the interface between the radio link and the communications hardware. They are used as the start and end point of the radio link for transmission and/or reception. An *isotropic antenna* radiating uniformly in all directions (uniformly into 4π solid angle of the sphere) is taken as the reference. Its power flux density at distance d is

$$\phi_i = \frac{P_t}{4\pi d^2} \qquad \text{in W/m}^2, \tag{3.1}$$

where P_t denotes the transmit power.

Radiator with Antenna Gain. A directional antenna focusses the power radiation, e.g. by using a reflector, and the power is mainly radiated into a solid angle less than 4π. Accordingly, only a limited part of the sphere is illuminated, and a power flux density $\phi > \phi_i$ is generated in this illumination area. The transmitted power P_t corresponds to the integral of the power flux density over a sphere with surface $4\pi d^2$. The antenna gain[1] is defined as

$$G_t \;=\; \frac{\phi}{\phi_i} \qquad \text{in linear form, or}$$

$$G_t \;=\; 10\log\frac{\phi}{\phi_i} \qquad \text{in dBi} \tag{3.2}$$

where the index t refers to a transmit antenna. Vice versa, for a given antenna gain, the power flux density produced at distance d is

$$\phi = G_t\phi_i = \frac{P_t G_t}{4\pi d^2}\ . \tag{3.3}$$

The *equivalent isotropic radiated power* (EIRP) is the main characteristic of a transmitter. It is defined as the power which must be radiated by an isotropic radiator in order to achieve the same power flux density as a given directional antenna:

---

[1] The antenna gain can be expressed in linear form or as a logarithmic measure, usually dBi, with "i" referring to the isotropic radiator. The function log denotes logarithm to the base 10, $\log_{10}$.

$$\text{EIRP} = P_t G_t \quad \text{in linear form, or}$$

$$= \frac{P_t}{\text{dBW}} + \frac{G_t}{\text{dBi}} \quad \text{in dBW .} \tag{3.4}$$

The same concept of antenna gain also holds for a directional receive antenna which preferably receives power from a limited range of directions.

Antennas for handheld terminals typically limit the radiation and reception to the upper hemisphere, and thus exhibit antenna gains of $G = 1\text{--}3$ dBi. They can be realized as dipole antennas or as helix antennas.

Antennas with a higher directivity are required on board the satellites, at fixed earth stations, in portable (briefcase or laptop) terminals, and in mobile terminals. Here, reflector antennas or array antennas can be used, commonly referred to as aperture antennas, see Chap. 8.

Gain of Aperture Antennas. Generally, the gain of an antenna is a function of the direction of radiation or reception, making up the antenna characteristic. Figure 3.2 shows the polar and Cartesian representations of an antenna characteristic.

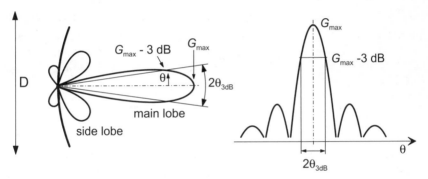

Fig. 3.2. Antenna characteristic of a rotational symmetric directional antenna; left figure: polar representation; right figure: Cartesian representation

The *boresight direction* of an antenna is the direction of the maximum gain. Usually this direction corresponds to the antenna symmetry axis. The maximum gain is proportional to the antenna area:

$$G_{\max} = \eta \left(\frac{\pi D}{\lambda} \right)^2 \quad \text{in linear units.} \tag{3.5}$$

Here, D is the diameter of the antenna aperture, i.e. the diameter of the reflector or of the array of radiating elements, respectively. Furthermore, the maximum gain is inversely proportional to the wavelength λ where $\lambda = c/f$ (with vacuum speed of light $c = 2.998 \cdot 10^8$ m/s and frequency f). The antenna efficiency η is typically in the range of 0.55–0.65. The antenna efficiency depends on the shape of the power distribution across the aperture: the power flux density may roll off towards the circumference of the aperture (tapering) in order to suppress antenna side lobes. For reflector antennas, the efficiency also depends on the amount of spill-over occurring when illuminating the reflector by a feed radiator.

The main parameter describing the directivity of an antenna is the beamwidth. It is the angular span in which the antenna gain is larger than $G_{max} - 3$ dB. For a typical tapered power flux density on the aperture (for instance, for parabolic antennas), this two-sided 3-dB (half-power) beamwidth can be approximated as

$$2\,\theta_{3\mathrm{dB}} \approx 70° \cdot \lambda/D \ , \tag{3.6}$$

whereas the approximation for uniformly illuminated apertures is

$$2\,\theta_{3\mathrm{dB}} \approx 58° \cdot \lambda/D \ . \tag{3.7}$$

For $\eta = 0.6$ the following relation between the gain and beamwidth of a tapered antenna can be derived:

$$G_{max} = 44.6\,\mathrm{dBi} - 20\log\left(2\,\theta_{3\mathrm{dB}}\right) \tag{3.8}$$

with $\theta_{3\mathrm{dB}}$ in degrees. Figure 3.3 shows the relationship between antenna diameter D, boresight antenna gain G_{max}, and half-power beamwidth $2\,\theta_{3\mathrm{dB}}$.

According to the former CCIR [CCI90], the antenna gain within the main lobe can be approximated by a quadratic decrease with off-boresight angle (off-axis angle) θ:

$$
\begin{aligned}
G(\theta) &= G_{max} - 3(\theta/\theta_{3\mathrm{dB}})^2 &&\text{in dBi} \\
G(\theta) &= G_{max} \cdot 10^{-0.3(\theta/\theta_{3\mathrm{dB}})^2} &&\text{in linear form.}
\end{aligned} \tag{3.9}
$$

Illumination Area of a Satellite Antenna. In Sect. 2.2.2 the geometric coverage area of a satellite was defined as the area within which the satellite can be seen with an elevation angle $\varepsilon \geq \varepsilon_{min}$. The coverage area with regard to service, however, is determined by the illumination area of the satellite antenna.

Often, the edge of coverage (EOC) is defined as the 3-dB contour of the illumination area on the earth's surface: $\vartheta_{max} = \theta_{3\mathrm{dB}}$. Then, from Eq. (3.6), the required diameter of the tapered satellite antenna can be derived as

$$D \approx \lambda \cdot \frac{35°}{\vartheta_{max}} \ . \tag{3.10}$$

Considering for example a GEO satellite with a global beam and minimum elevation angle $\varepsilon_{min} = 5°$, the antenna beamwidth must be $2\,\vartheta_{max} = 17.3°$ (cf. Eq. (2.24)). For a frequency of $f = 1.5$ GHz, the required antenna diameter is $D = 0.8$ m, and the antenna gain is $G_{max} = 20$ dBi.

For a constant diameter of the illumination area, the antenna beamwidth decreases with increasing orbit altitude. Therefore, the satellite antenna must be larger for higher-altitude satellites.

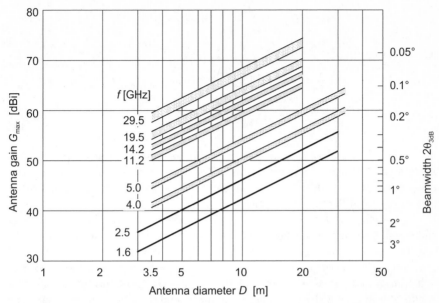

Fig. 3.3. Maximum antenna gain G_{max} and 3-dB (half-power) beamwidth $2\theta_{3\mathrm{dB}}$ as a function of antenna diameter D for various frequencies. Shaded areas correspond to ranges of antenna efficiency. Source: ITU

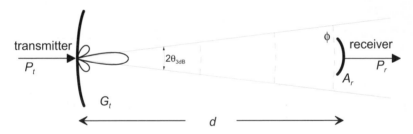

Fig. 3.4. Radio link with directional antennas

3.1.2 Free Space Loss and Received Power

Figure 3.4 shows a radio link with two directional reflector antennas. From the power radiated by the transmit antenna, the receive antenna picks up an amount that depends on the power flux density ϕ at the receive antenna, on the reflector area A_r of the receive antenna, and on the efficiency η_r of the receive antenna. With these characteristics, the received power is

$$P_r = \eta_r A_r \phi = \eta_r A_r \frac{P_t G_t}{4\pi d^2} \ . \tag{3.11}$$

Substituting the term $A_r = \pi D^2/4$ with Eq. (3.5) and taking G_r as the gain of the receive antenna, we get

$$P_r = \frac{P_t G_t G_r}{(4\pi d/\lambda)^2} = \frac{P_t G_t G_r}{L_0} \,. \tag{3.12}$$

The variable

$$
\begin{aligned}
L_0 &= \left(\frac{4\pi d}{\lambda}\right)^2 \quad \text{in linear form} \\
&= 20 \log\left(\frac{4\pi d}{\lambda}\right) \quad \text{in dB} \tag{3.13}
\end{aligned}
$$

is called the *free space loss*. It describes the decrease of received power with increasing distance d between transmitter and receiver, due to the spatial dissipation of the radiated power. Figure 3.5 is a graphical representation of Eq. (3.13) for different frequencies.

In addition to free space loss, other kinds of attenuation occur when the signal propagates through the troposphere and the ionosphere (see e.g. [Ric99]): gaseous absorption, rain attenuation, attenuation from clouds and fog, and tropospheric and ionospheric scintillation. Rain attenuation becomes substantial at frequencies above 10 GHz and is the most important effect of them. Summarizing these effects by an additional signal attenuation ΔL, the total signal attenuation results in

$$L = L_0 + \Delta L \quad \text{in dB.} \tag{3.14}$$

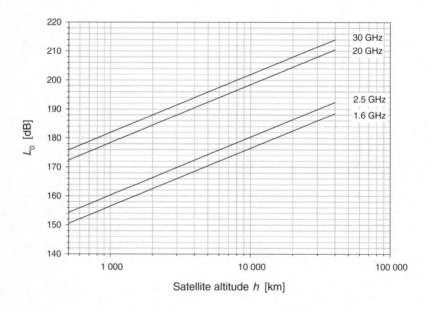

Fig. 3.5. Free space attenuation as a function of distance (satellite altitude), for different frequencies

3.1.3 Link Budget

Since signal propagation does not depend on direction, the following considerations can be applied for both an uplink and a downlink. Designating the gain of the receive antenna by G, the received power can be written as

$$C = P_t \frac{G_t G}{L} = \text{EIRP} \, \frac{G}{L} \, . \tag{3.15}$$

In logarithmic terms, the received power can be represented as

$$\begin{aligned} C &= P_t + G_t - L + G \quad \text{in dBW} \\ &= \text{EIRP} - L + G \quad \text{in dBW.} \end{aligned} \tag{3.16}$$

These expressions are commonly known as *transmission equations* [Ric99].

Signal-to-Noise Ratio C/N. In order to derive the signal-to-noise ratio at the receiver, we have to consider thermal noise, which is superimposed on the received signal. Thermal noise is characterized by the one-sided *thermal noise power spectral density* $N_0 = kT$ which is constant over frequency. Here, $k = 1.38 \cdot 10^{-23}$ Ws/K $= -228.6$ dBWs/K is the Boltzmann constant, and T is the effective noise temperature. The two main contributions are the noise picked up by the receive antenna and the noise occurring in the first stage of the receiver low-noise amplifier (LNA).[2] Accordingly, the effective noise temperature T can be taken as the sum of the noise temperature of the antenna, T_a, and the effective noise temperature of the receiver, T_r:

$$T = T_a + T_r. \tag{3.17}$$

Typical antenna noise temperatures are given in Tab. 3.1.

Table 3.1. Noise temperatures in antennas

Antenna		Noise temperature
Directional satellite antenna	earth from space	290 K
Directional terminal antenna	space from earth at 90° elev.	3–10 K
	space from earth at 10° elev.	≈ 80 K
	sun $(1 \ldots 10$ GHz)	$10^5 \ldots 10^4$ K
Hemispherical terminal antenna	at night	290 K
	cloudy sky	360 K
	clear sky with sunshine	400 K

The receiver noise temperature is often expressed by the receiver's noise figure N_r in dB:

---

[2] For simplicity we neglect the influence of the connection cable between the receive antenna and the LNA.

$$T_r = \left(10^{N_r/10} - 1\right) \cdot 290 \text{ K} . \tag{3.18}$$

Typical noise figures of LNAs are in the range $N_r \approx 0.7$–2 dB.

From Eq. (3.15), the ratio between signal power and noise power spectral density follows as (all terms linear)

$$C/N_0 = \frac{C}{kT} = \frac{P_t G_t}{L} \cdot \frac{G}{kT} = \frac{\text{EIRP}}{L} \cdot \frac{G/T}{k} . \tag{3.19}$$

The term G/T is commonly called the *figure of merit* of the receiver. It characterizes the quality of the receiver with regard to signal reception, being determined by its antenna gain G and its overall noise temperature T.

The logarithmic representation of this equation (all terms in dB) is commonly called the *link budget*:

$$\begin{aligned} C/N_0 &= P_t + G_t - L + G - k - T \\ &= \text{EIRP} - L + G/T - k \quad \text{in dBHz.} \end{aligned} \tag{3.20}$$

Considering a receive filter with noise equivalent bandwidth B_r, the noise power is $N = N_0 B_r$, and the *signal-to-noise ratio* C/N becomes

$$C/N = C/N_0 \cdot \frac{1}{B_r} . \tag{3.21}$$

In real system scenarios, in addition to thermal noise, several kinds of interference should be taken into account, such as co-channel interference (Sect. 6.2.1), adjacent channel interference, and interference caused by other systems.

Symbol Energy-to-Noise Power Density E_s/N_0. In Chap. 4, we will see that the bit error rate for digital transmission can conveniently be expressed as a function of the ratio between the signal energy contained in a data symbol, E_s, and the noise power spectral density N_0. Although not a power ratio, we will designate the term E_s/N_0 as the *signal-to-noise ratio per symbol*.

For a transmission rate R_s in symbols/sec and symbol duration $T_s = 1/R_s$, the symbol energy is $E_s = CT_s = E_s/R_s$, and the signal-to-noise ratio per symbol becomes (all terms linear)

$$E_s/N_0 = C/N_0 \cdot T_s = \frac{C/N_0}{R_s} . \tag{3.22}$$

In logarithmic representation we have (all terms in dB)

$$\begin{aligned} E_s/N_0 &= C/N_0 - R_s \\ E_s/N_0 &= \text{EIRP} - L + G/T - k - R_s \quad \text{in dB.} \end{aligned} \tag{3.23}$$

The E_s/N_0 can be used to determine the bit error rate of a digital transmission scheme, or, vice versa, the characteristics of a satellite link can be determined such that a given value for the signal-to-noise ratio E_s/N_0 is achieved, guaranteeing a sufficiently low bit error rate.

Link Budget Example. In this example we show that it is technically feasible to communicate from a handheld terminal with low transmit power to a MEO satellite more than 10 000 km away. Table 3.2 lists the main elements of the link budget. The satellite is assumed to be regenerative, i.e. it demodulates and decodes the received signal. Then, the uplink can be analyzed independently of the succeeding downlink. For this example the signal-to-noise ratio results in

$$E_s/N_0 = \text{EIRP} - L_0 + G/T - k - R_s = 9.1 \text{ in dB}. \tag{3.24}$$

Table 3.2. Example for a MEO uplink budget, assuming a handheld terminal and a regenerative satellite

Transmit power of handheld terminal	$P_t = 0.25$ W $\hat{=}$ –6 dBW
Terminal antenna gain (hemispher. characteristic)	$G_t = 3$ dBi
Terminal EIRP	$P_t + G_t = -3$ dBW
Free space attenuation ($f = 2$ GHz, 10 354 km orbit altitude, 10° satellite elevation)	$L_0 = 181.6$ dB
Gain of satellite receive antenna (3 m aperture diameter, gain at edge of coverage)	$G = 33.7$ dBi – 3 dB $= 30.7$ dBi
Antenna noise temperature	290 K
LNA noise temperature (noise figure $N_r = 1$ dB)	75 K
Effective noise temperature	$T = 365$ K $\hat{=}$ 25.6 dBK
Satellite G/T	$G/T = 5.1$ dB/K
Boltzmann constant	$k = -228.6$ dBWs/K
Transmission rate	$R_s = 10$ kb/s $\hat{=}$ 40 dBHz

From Fig. 4.15 we can see that for coherent BPSK transmission with rate-1/2 convolutional coding a signal-to-noise ratio $E_s/N_0 = 2$ dB is sufficient to achieve a bit error rate below 10^{-6} in a non-fading channel. This means that we actually have a contingency of 7.1 dB. This kind of contingency is called the *link margin*; it can be used to account for implementation losses, additional signal attenuation as considered in Eq. (3.14), and signal power variations due to multipath fading, see Sect. 3.3. Table 3.5 in Sect. 3.4 shows the link margins of some actual systems.

Link Budget for a Full Satellite Hop (Uplink + Downlink). In order to assess the quality of a full satellite hop, the series connection of uplink and downlink must be considered.

In a *transparent satellite* the received uplink signal is amplified and shifted in frequency before it is retransmitted into the downlink. Basically, the uplink noise at the satellite adds to the downlink noise at the receiving earth station. Therefore, the following relation holds for the total signal-to-noise ratio $(E_s/N_0)_{\text{tot}}$:

$$\frac{1}{(E_s/N_0)_{\text{tot}}} = \frac{1}{(E_s/N_0)_u} + \frac{1}{(E_s/N_0)_d} \quad \text{in linear units} \qquad (3.25)$$

with index u referring to the uplink and index d referring to the downlink.

A *regenerative satellite* decides on the received data symbols by demodulating and decoding the received signal. The data are then re-coded and re-modulated for transmission on the downlink. Thus, uplink and downlink can be analyzed separately, and the respective bit error rates approximately add up:

$$p_{b,\text{tot}} \approx p_{b,u} + p_{b,d}. \qquad (3.26)$$

The advantages of a regenerative satellite are some savings in transmit power and the possibility to match the uplink and downlink transmission schemes to different requirements.

3.1.4 Spot Beam Concept

Generally, for satellite communications with handheld terminals the link budget requires a large gain of the satellite antenna. For example, in Table 3.2 a boresight gain of 33.7 dBi was assumed. According to Eq. (3.8), the corresponding half-power beamwidth is $2\theta_{3\text{dB}} = 3.5°$. This represents a highly focussed antenna beam, called a *spot beam*, with maximum off-boresight angle $\theta_{\text{spot}} = 1.75°$.

However, if the footprint of the satellite should extend up to a minimum elevation of $\varepsilon_{\min} = 10°$, Eq. (2.23) yields a maximum nadir angle of $\vartheta_{\max} = 22°$. This means that the illumination area of the spot beam antenna (constituting a radio cell) is substantially smaller than the desired coverage area of the satellite. Hence, the coverage area must be "filled up" with several spot beams, as illustrated in Fig. 3.6. Several tiers of spot beams are arranged in a circular way (this is discussed in more detail in App. A.2). From a technological point of view, spot beams can be produced by multifeed reflector antennas or with phased-array antennas, see Chap. 8.

The signals in overlapping spot beams can be separated by using different frequency bands, see Chap. 6. Alternatively, the separation of the spot beams can be achieved by using different orthogonal signal codes, orthogonal polarizations, or through beam scanning (orthogonal time periods).

Determination of the Required Number of Spot Beams. The solid angle covered by an antenna is

$$\Omega = 2\pi(1 - \cos\theta_{\text{spot}}). \qquad (3.27)$$

With 21% overlapping of the radio cells (assuming that circles circumscribe hexagons in a regular pattern), the required number of spot beams follows as

$$n = 1.21 \frac{1 - \cos\vartheta_{\max}}{1 - \cos\theta_{\text{spot}}} . \qquad (3.28)$$

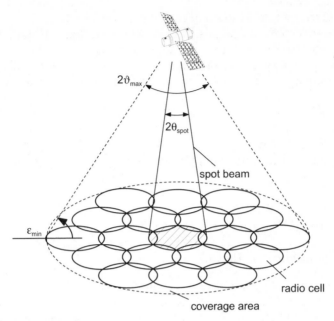

Fig. 3.6. Spot beam concept

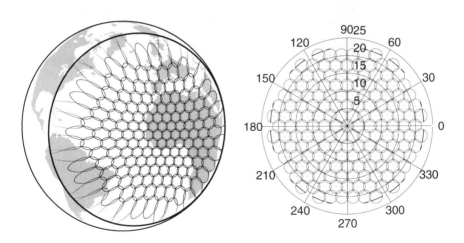

Fig. 3.7. Spot beam pattern for the ICO system (163 spot beams)

For the above example we get $n = 188$. The similar ICO system uses 163 spot beams, Fig. 3.7.

For a given spot beam transmit power and a given extent of the radio cell, the produced power flux density is independent of orbit altitude. For ε_{min} fixed, the coverage area of the satellite and therefore the required number of spot beams increases with orbit altitude. Furthermore, for a constant cell area the diameter of the satellite spot beam antenna increases with orbit altitude.

The LEO Globalstar system uses 16 spot beams per satellite; Iridium uses 48 spot beams per satellite. The MEO ICO system uses 163 spot beams per satellite, Fig. 3.7, indicating the increase of the number of spot beams with increasing satellite altitude. Spill-over of the earth's horizon occurs for some of the outermost cells.

3.2 Peculiarities of Satellite Links

3.2.1 Dependence on Elevation

Slant Range. Eqs. (2.25) and (2.26) express the distance d to the satellite (slant range) as a function of satellite elevation ε. For $\varepsilon = 90°$, the distance equals the satellite altitude $d = h$; for $\varepsilon < 90°$, the slant range is larger than the satellite altitude. In the latter case, an additional free space loss

$$\Delta L = 20 \log \frac{d}{h} \tag{3.29}$$

occurs, Fig. 3.8, which must be included in the link budget. For the example in Table 3.2 the additional free space loss due to the slant range amounts to 2.9 dB.

Round-Trip Delay. Also the round-trip delay Δ (for one hop) depends on the satellite elevation, cf. Fig. 3.9:

$$\Delta = 2d/c. \tag{3.30}$$

Additional delay occurs in the terrestrial network, for which an ITU estimation is 12 ms + (4 ms/1000 km). For telephony, the total delay should not exceed 400 ms. For a delay larger than 150 ms the use of echo compensation is suggested.

3.2.2 Time Dependence of Satellite Links

Time dependences are caused by the motion of non-geostationary satellites with regard to a user on the earth's surface. At time $t = 0$ the satellite will pass over the user. Then, neglecting the earth's rotation, the time-varying earth central angle is

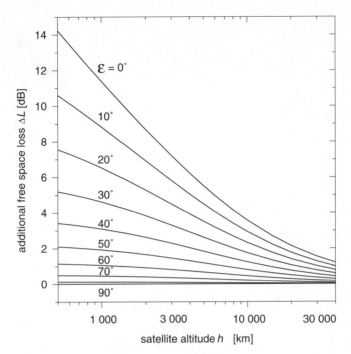

Fig. 3.8. Additional free space loss due to the slant range

$$\psi(t) = 360° \cdot t/T. \tag{3.31}$$

From Eq. (2.26) the time-varying distance $d(t)$ is obtained, and Eq. (3.29) gives the time-varying (additional) signal attenuation ΔL. Similarly, the time-varying round-trip delay Δ can be derived.

Doppler Shift. The time-varying signal delay causes a time-varying Doppler shift, Fig. 3.10:

$$\Delta f(t) = -\frac{d'(t)}{\lambda} = -\frac{2\pi f}{cT} r R_e \frac{\sin \psi(t)}{d(t)} \tag{3.32}$$

with wavelength λ, carrier frequency f, and speed of light c. For a worst-case approximation, the Doppler shift caused by the earth's rotation can be added:

$$\Delta f_e = \frac{f}{c} \cdot \frac{2\pi R_e}{T_e} \cos \varepsilon_{\min} = 1.55 \cdot 10^{-6} f \cos \varepsilon_{\min} . \tag{3.33}$$

The time-dependent elevation angle was discussed in Sect. 2.2.1.

Time-Variant Antenna Gain. Time-variant elevation and azimuth angles of satellites cause time-variant antenna gains corresponding to the characteristics of the satellite and terminal antennas. Also, the movement of mobile/personal users cause time-variant antenna gains.

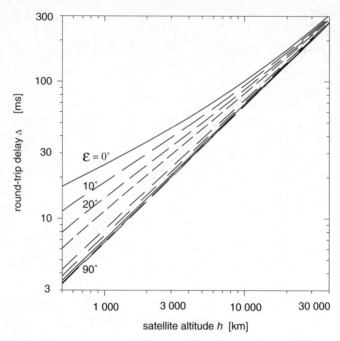

Fig. 3.9. Round-trip delay as a function of satellite altitude for various elevation angles ε

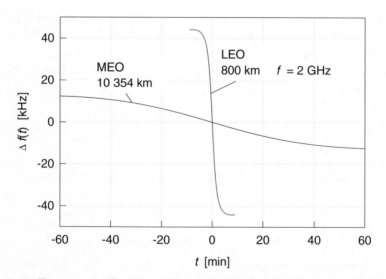

Fig. 3.10. Time-varying Doppler shift during a LEO or MEO passage. The maximum additional Doppler shift caused by the earth's rotation is 3.1 kHz

3.2.3 Faraday Rotation

The ionosphere introduces a rotation of the plane of a linearly polarized wave [MB93]. The angle of the rotation is inversely proportional to the square of the frequency. The rotation depends on the ion density in the ionosphere and is thus a function of time, season, and solar activity. Typical values for the Faraday rotation that can be experienced from GEO satellites to temperate latitudes are $108°$, $9°$, $4°$, and $1°$ for carrier frequencies of 1 GHz, 4 GHz, 6 GHz, and 12 GHz, respectively. Furthermore, for mobile handsets the antenna orientation is not steady and the polarization planes are likely to vary with time. By contrast, circular polarization will keep the signal level constant regardless of these anomalies. Therefore, S-PCN systems working at L or S band are using circular polarized antennas whereas K/Ka band systems with fixed or portable terminals typically have linear polarization.

3.3 Signal Shadowing and Multipath Fading

The availability and quality of the service that satellite PCN can offer is crucially influenced by the particular characteristics of signal propagation in the link between the mobile or personal user and the satellite. In order to investigate this subject, a number of propagation measurements have been performed, and several channel models have been derived, describing the transmission path between a mobile/personal user and a GEO or non-GEO satellite, e.g. [Loo85, LCD$^+$91, JSBL95b].

In the mobile satellite link, multipath fading occurs because the received signal contains not only the direct signal but also echo components being reflected from objects in the surroundings. The received total power of the echoes mainly depends on the type of user environment (urban, suburban, rural, etc.) and on the antenna characteristic of the user terminal. Antennas with wide-angle patterns tend to gather more echo power than directive antennas. In contrast to antennas mounted on top of a vehicle, handheld terminal antennas may pick up strong specular reflections from the ground [JL94]. Variation of the received power with time is caused by movement of the user, of the (non-geostationary) satellite, or of reflecting objects.

Shadowing of the satellite signal is caused by obstacles in the propagation path, such as buildings, bridges, and trees. Shadowed signals will suffer deep fading with substantial signal attenuation. The percentage of shadowed areas on the ground, as well as their geometric structure, strongly depend on the type of environment. For low satellite elevation the shadowed areas are larger than for high elevation. Especially for streets in urban and suburban areas, the percentage of signal shadowing also depends on the azimuth angle of the satellite.

Due to the movement of non-geostationary satellites, the geometric pattern of shadowed areas is changing with time. Similarly, the movement of a mobile/personal user translates the geometric pattern of shadowed areas into a time series of good and bad channel states. The mean durations of the good and bad state, respectively, depend on the type of environment, satellite elevation, and mobile user speed.

3.3.1 Narrowband Model for the Land Mobile Satellite Channel

Several institutions have performed measurements of the land mobile satellite channel (Davarian offers a good survey [Dav94]), mainly the Jet Propulsion Laboratory (JPL) [RSHP96], the European Space Agency (ESA) [SBM93], the University of Texas [GV94], the University of Surrey [BER92], and the German Aerospace Center DLR [LJWB96]. The latter performed several campaigns with narrowband and wideband signals from L band to EHF band (40/50 GHz) [LCD+91, JSBL95a, JL97]. A handheld mock-up with an RHCP drooping dipole antenna as well as a car-roof-mounted RHCP antenna have been used. The measurement setup is described in [JL94, CJLV95] for the L band campaigns and in [JL96] for the EHF band. Small aircraft simulated different satellite orbit characteristics. Due to the limited flight altitude the experiment geometry is different from the satellite case and thus the signal propagation is affected (e.g., the free space path loss of echoes with respect to the attenuation of the direct line-of-sight signal is smaller for the aircraft scenario than for the satellite scenario). The related measurement errors and the error processing requirements are discussed in [HJW+97].

Figure 3.11 shows the received power level for a quasi-fixed (personal) user with a handheld terminal. The 0-dB value of the power plot corresponds to the expected mean power of a direct signal without multipath in line-of-sight (LOS) condition. The handheld terminal mainly suffers from tree shadowing and two-path fading caused by specular reflections from the ground. During tree shadowing the received signal is attenuated by 10–20 dB. Also multipath contribution is visible. Earlier narrowband measurements have been carried out at 1.54 GHz, using the geostationary MARECS satellite and a car-roof-mounted receive antenna. Figure 3.12 shows an example for the received power level in a suburban area. Again, states with good signal levels when the satellite is in line-of-sight and bad states with severe losses due to shadowing are visible.

A two-state narrowband model for the land mobile satellite channel was derived from the measurements [LCD+91], Fig. 3.13. Narrowband models assume that the signal bandwidth is smaller than the coherence bandwidth of the transmission channel. This assumption is valid for sinusoidal signals or for signals with data rates typically up to 1 Mbps. The model is based on the physical effects of signal shadowing and multipath fading. It distinguishes "good" channel states with the satellite being visible from "bad" channel states with the satellite being shadowed by an obstacle.

In both of these cases, the satellite signal is reflected from a large number of objects in the surroundings of the mobile user. These signal components are

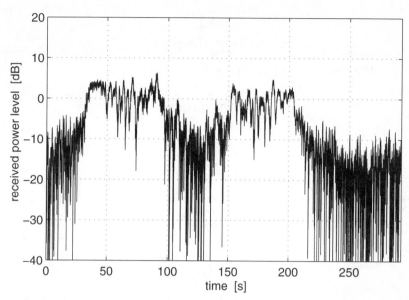

Fig. 3.11. Received power level from narrowband measurements. Randomly moving user with handheld terminal in rural (tree-shadowed) environment, 25° elevation, measurement frequency 1.82 GHz

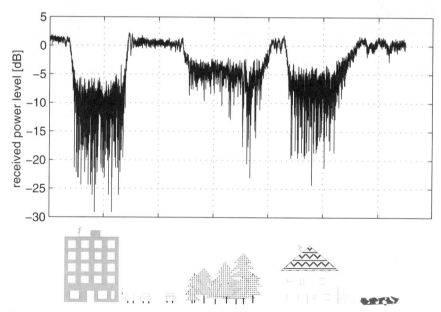

Fig. 3.12. Measured signal power from mobile reception of satellite beacon. Satellite elevation 21°, measurement frequency 1.54 GHz

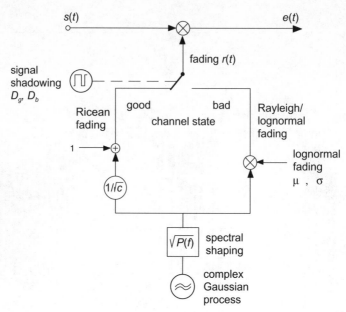

Fig. 3.13. Narrowband model of the land mobile satellite channel [LCD+91]

received with independently time-varying amplitudes and phases; according to the central limit theorem, these components add up to a complex Gaussian process. When no shadowing is present, this multipath signal is superimposed on the direct satellite signal, with the total received signal forming a Ricean process. The momentary received power S obeys a Ricean probability density:

$$p_{\text{Rice}}(S) = c\,e^{-c(S+1)}\,I_0(2c\sqrt{S}).\tag{3.34}$$

Here, c is the direct-to-multipath signal power ratio (Rice factor) and I_0 is the modified Bessel function of order zero. The power of the unfaded satellite link is normalized to unity. If no shadowing is present, the mean received total power is $\text{E}\{S|\text{no shadowing}\} = 1 + 1/c$.

When shadowing is present, it is assumed that no direct signal path exists and that the multipath fading has a Rayleigh characteristic with short-term mean received power S_0. The probability density function of the received power conditioned on mean power S_0 is

$$p_{\text{Rayl.}}(S|S_0) = \frac{1}{S_0}\exp(-S/S_0).\tag{3.35}$$

The slow shadowing process results in a time-varying short-term mean received power S_0 for which a lognormal distribution is assumed:

$$p_{\text{LN}}(S_0) = \frac{10}{\sqrt{2\pi}\,\sigma_{\text{dB}}\ln 10}\cdot\frac{1}{S_0}\exp\left[-\frac{(10\log S_0 - \mu_{\text{dB}})^2}{2\sigma_{\text{dB}}^2}\right].\tag{3.36}$$

Here μ_{dB} is the mean power-level decrease (in dB) and σ_{dB}^2 (often called *dB-spread*, in dB^2) is the variance of the power level due to shadowing. From Eqs. (3.35) and (3.36), the received power is described by a Rayleigh/lognormal distribution which is also often used to model the terrestrial land mobile channel [HM77].

In order to derive the resulting probability density function of the received signal power, the densities (3.34) – (3.36) must be properly combined. To this end, the time-share of shadowing, A, is defined, and the resulting probability density function becomes

$$p(S) = (1 - A) \cdot p_{Rice}(S) + A \cdot \int_0^\infty p_{Rayl.}(S|S_0) p_{LN}(S_0) \, dS_0. \qquad (3.37)$$

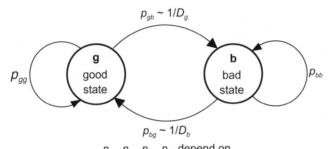

Fig. 3.14. Two-state Markov model for the good and bad channel states of the land mobile satellite channel

In the channel model, Fig. 3.13, the switching process between Ricean and Rayleigh/lognormal fading is modeled by a two-state Markov chain and is characterized by the mean durations D_g and D_b for which the channel remains in the good or bad state, respectively, cf. Fig. 3.14. For personal communications of pedestrians the durations D_g and D_b may be given in meters; for mobile applications with a certain speed they can be translated into bit durations. The time-share of shadowing, A, representing the percentage of time when the channel is in the bad state, is given by

$$A = \frac{D_b}{D_g + D_b}. \qquad (3.38)$$

The parameters of the channel model were fitted to measurement results for different environments, antenna types, and elevation angles. Figure 3.15 especially shows the elevation dependence of the mean duration of the good and bad channel states. Values for the channel parameters are published in [LCD$^+$91] and [BWL96] for different environments, satellite elevations, and

(a)

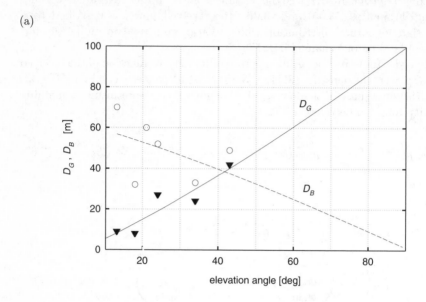

(b)

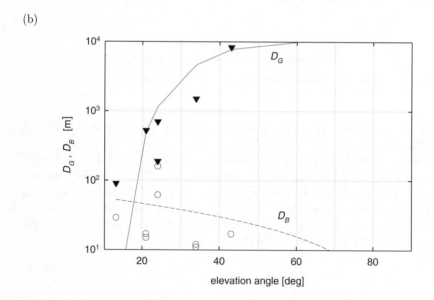

Fig. 3.15. Dependence of mean durations of good and bad channel states, D_g and D_b, on the elevation angle [LJWB96]: (a) urban area, (b) highway

terminal types. Typical parameter values are compiled in Table 3.3. Moreover, detailed parameter tables for several environments, elevation angles, and frequency bands are given in App. B. Compared to mobile terminals, handheld terminals tend to exhibit a somewhat larger time-share of shadowing and a lower Rice factor. In general, the results of the propagation measurements indicate that the channel behavior is dominated by the effect of signal shadowing. With increasing satellite elevation the Rice factor in LOS situations tends to increase and the time-share of shadowing decreases. The signal attenuation in shadowed situations is usually more severe at higher elevation angles.

Table 3.3. Typical parameter values for the narrowband characterization of the personal and land mobile satellite channel, for satellite elevation angles $10° \ldots 30° \ldots 50°$ [Lut98]

Environment	City	Highway
Mean duration of good states, D_g	7 ... 25 ... 50 m	$10^2 \ldots 10^3 \ldots 10^4$ m
Rice factor for LOS situations, c	5 ... 9 ... 10 dB	10 ... 14 ... 18 dB
Mean duration of bad states, D_b	70 ... 50 ... 35 m	50 ... 30 ... 20 m
Time-share of shadowing, A	0.9 ... 0.7 ... 0.4	0.3 ... 0.2 ... 0
Mean attenuation at shadowing, $-\mu_{dB}$	12 ... 12 ... 15 dB	9 ... 10 ... 14 dB
Standard deviation of power level for shadowed situations, σ_{dB}	4 dB	4 dB

In [Loo85], a different approach suitable for foliage attenuation is described, assuming a lognormally distributed direct signal plus a Rayleigh distributed multipath signal with constant power. Vatalaro and Corazza [CV94, CJLV95] have similarly modeled the elevation dependence of the mobile satellite channel in a single environment by a lognormal distribution for the direct signal and the multipath.

3.3.2 Satellite Channels at Higher Frequencies

A measurement campaign at EHF band was performed by DLR to characterize the channel statistics for mobile multimedia services at millimeter frequencies [JL97]. For this purpose a measurement van was equipped with a steered antenna platform. An RHCP horn antenna with a 3-dB half-beamwidth ($\theta_{3dB} = 25°$) was used. The antenna pointing employed GPS data and high-precision inertial sensors. The antenna deviation from boresight was proven to be less than $0.7°$. The measurement setup and a detailed description of the environments and scenarios are given in [JL96].

A typical narrowband power plot in an urban environment is shown in Fig. 3.16. The channel is characterized by long states with good and shadowed power levels. Compared to L band measurements, the EHF channel

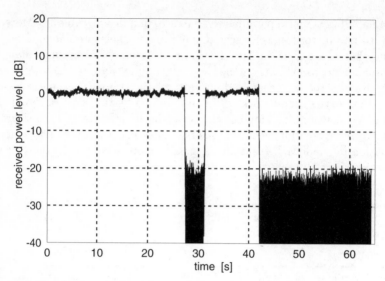

Fig. 3.16. Received power level for narrowband measurements at EHF band, moving van, steered high-gain antenna; elevation: 35°; carrier frequency: 40.1 GHz, [JH98] © 1998 IEEE.

exhibits stronger shadowing attenuation but less multipath fading in LOS states. The parameters of the two-state model are listed in App. B.2 for several elevation angles. Compared to the L band parameters higher values for the Rice factor arise, e.g. $c = 21.5$ dB for rural roads or $c = 15.7$ dB for the urban environment. The stronger shadowing attenuation in the higher-frequency bands is reflected by the lower mean power μ_{dB} (−22 to −27 dB) of the lognormal distribution. Parameters at K band frequencies (20 GHz) with directive antennas are reported in [RSHP96].

A measurement campaign to investigate the wideband effects at millimeter wavelength was performed in 1996 [JL97]. Although the satellite systems use directed terminal antennas a hemispherical antenna was used in order to receive echoes from all directions. Only a few significant echoes (approximately. two to three) were observed with strong echo attenuation (−22 to −27 dB). The EHF channel can therefore be considered frequency non-selective for a bandwidth up to 30 MHz.

3.3.3 Wideband Model for the Land Mobile Satellite Channel

In the following, we summarize the wideband characteristics of the land mobile satellite channel that have been obtained by several measurement studies of DLR. These were performed at L band at a carrier frequency of 1.82 GHz. The test signal was an m-sequence of length 511 with a bandwidth of 30 MHz. The receiver was designed using the pulse compression scheme through time-delay correlation after Parsons [Par91]. This yielded a time resolution

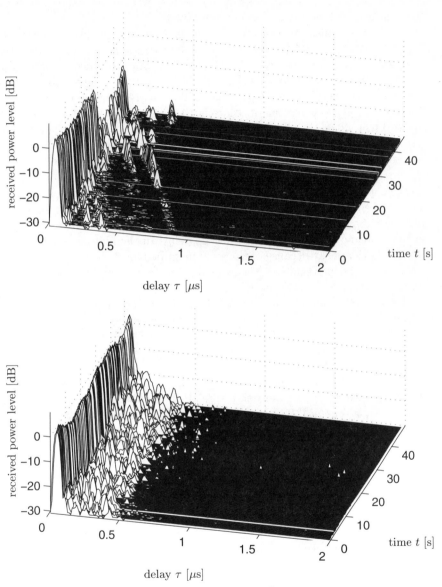

Fig. 3.17. Measured power delay profiles $|h(t, \tau)|^2$ for car-roof-mounted receive antenna in move [JSBL95a]; upper figure: urban environment, $15°$ elevation; lower figure: highway environment, $65°$ elevation

of 30 ns for the echoes corresponding to a location resolution of 10 m. The measurement rate was 15.6 impulse profiles per second with a maximum delay of 17 μs. The dynamic range of the receiver was 40 dB. More details on the measurement setup can be found in [JL94].

Typical examples of measured power delay profiles in dB are shown in Fig. 3.17. The power delay profiles correspond to the power $|h(t, \tau)|^2$ of the complex channel impulse response $h(t, \tau)$. The time axis refers to the measurement time t, the delay axis shows the echo delay[3] τ. Signal echoes occur typically with short delays smaller than 600 ns (corresponding to detours below 180 m). The attenuation of echoes is 10–30 dB with respect to the power of the direct signal. In addition to near echoes, usually some far echoes occur with delays ranging to a few μs. This behavior is typical for all environments and scenarios. The environments mainly differ in the extent of shadowing of the direct signal at $\tau = 0$ and in the number of echoes. In the urban environment (upper figure) the LOS state changes after 25 s to the shadowed state. Here, the signal is attenuated by approximately 15–20 dB. Additionally, the influence of a single reflector being approached by the measurement vehicle is visible, causing a steady shift of the echo delay. In the highway environment many echoes appear because of the good metallic reflectors of the car bodies. An exponential power decay versus delay time can be noticed here. Similar results are reported by Vogel [KV92].

For a fixed measurement time t_0, the delay spread Δ is defined as

$$\Delta = \sqrt{\frac{\int_{-\infty}^{\infty} (\tau - m_\tau)^2 \cdot |h(t_0, \tau)|^2 \, d\tau}{\int_{-\infty}^{\infty} |h(t_0, \tau)|^2 \, d\tau}} \tag{3.39}$$

where $m_\tau = \int_{-\infty}^{\infty} \tau \cdot |h(t_0, \tau)|^2 \, d\tau / \int_{-\infty}^{\infty} |h(t_0, \tau)|^2 \, d\tau$ denotes the mean echo delay. Typical values of the delay spread Δ for handheld terminals range from 15 to 500 ns [BJL96a] if, according to CCIR, echoes below –25 dB are neglected.

Assuming an exponential power delay profile and determining the coherence bandwidth as the bandwidth in which two fading signal envelopes have a correlation greater than 0.5, the coherence bandwidth of the channel (with regard to the signal amplitude) can be derived as [Lee93]

$$B_{\text{coh}} = \frac{1}{2\pi\Delta}. \tag{3.40}$$

The time-variant channel transfer function can be determined from the measured impulse responses $h(t, \tau)$ by Fourier transformation. Figure 3.18 shows typical examples for several environments. The transfer function remains nearly constant versus frequency in the open environment for LOS and in environments with flat background since echoes arise rarely. This corresponds to an ideal channel transfer function. If echoes appear frequency-selective behavior is exhibited. Echoes with long delays cause fluctuations

---

[3] A delay $\tau = 0$ corresponds to the propagation delay of the direct path.

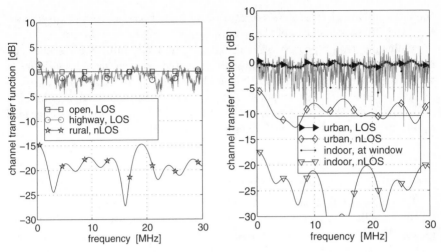

Fig. 3.18. Frequency selective fading in several environments

Table 3.4. Coherence bandwidth B_{coh} for several environments

Environment	Coherence bandwidth in MHz
Highway, LOS	11.6
Rural, nLOS	8.7
Urban, LOS	11.3
Urban, nLOS	7.4
Indoor, LOS	0.23
Indoor, nLOS	4.7

of the transfer function in short frequency intervals[4] whereas echoes with short delays cause wide frequency variations. The amplitude of the variation corresponds to the echo power. The biggest impact reflects the shadowing situation. In LOS states the transfer function varies little around the 0-dB line since the weak echoes hardly have an impact on the strong direct signal. However, in nLOS states, the channel attenuation is in the range from –10 to –30 dB and the transfer function shows a strong variation versus frequency since the echoes now have adequate power compared to the direct signal. The coherence bandwidth B_{coh} of the channel can also be derived from the correlation of the transfer function (50%-bandwidth of the correlation). Table 3.4 shows values of B_{coh} for the environments given in Fig. 3.18. The values for B_{coh} in shadowed conditions range from 4.7 to 8.7 MHz. In LOS conditions, the coherence bandwidth is far higher with values over 10 MHz except for

---

[4] In the indoor environment at a window an echo with long delay occured from outside, causing narrowband fluctuations of the transfer function.

the indoor channel. The values of B_{coh} seem higher as one would expect from Fig. 3.18. This is because of the averaging effect of the correlation.

Following the approach of Bello [Bel63] and assuming that echoes arriving from different distances are uncorrelated, the channel is wide-sense stationary with uncorrelated scattering (WSSUS). Accordingly, the complex-valued channel impulse response $h(t, \tau)$ can be expressed as a sum of a direct path $h_1(t) \cdot \delta(\tau - \tau_1(t))$ with delay $\tau_1(t)$ and a number of echoes $h_k(t) \cdot \delta(\tau - \tau_k(t)), k \geq 2$, with delays $\tau_k(t) = \tau_1(t) + \Delta_k(t)$:

$$h(t, \tau) = \sum_{k \geq 1} h_k(t) \cdot \delta(\tau - \tau_k(t)). \tag{3.41}$$

In order to adapt the channel model to the characteristics of the land mobile satellite channel the impulse response $h(t, \tau)$ is divided into three regions: the *direct path*, *near echoes*, and *far echoes*, cf. Fig. 3.19. For each region the distribution and channel coefficients will be determined. This procedure seems appropriate due to the properties of the impulse response as shown in Fig. 3.17.

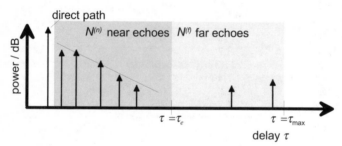

Fig. 3.19. Regions of the wideband impulse response

With this approach, the wideband satellite channel can be modeled as a tapped delay-line as shown in Fig. 3.20 [JSBL95b, JBH96]. The power delay profile is related to the impulse response by

$$p(t, \tau) = |h(t, \tau)|^2. \tag{3.42}$$

The signals in the three regions in Figs. 3.19 and 3.20 have the following characteristics:

1. **Direct path** h_1. The amplitude distribution depends on the shadowing state. For LOS, a Rice distribution using Eq. (3.34) describes the probability density function (PDF) for $|h_1(t)|$. For non-line-of-sight (nLOS), the PDF of $|h_1(t)|$ follows a Rayleigh distribution (Eq. (3.35)) with lognormally distributed mean power (cf. Eq. (3.36)).

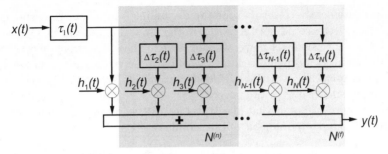

Fig. 3.20. Wideband model of the land mobile satellite channel

2. The region of **near echoes**. The number $N^{(n)}$ of near echoes in the vicinity of the receiver follows a Poisson distribution:

$$p_{\text{Poisson}}(N) = \frac{\lambda^N}{N!} e^{-\lambda} \qquad (3.43)$$

with the parameter λ being the mean number of echoes.
The delay times lie in an interval $0 < \Delta\tau_i^{(n)} \leq \tau_e$, $\tau_e \approx 400$–600 ns, corresponding to a detour of less than 120–180 m. The distribution of the delays $\Delta\tau_i^{(n)}$ of the near echoes follows an exponential distribution

$$p_{\exp}(\Delta\tau_i^{(n)}) = \frac{1}{b} e^{-\frac{\Delta\tau_i^{(n)}}{b}} . \qquad (3.44)$$

Most of the echoes in the mobile satellite channel appear in the near region. The mean power (or the delay power spectral density) $P_h(\tau)$ of the near echoes decreases exponentially

$$P_h(\tau) = P_{h,0} \cdot e^{-\delta\tau} \qquad (3.45)$$

which can be expressed in logarithmic units:

$$\frac{P_h(\tau)}{\text{dB}} = \frac{P_{h,0}}{\text{dB}} - \frac{d}{\text{dB}}\tau , \qquad (3.46)$$

with $d = 10\log\delta/(10\log e)$. For a given echo power $P_h(\tau)$ and a fixed delay τ, the echo amplitude $|h_i|^{(n)}$ of the near echoes varies around this mean value with Rayleigh-distributed PDF and variance $2\sigma^2 = P_h(\tau)$, cf. Eq. (3.35).
The reader should note that the statistics of the near echoes depend on the near environment of the mobile receiver.

3. The region of **far echoes**. The number of far echoes $N^{(f)} = N - N^{(n)} - 1$ follows again a Poisson distribution according to Eq. (3.43). The echoes appear uniformly distributed in an interval $\tau_e < \Delta\tau_i^{(f)} \leq \tau_{\max}$ though only few echoes can be observed. Because of the limited dynamic range of the

measurement receiver and the rapidly falling echo power the exponential power decay could not be proven in the far region beyond τ_e. Only single echoes with stronger power were observed. The echoes were reflected from pronounced morphological obstacles (such as water towers, power lines, skyscrapers, etc.). A Rayleigh distribution (cf. Eq. (3.35)) is used for the amplitudes $|h_i|^{(f)}$ of the far echoes.

It should be noted that the statistics of the far echoes depend on the background environment of the mobile receiver.

For the Doppler power spectral densities of the echoes a Ricean Doppler power spectral density can be taken for the direct path under LOS conditions, otherwise the Jakes power spectral density can be used. The initial phase of the echoes can be assumed to be uncorrelated and uniformly distributed in $[0, 2\pi)$ because of the assumption of uncorrelated scattering.

From the measured wideband data the parameters of the wideband model at L band were determined [JBH96] using the Gauss–Newton algorithm for minimization of the least-square root error. The parameters of the model are listed in App. B.3 for several environments. They have been submitted for standardization to the European Telecommunications Standards Institute ETSI [BJL96a]. The model and its parameters were also verified and confirmed by Parks et al. [PEBB96] and compared to results from their measurements.

ITU Wideband Model. For the sake of simplicity, three parameter sets with fixed number of taps (typically 2-4) and fixed delays have also been derived from the measured data. The chosen environments *urban*, *rural*, and *suburban* correspond to the 90%, 50%, and 10% percentiles of the delay spread. The model has been submitted to the standardization bodies of ETSI and ITU and was approved as a draft standard [BJL96b, BJL96c]. The parameter sets are listed in App. B.3.

3.4 Link Availability and Satellite Diversity

In Sect. 3.3 it was shown that the mobile satellite channel suffers signal shadowing and multipath fading. Such channels require high link margin or power control to compensate for signal fades. Satellite diversity is a concept to avoid shadowing and thus to improve the service quality with moderate link margins.

3.4.1 Concept of Satellite Diversity

Diversity uses the geometric correlation of shadowing obstacles to reduce the probability of shadowing. Systems such as Globalstar or ICO essentially provide double coverage of the earth. This feature enables the application of

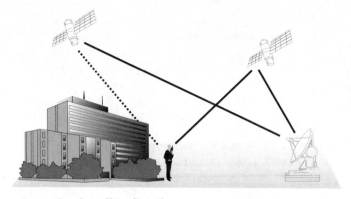

Fig. 3.21. Principle of satellite diversity

satellite diversity, i.e. the simultaneous communication with a user via two or more satellites. If one of these satellites is shadowed, there is some chance of another satellite being still in view to the user and maintaining the service, Fig. 3.21. In this way, satellite diversity can substantially improve service availability (the percentage of time when the service is available) and reduce the required link power margin.

The required link margins for PCN systems have been determined by measurement for many environments with different shadowing [Jah94]. To permit an investigation of diversity, two transmitters have been used with different elevation and azimuth angles. The required values for the link margins with and without diversity are shown in Fig. 3.22 versus the time-share of shadowing A for a link availability of 95%. Selection combining was assumed for diversity combination. The measured points correspond to low and medium elevation angles $\varepsilon = 15° - 45°$. In all environments a substantial improvement can be achieved by diversity. Even in environments with very strong shadowing (such as forest or urban areas) realistic link margins around 10 dB are sufficient for a good quality of service. For comparison, Table 3.5 lists the link margin available in some LEO and MEO satellite systems for personal communications.

Table 3.5. Link margins of S-PCN LEO/MEO systems

Iridium:	16 dB
Globalstar:	3 dB + power control + satellite diversity
ICO:	10 dB + satellite diversity.

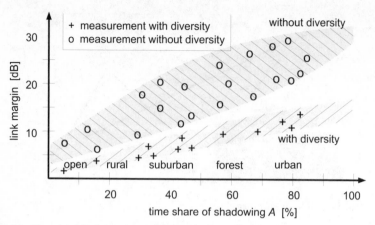

Fig. 3.22. Required link margins at L band for different shadowing with handset reception; required link availability 95%

3.4.2 Correlation of Channels

Of course, a gain in service availability can only be achieved if the considered satellite channels behave *differently*. Therefore, any dependence between the channels influences the benefit of satellite diversity. In [Lut96] a concept for modeling two statistically dependent satellite channels was developed. The two-state Markov channel model characterizing the shadowing process (i.e. the switching between good and bad channel states, cf. Fig. 3.14) was extended to two independent channels, see Fig. 3.23. In order to take into account the correlation between the shadowing processes of the channels, the transition probabilities of the combined Markov model were modified in such a way that the initial channel models are maintained but a certain correlation coefficient ρ is produced.

For the definition of the correlation coefficient we consider the amplitude $a_i(t)$ of channel $i = 1, 2$ as a stochastic process which is 0 for the shadowed channel state and 1 for the LOS condition:

$$a_i(t) = \begin{cases} 0 & \text{bad channel state} \\ 1 & \text{good channel state.} \end{cases} \tag{3.47}$$

The mean value and variance of the channel amplitude are

$$\mathrm{E}\{a_i(t)\} = \overline{a_i} = A_i \tag{3.48}$$

$$\mathrm{E}\{(a_i(t) - \overline{a_i})^2\} = \sigma_i^2 = A_i(1 - A_i) \quad , \tag{3.49}$$

with A_i denoting the time-share of shadowing of channel i. Using (3.48), the correlation coefficient can be defined as the time average

$$\rho = \frac{\mathrm{E}\{(a_1(t) - \overline{a_1})(a_2(t) - \overline{a_2})\}}{\sigma_1 \sigma_2} \quad . \tag{3.50}$$

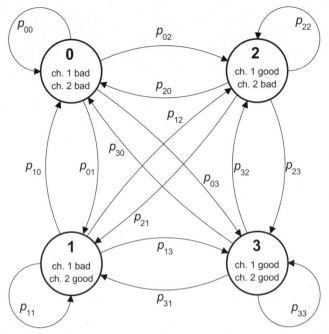

Fig. 3.23. Four-state Markov model for two land mobile satellite channels [Lut96]
© John Wiley & Sons Limited, reproduced with permission

According to (3.50), the correlation coefficient can be evaluated from pairs of (time-synchronized) channel measurements with regard to the same mobile terminal. The correlation depends on the user environment, as well as the elevation and azimuth angles of the channels. As shown in [REE92], the dependence on the azimuth angles may approximately be described as a function of their difference $\Delta\varphi$. With this simplifying limitation, ρ can also be estimated from circular measurements at constant elevation angles or from "fish-eye" photos [AV95] for a single fixed user position, according to

$$\rho = \frac{\mathrm{E}\{(a_1(\varphi) - \overline{a_1})(a_2(\varphi + \Delta\varphi) - \overline{a_2})\}}{\sigma_1\sigma_2} \quad . \tag{3.51}$$

Here, ρ depends on the user environment, the chosen pair of elevation angles, and the azimuth separation $\Delta\varphi$.

Two examples for the azimuth correlation of shadowing in an urban area are given in Fig. 3.24 for two channels at 35° and 45° elevation. The correlation decreases with increasing azimuth separation of the satellites, and is smaller if the satellites have different elevation angles. Furthermore, for some values of $\Delta\varphi$ the channels are counter-correlated.

The correlation coefficient ρ can be used to describe the state probabilities p_0, p_1, p_2, p_3 and transition probabilities $p_{ij}, i, j = 0, \ldots, 3$ for the four-state model in Fig. 3.23. In [Lut96] it is shown for correlated channels that

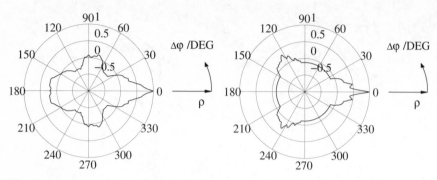

Fig. 3.24. Azimuth correlation of shadowing in urban environment at $35°$ (left) and $45°$ elevation (right) [BWL96]

$$
\begin{aligned}
p_0 &= \rho\sigma_1\sigma_2 + A_1 A_2 \\
p_1 &= A_1 - p_0 \\
p_2 &= A_2 - p_0 \\
p_3 &= 1 + p_0 - A_1 - A_2 \ .
\end{aligned} \tag{3.52}
$$

The transition probabilities can be derived from the probabilities $\mathbf{P} = (p_{ij,\text{uncorr.}})$ of two uncorrelated channels

$$
\mathbf{P} = \begin{pmatrix}
(1-g_1)(1-g_2) & (1-g_1)g_2 & g_1(1-g_2) & g_1 g_2 \\
(1-g_1)b_2 & (1-g_1)(1-b_2) & g_1 b_2 & g_1(1-b_2) \\
b_1(1-g_2) & b_1 g_2 & (1-b_1)(1-g_2) & (1-b_1)g_2 \\
b_1 b_2 & b_1(1-b_2) & (1-b_1)b_2 & (1-b_1)(1-b_2)
\end{pmatrix} \tag{3.53}
$$

(where $g_i = 1/D_{b,i}, b_i = 1/D_{g,i}$ denote the transition probabilities of the two-state model of channel i) by applying some adjustments that reflect the correlation properties

$$
(p_{ij,\text{corr.}}) = \mathbf{P} + c \cdot \begin{pmatrix}
x & -x & -x & x \\
y & -y & -y & y \\
v & -v & -v & v \\
w & -w & -w & w
\end{pmatrix} , \quad 0 \le c \le 1 , \tag{3.54}
$$

where

$$
\begin{aligned}
c &= \frac{\rho\,K Q}{R_0 - \rho\,S_0 Q} \quad \text{and} \\
K &= 1 - (1 - g_1 - b_1)(1 - g_2 - b_2) \\
Q &= \sqrt{g_1 b_1 g_2 b_2} \\
R_0 &= g_1 g_2 w + g_1 b_2 v + b_1 g_2 y + b_1 b_2 x \\
S_0 &= y - x + v - w \ .
\end{aligned} \tag{3.55}
$$

The introduced variables x, y, v, and w depend on the correlation

$$x, y, v, z = \begin{cases} x_{max}, y_{max}, v_{max}, w_{max} & \rho \geq 0 \\ x_{min}, y_{min}, v_{min}, w_{min} & \rho < 0 \end{cases} \qquad (3.56)$$

where the useful range of the parameters is limited by the requirement $0 \leq p_{ij} \leq 1$:

$$
\begin{aligned}
x_{min} &= -g_1 g_2 & x_{max} &= \min\{g_1; g_2\} - g_1 g_2 \\
y_{min} &= g_1 b_2 - \min\{g_1; b_2\} & y_{max} &= g_1 b_2 \\
v_{min} &= b_1 g_2 - \min\{b_1; g_2\} & v_{max} &= b_1 g_2 \\
w_{min} &= -b_1 b_2 & w_{max} &= \min\{b_1; b_2\} - b_1 b_2 \; .
\end{aligned} \qquad (3.57)
$$

3.4.3 Link Availability and Satellite Diversity Service Area

To exploit the potential of satellite diversity, the mobile subscriber must be served by at least two satellites. Moreover, the satellites must also be connected to the same gateway. Extensive simulations have been performed in [BWL96, BW97] for a gateway at (100°W,40°N) and 16 user positions equally distributed in a circle around the gateway, where it was assumed that the gateway always selects the two satellites with the highest elevation angles. Figure 3.25 shows the service availability for the Globalstar constellation. A link margin of 7 dB was assumed, and the elevation-dependent channel parameters were taken from measurements. When the distance between user and gateway increases, the probability of seeing two satellites connected to the same gateway diminishes and the elevation angles of the satellites visible to both the gateway and the subscriber are lower. Thus, the link availability is reduced.

Satellite diversity results in a significant improvement of service availability. The simulations show that in the city environment a user has two visible satellites for 80% of the time. Comparing the highway environment with the city environment it can be concluded that S-PCNs can provide high service quality in highway or rural areas. In urban areas satellite diversity can substantially improve the service availability.

Exaggerating the two-state channel behavior of the mobile user link to a pure on/off channel, the concept of satellite diversity would be equivalent to a system initiating a seamless satellite handover each time the currently used satellite becomes shadowed. A requested handover is only successful, however, if an alternative satellite is available to the user. These handovers would occur in addition to the handovers required due to time-limited satellite visibility, cf. Sect. 7.7.

3.5 System Implications

Signal Shadowing. The most detrimental propagation impairment is signal shadowing. Light shadowing, e.g. caused by a single tree, or head shadowing

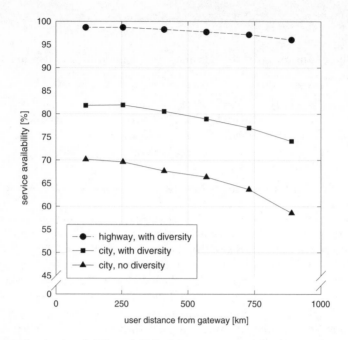

Fig. 3.25. Service availability of Globalstar versus user distance from gateway

may be compensated by power control. In the presence of heavy shadowing (blockage) the link is interrupted.

For real-time services, such as speech, the influence of signal shadowing can be reduced by satellite diversity, see Sect. 3.4. The better link availability of satellite diversity systems has to be paid for by increased (e.g. doubled) transmit power and spectrum usage.

For data services and signaling, the application of ARQ (automatic repeat request) and related transmission protocols is suitable, also without satellite diversity, see Sect. 4.4.

Frequency Non-Selective Multipath Fading. Multipath fading appears to be frequency non-selective (flat) if the coherence bandwidth of the channel is larger than the signal bandwidth. Here, any kind of diversity can improve the signal transmission. The most common type of diversity is receive antenna diversity, which improves the link quality without increasing the transmitted power or bandwidth. For personal communication terminals, antenna diversity is not favorable, however. Of course, also (microscopic or macroscopic) transmit antenna diversity decreases the influence of fading.

Frequency diversity reduces frequency non-selective fading by periodically changing the transmit frequency with steps larger than the coherence bandwidth of the channel (frequency hopping). The hopping sequence may be slow or fast compared to a bit duration. Packet retransmission or channel coding can cope with the transmission errors occurring during a hop within a fade.

A similar effect is achieved by time diversity. Interleaving the transmitted bits and de-interleaving the received bits time-spreads the error bursts caused by fades. Channel coding can cope much better with such randomized errors than with burst errors.

For continuous transmission with a given net rate, channel coding increases the bandwidth (except if trellis-coded modulation is used) but achieves a net reduction in transmit power, compared to uncoded transmission, see Sect. 4.3. For data transmission with a given transmission rate, the redundancy bits of channel coding lower the information throughput; however, the reliability of the transmitted data is increased.

Frequency-Selective Multipath Fading. If the signal bandwidth exceeds the coherence bandwidth of the channel, the duration of the impulse response is noticeable compared to the bit duration. The transmitted pulses are spread in time, and intersymbol interference (ISI) arises. This filtering effect of the channel can be compensated for by an inverse filter at the receiver, known as an equalizer. Usually, the channel is unknown and time-varying. This means that the equalizer must be adaptive. With an equalizer, the link quality is improved without increasing the transmitted power or bandwidth. Since TDMA systems exhibit a high burst rate compared to the user bit rate (see Sect. 5.6), such systems often require adaptive equalization, adding to the complexity of the receiver.

In CDMA systems, the chip rate is usually larger than the coherence bandwidth of the channel. However, due to the pseudo-noise feature of signature sequences there is very low correlation between successive chips. Moreover, the interchip interference is averaged out by the correlator within one bit duration. Therefore, no equalization is necessary.

If the delay of a multipath echo is larger than a chip duration (but shorter than a bit duration), and if another correlator is tuned to the relevant delay, then the echo signal can be separately detected. Finally, it can be combined with the time-delayed version of the original signal. A Rake receiver collects the time-shifted versions of the original signal by providing a separate correlation receiver for each of the multipath signals. In this way, the signal energy contained in the echo signals is exploited for the bit decision, in addition to the energy of the directly received signal.

In the mobile satellite channel, only relatively weak echoes are to be expected, cf. Fig. 3.17. Therefore, not much gain can be achieved by a Rake receiver. A Rake receiver can, however, be used in CDMA systems to implement satellite diversity and seamless satellite handover. In these cases, the signal is simultaneously transmitted over more than one satellite. Thus, artificial multipath is produced, which can be efficiently exploited by the Rake receiver.

None the less, CDMA with Rake reception is not the only way to implement satellite diversity or seamless handover. Diversity can be achieved with TDMA systems as well, if the signal is received from different satellites at different frequencies or within different time slots. The combining or selection can be done at the bit or packet level.

4. Signal Transmission

This chapter is not intended to provide a thorough treatment of signal transmission techniques, rather we review the basic concepts of speech coding, modulation, channel coding, and packet transmission schemes as far as they are relevant for the following chapters. For more detailed information the reader is referred to the cited literature.

4.1 Speech Coding

The basic goal of speech coding is the representation of an (analog) speech signal by a digital signal with a bit rate as low as possible. This means that during the process of digitizing the speech signal as much redundancy as possible should be removed. However, some constraints must be taken into account:

- The speech quality should be maintained to a maximum extent.
- The complexity of the speech coding process should be limited.
- The time duration of the coding/decoding process represents additional delay and should be restricted, e.g. to <50 ms.
- For mobile communications, the coded speech should be robust with regard to acoustic background noise and error-prone transmission.

The fundamental procedure of speech coding consists of the following steps:

- Band-limiting of the analog speech signal to 300–3400 Hz.
- Sampling of the analog speech signal, typically with 8000 samples/s.
- Analog-to-digital conversion of the samples. Typically, each speech sample is quantized with 8 bits, such that a raw bit rate of 64 kb/s results for the digital speech signal.
- Further processing and coding of the digital samples. In this step, the number of bits used for the representation of the speech signal is reduced and a lower bit rate for the digital speech signal is obtained.

4.1.1 Quality of Coded Speech

In order to evaluate the quality of a coded speech signal, the speech may be assessed by a number of listeners, according to an agreed scale of speech quality and speech degradation, the mean opinion score (MOS), see Table 4.1. Table 4.2 lists different requirements for the quality of telephone speech, depending on the kind of application.

Table 4.1. Mean opinion score (MOS) for speech quality

MOS	Quality	Degradation
5	excellent	imperceptible
4	good	just perceptible
3	fair	perceptible, slightly annoying
2	poor	annoying but not objectionable
1	bad	very annoying and objectionable

Table 4.2. Requirements for speech quality

Speech quality	MOS	Application
network quality, toll quality	4	transmission in fixed networks, teleconferencing
communications quality	3.5	e.g. mobile radio communications
synthetic quality	3	

Typical values for the bit rates of digital speech signals are:

– speech with network quality: 16 kb/s, 32 kb/s, 64 kb/s (these bit rates are international standards)
– speech in mobile radio communications: 2.4 ... 13 kb/s.

4.1.2 Overview of Speech Coding Schemes

Figure 4.1 gives an overview of various types of speech coders. Whereas waveform coders try to reproduce the speech signal in the time domain or its spectrum in the frequency domain, voice coders try to model the physiological process of voice production.

PCM (pulse code modulation) is a standard speech coding scheme for fixed networks. It is based on a logarithmic quantization of speech samples with 8 bits per sample. With a sampling rate of 8000 samples/s a speech bit rate of 64 kb/s results. PCM exhibits a very high voice quality with an MOS of 4.3.

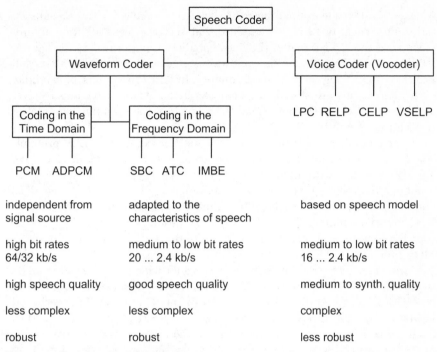

Fig. 4.1. Speech coders for telephony

Important Coding Schemes for Mobile Communications. ADPCM (adaptive differential PCM) uses a differential PCM quantization, which is adapted to the dynamics of the speech signal. A typical bit rate for ADPCM is 32 kb/s (without channel coding), and an MOS of 4.1 can be achieved. ADPCM with its relatively high bit rate is applied in the cordless DECT telephone system, e.g. where bandwidth efficiency is not of primary importance.

LPC (linear predictive coding) is the basic vocoder scheme making use of a speech generation model. The vocal tract is modeled by an adaptive filter (the LPC synthesis filter) fed by an excitation signal. In the LPC coder the filter coefficients along with other parameters such as amplitude, pitch, and voiced/unvoiced state of the speech, are determined for successive frames of the input analog speech signal. These parameters are transmitted to the receiver, which uses a synthesis filter to reproduce the initial voice signal. The synthesis filter is tuned by the received filter coefficients and is excited by periodic pulses at the pitch frequency or noise, respectively, depending on the voiced/unvoiced state. Also, the voice amplitude is reproduced. This fundamental speech coding scheme can work with a remarkably low bit rate of 2.4 kb/s [Tre82] but can provide only synthetic speech quality with an MOS around 2.5.

The CELP (code excited linear prediction) scheme constitutes a substantial improvement of LPC, in the sense that the receiver is additionally provided with the error signal of the linear prediction performed at the encoder (the residual). This is used as excitation signal of the synthesis filter, substantially improving output speech quality. In CELP a number of candidate excitation signals are stored in codebooks at the transmitter and receiver, and instead of the excitation signal the address of the optimum candidate is transmitted [SA85].

A large number of low-rate speech coders is based on the CELP principle. The 16-kb/s low-delay CELP coder uses a single codebook and provides MOS 4.0. CELP coders with 4.8 kb/s net rate [CTW91] typically use an adaptive codebook representing the long-term periodicity (pitch) of the speech signal and being updated with each coding frame as well as a stochastic codebook matching the remainig signal. The total excitation is a weighted sum of the contributions of both codebooks. The 4.8-kb/s CELP version achieves an MOS around 3.5. The Qualcomm variable-rate version of CELP is used in the American terrestrial cellular network IS-95 with net bit rates from 1.2 to 9.6 kb/s. Also, the Globalstar satellite network (supported by Qualcomm) uses CELP speech coding. Finally, adaptive multirate (AMR) speech coding based on multirate CELP is envisaged for GSM phase 2+ and UMTS. Such a scheme allows to adaptively choose the ratio between source bit rate and error-protecting redundancy according to the channel condition. Thus, a trade-off between speech quality and error robustness is achieved.

VSELP (vector sum excited linear prediction) is an extension of CELP, using more than one stochastic codebooks in addition to an adaptive long term (pitch) predictor. An 8-kb/s VSELP coder using two stochastic codebooks and achieving MOS 3.5 was selected as standard for North American digital cellular telephone systems [GJ90]. Also the 5.6-kb/s GSM half rate codec is based on VSELP and achieves MOS 3.3.

Whereas CELP belongs to the family of voice coders, some present and future satellite networks use IMBE (improved multiband excitation). With this scheme, the frequency band of speech is divided into subbands, and for each subband a decision is made between voiced and unvoiced state of speech. The relevant parameters are the fundamental frequency, a set of voiced/unvoiced decisions, and a set of spectral amplitudes. These are transmitted to the receiver, which excites the according subbands with a sine signal or with noise, respectively. The more sensitive bits are channel-coded and the voiced/unvoiced decisions are smoothed at the receiver. IMBE is robust against channel impairments and presence of background noise and therefore is well suited for mobile communications. IMBE speech coding with 6.4 kb/s achieves MOS 3.4 for error-free transmission as well as for a channel with 1% bit error rate [HL91]. IMBE is used by Inmarsat-M with a net bit rate of 4.8 kb/s and by ICO with 3.6 kb/s. Iridium uses a modified version of IMBE, called advanced multiband excitation, with a net bit rate of 2.4 kb/s.

Wideband Speech and Audio. As an extension to telephone speech coding, wideband speech coding deals with a speech bandwidth of 50–7000 Hz (16 kHz sampling rate) and is used for ISDN teleconferencing and video conferencing. Audio coding is used for the transmission and storage of high-fidelity sounds, e.g. using an ISO/MPEG audio coding standard [Nol97].

4.2 Modulation

In order to transmit the digital (speech) signal via radio, this signal is used to modulate the amplitude, the phase, or the frequency of a radio carrier. During this process, the digital signal may also undergo suitable filtering to define the radio spectrum of the transmitted signal. The main evaluation criteria of modulation schemes are their bandwidth requirement and their performance with regard to transmission errors.

4.2.1 Modulation Schemes for Mobile Satellite Communications

In mobile satellite communications, signal transmission is deteriorated by multipath fading, Doppler shift, phase variations, and nonlinearities of the power amplifier of the transmitter or of the satellite transponder. Therefore, modulation schemes for mobile satellite communications must be robust. But equally important, a high bandwidth efficiency is required. In the following, we concentrate on phase modulation schemes which because of their high bandwidth efficiency are widely used in satellite communications.

Generally, a phase-modulated carrier signal can be represented by

$$
\begin{aligned}
s(t) &= A\cos\left(\omega_c t + \phi(t)\right) \\
&= A\cos\phi(t)\cdot\cos\omega_c t - A\sin\phi(t)\cdot\sin\omega_c t.
\end{aligned}
\tag{4.1}
$$

The baseband signal modulates the phase ϕ of the carrier. $\cos\phi(t) = I$ is denoted as in-phase component (in phase with the cosine-carrier) and $\sin\phi(t) = Q$ is denoted as quadrature component of the modulating signal.

Binary Phase-Shift Keying, BPSK. This is the basic binary modulation scheme. Depending on the value of a bit of the digital signal to be transmitted, the phase of a radio carrier is set to 0 or π, respectively, Fig. 4.3. Each bit of the digital signal produces a transmit symbol with duration T_s, being identical to the bit duration T_b.

For an NRZ baseband data signal, the power spectral density of a BPSK-modulated signal is given by

$$
L(f) = T_s \left(\frac{\sin(\pi f T_s)}{\pi f T_s} \right)^2 .
\tag{4.2}
$$

Here, f represents the deviation of the frequency from the carrier frequency (equivalent baseband representation).

In order to recover the transmitted information at the receiver, the PSK-modulated signal is coherently demodulated. The demodulator comprises an input bandpass filter, a carrier recovery circuit, a multiplier, and a bit timing recovery circuit, Fig. 4.2.

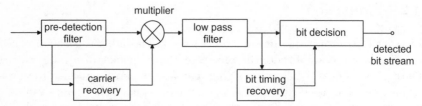

Fig. 4.2. Simplified block diagram of a coherent BPSK demodulator

The ideal received signal can be written as $a(t)\cos\omega_c t$, with $a(t)$ representing the bipolar binary data signal. Multiplying the received signal by a locally generated carrier results in

$$a(t)\cos^2\omega_c t = a(t)\cdot\frac{1}{2}(1+\cos 2\omega_c t). \qquad (4.3)$$

The desired data signal $a(t)$ is recovered by lowpass filtering and deciding this signal according to the recovered bit timing.

The recovered carrier phase is ambiguous with regard to the transmitted bipolar data. This ambiguity can be removed by a differential encoding of the bits, resulting in DPSK (differential phase-shift keying), or DEPSK (differentially encoded phase-shift keying) when demodulated coherently. For transmitting a 1, the carrier phase is maintained; for transmitting a 0, the carrier phase is changed by 180°.

Quadrature Phase-Shift Keying, QPSK. In this modulation scheme two bits at a time are combined, and the radio carrier is phase-modulated according to the four possible patterns of the two bits, Fig. 4.3. All possible state transitions of the carrier phase are allowed.

From the principle of QPSK it is clear that a transmit symbol lasts twice as long as a bit and contains twice the energy E_b corresponding to a bit:

$$\begin{aligned} T_s &= 2T_b \\ E_s &= 2E_b. \end{aligned} \qquad (4.4)$$

The power spectral density of a QPSK signal is

$$L(f) = T_s\left(\frac{\sin(\pi f T_s)}{\pi f T_s}\right)^2 = 2T_b\left(\frac{\sin(2\pi f T_b)}{2\pi f T_b}\right)^2. \qquad (4.5)$$

QPSK is the standard modulation scheme for satellite transmission. For equal bit rates, QPSK requires only half the bandwidth of BPSK.

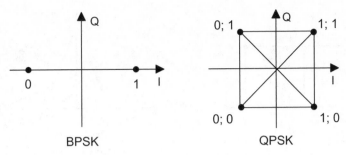

Fig. 4.3. Representation of possible phases and phase transitions of BPSK and QPSK in the complex signal plane

Offset-QPSK, O-QPSK. The O-QPSK scheme is similar to QPSK, how-ever, a $T_s/2$-shift is introduced between the in-phase and quadrature compo-nents of the signal. Accordingly, the carrier phase may change after each bit duration, but only one signal component can change at a time. Therefore, no 180° phase transitions exist during which the signal would cross the origin (the signal having zero amplitude), Fig. 4.4.

This feature of O-QPSK has the effect that, after a band-limiting filter (in the transmitter), the transmit signal exhibits less amplitude variations than a QPSK signal. This leads to less broadening of the signal spectrum due to the nonlinearity of the satellite transponder amplifying the signal. In turn, the satellite transponder can be operated at a higher power level.

O-QPSK has the same power spectral density as QPSK. Unlike QPSK, no fixed bit patterns can be assigned to the phase states of an O-QPSK signal.

$\pi/4$-QPSK. $\pi/4$-QPSK uses two QPSK phase constellations which are phase-shifted by $\pi/4$. The transmit signal points are selected in turn from the two constellations. Therefore, after each symbol of the transmit signal a phase jump of $\pm45°$ or $\pm135°$ appears, Fig. 4.4. As for O-QPSK, no zero crossings appear, and a suitably filtered $\pi/4$-QPSK signal suffers less spectral spreading when passed through a nonlinear amplifier. Moreover, the phase jumps after each transmit symbol ease the symbol synchronization in the receiver.

Continuous-Phase Frequency-Shift Keying, CPFSK. Frequency-shift keying is another simple modulation scheme, where during each transmit symbol interval one of two signal tones is sent, depending on the information. Continuous-phase FSK (CPFSK) can be achieved by frequency-shifting a carrier by a fixed deviation $\pm f_d$, according to the sign of the current bit. The frequency deviation is determined by the modulation index $h = 2f_dT_s$, which can be chosen as a parameter. The transmit symbol duration T_s is equal to the bit duration.

During the time interval of a transmit symbol, the CPFSK signal exhibits a linear increase or decrease of the phase. No phase jumps occur at the symbol transitions. Therefore, CPFSK signals are highly bandwidth efficient.

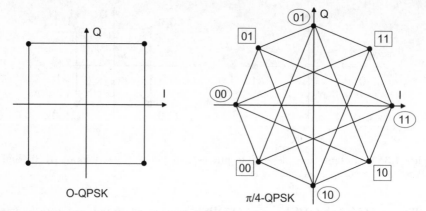

Fig. 4.4. Possible carrier phases and phase transitions for O-QPSK and $\pi/4$-QPSK

The complex modulated signal has constant amplitude. Therefore, CPFSK is robust against nonlinearities of a satellite transponder.

The smallest value of h for which the signals representing 0 and 1 are orthogonal is $h = 1/2$. This choice leads to the following scheme.

Minimum-Shift Keying, MSK. Figure 4.5 shows the trace of an MSK-modulated signal in the complex plane. The points mark the carrier phase at the end of a transmit symbol. An example for the transition of the signal phase with time is illustrated in Fig. 4.6. MSK can also be considered as O-QPSK with the baseband pulses shaped according to a sinusoidal function to effect a linear change of phase.

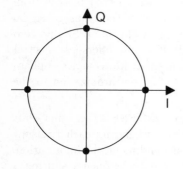

Fig. 4.5. Phase diagram of an MSK modulated signal

Figure 4.7 compares the power spectral densities for different modulation schemes. It can be seen that the main lobe of QPSK and O-QPSK signals occupies only half the bandwidth of the main lobe for BPSK. Also, the main lobe for MSK signals is narrower than the BPSK main lobe. Moreover, the side lobes of an MSK-modulated signal decrease substantially faster than for BPSK.

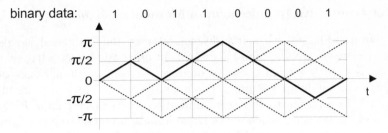

Fig. 4.6. Phase transition of an MSK modulated signal

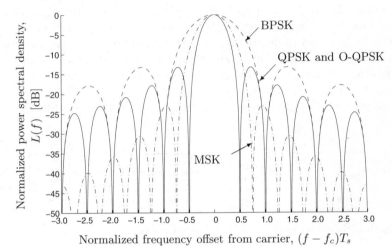

Normalized frequency offset from carrier, $(f - f_c)T_s$

Fig. 4.7. Power spectral densities for BPSK, QPSK, O-QPSK, and MSK

Gaussian-Filtered MSK, GMSK. The MSK spectrum can made even more compact by lowpass filtering the data stream controlling the FM modulator. This technique provides a smoothing of the phase transitions at the symbol boundaries and still preserves the constant-envelope property of the signal. For GMSK, a Gaussian lowpass filter is used, characterized by its normalized 3-dB bandwidth BT_s [MH81]. The choice of this parameter provides a trade-off between spectral confinement and performance loss due to inter-symbol interference. A typical value is $BT_s = 0.3$ which was chosen for the GSM standard.

The following modulation schemes are used in satellite systems for mobile communications:

Ellipso:	BPSK	Iridium:	QPSK
Inmarsat-M:	O-QPSK	Teledesic:	trellis-coded 8-PSK
Euteltracs:	MSK	ICO:	downlink: BPSK, QPSK
			uplink: GMSK $(BT_s = 0.4)$.

4.2.2 Bandwidth Requirement of Modulated Signals

As shown in Fig. 4.7, without additional filtering, the modulated signals discussed above exhibit infinite bandwidth. In order to protect adjacent radio
channels from interference and to improve the bandwidth efficiency of the
system, the modulated signal is bandpass filtered before it is transmitted,
Fig. 4.8. In addition to this band limitation, the transmit filter $P(f)$ also
includes the transformation of the Dirac impulses representing the data bits
into rectangular pulses used for BPSK or QPSK, e.g. Further on, the spectrum of the transmitted signal is modified by the channel frequency response
$G(f)$ and the receive filter $R(f)$. Thus, the shape of the detection pulse $h(t)$
at the receiver side is determined by the series connection of $P(f)$, $G(f)$, and
$R(f)$. In order not to deteriorate the signal detection in the receiver, this total filter process should have no influence on the signal phase at the sampling
times (representing the information). Also, neighboring symbols should have
no mutual influence (intersymbol interference) at the sampling times. A type
of filtering which fulfills these requirements is called Nyquist filtering.

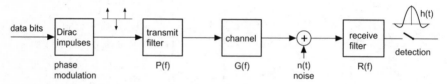

Fig. 4.8. Baseband model of a transmission system

A widely used kind of Nyquist filter is one with cosine roll-off. In the
baseband, the frequency response of this Nyquist filter is given by [Pro89]

$$H(f) \;=\; P(f)G(f)R(f) \tag{4.6}$$

$$= \begin{cases} T_s & \text{for } 0 \le |f| \le \frac{1-\beta}{2T_s} \\[2mm] \frac{T_s}{2}\left\{1 + \cos\left[\frac{\pi T}{\beta}\left(|f| - \frac{1-\beta}{2T_s}\right)\right]\right\} & \text{for } \frac{1-\beta}{2T_s} < |f| < \frac{1+\beta}{2T_s} \\[2mm] 0 & \text{for } |f| \ge \frac{1+\beta}{2T_s} \end{cases}$$

with duration T_s of a transmit symbol and roll-off factor β. As shown in Fig.
4.9, β determines the steepness of the filter slope. Common values for the
roll-off factor range from $\beta = 0.2$ to 0.5. For example, Iridium uses $\beta = 0.26$.

The detection impulse corresponds to the impulse response of the Nyquist
filter with cosine roll-off:

$$h(t) = \frac{\cos(\beta\pi t/T_s)}{1 - (2\beta t/T_s)^2} \cdot \frac{\sin(\pi t/T_s)}{\pi t/T_s} \tag{4.7}$$

with

$$h(nT_s) = \begin{cases} 1 & n = 0 \\ 0 & n \neq 0. \end{cases} \qquad (4.8)$$

Equation (4.8) states that the Nyquist impulse causes no intersymbol interference at the sampling times. From Fig. 4.9 it can be seen that a low value of β means a low bandwidth requirement, but due to the pronounced tails of the Nyquist pulse, the detection is more sensitive to timing errors.

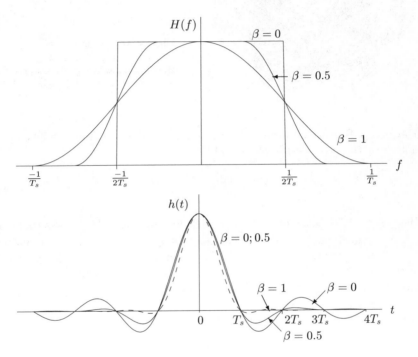

Fig. 4.9. Cosine Nyquist spectrum and corresponding impulse response [Pro89]

The distribution of $H(f)/G(f)$ between the transmit and receive filters has no influence on the shape of $h(t)$. For a frequency-flat channel with $G(f) \equiv 1$ and white noise $n(t)$, the best signal-to-noise ratio is achieved with

$$P(f) = R(f) = \sqrt{H(f)}. \qquad (4.9)$$

Due to the non-rectangular shape of the detection pulse, the representation of the signal in the complex plane no longer looks like Fig. 4.4 for $\pi/4$-QPSK, e.g. An example for square root Nyquist-filtered $\pi/4$-QPSK is shown in Fig. 4.10.

Fig. 4.10. Square root Nyquist-filtered $\pi/4$-QPSK with $\beta = 0.5$. Reprinted with permission from [Gib96]. ©Lewis Publishers, CRC Press

Required Signal Bandwidth. According to Fig. 4.9, the bandwidth required by the Nyquist-filtered signal is

$$B = \frac{1 + \beta}{T_s} = \frac{(1 + \beta)R_b}{\log_2 M}. \tag{4.10}$$

Here, $R_b = 1/T_b$ is the bit rate of the digital signal, and M denotes the number of signal states (levels) of the modulation scheme. For binary modulation $M = 2$, and for quaternary modulation schemes $M = 4$. Thus, the transmit symbol duration is $T_s = T_b \cdot \log_2 M = 1/R_b \cdot \log_2 M$. Accordingly, the symbol energy E_s and the bit energy E_b are related by

$$E_s = E_b \cdot \log_2 M. \tag{4.11}$$

The bandwidth efficiency of a modulation scheme can be defined as the number of bits per second which can be transported within a given bandwidth:

$$\eta_b = \frac{\text{bit rate}}{\text{bandwidth}} = \frac{\log_2 M}{1 + \beta} \quad \text{in} \quad \frac{\text{b/s}}{\text{Hz}} . \tag{4.12}$$

4.2.3 Bit Error Rate in the Gaussian Channel

In Fig. 4.8, the transmission channel was characterized by its frequency response $G(f)$ and the addition of noise $n(t)$. A basic channel model is the Gaussian channel, which has a flat frequency response and adds white Gaussian noise to the signal at the input of the receiver.

BPSK. In [Pro89] the bit error rate of binary antipodal transmission (BPSK) over the Gaussian channel and optimum (e.g. matched-filter) detection is derived as

$$p_b = \frac{1}{2}\mathrm{erfc}\,\sqrt{E_b/N_0} = \frac{1}{2}\mathrm{erfc}\,\sqrt{E_s/N_0} \qquad (4.13)$$

(for BPSK the bit energy E_b is equal to the symbol energy E_s). E_b/N_0 is denoted as signal-to-noise ratio per bit, which is often expressed by the logarithmic measure $10\log E_b/N_0$ in dB, and

$$\mathrm{erfc}\,x = \frac{2}{\sqrt{\pi}}\int_x^\infty e^{-t^2}\,dt \qquad (4.14)$$

is the complementary error function [Pro89].

QPSK. A QPSK signal can be considered as a combination of two orthogonal BPSK signals with half the power of the QPSK signal but symbol duration $2T_b$. Thus, the bit error rate of QPSK related to bit energy E_b is the same as for BPSK:

$$p_b = \frac{1}{2}\mathrm{erfc}\,\sqrt{E_b/N_0} = \frac{1}{2}\mathrm{erfc}\,\sqrt{E_s/2N_0} \qquad (4.15)$$

with $E_s = 2E_b$ being the symbol energy of the QPSK signal, cf. Eq. (4.4).

Other Modulation Schemes. The bit error rate for MSK with coherent FSK demodulation is given by [PL95]

$$p_b = \frac{1}{2}\mathrm{erfc}\,\sqrt{E_b/N_0}. \qquad (4.16)$$

The bit error rate for differential detection of PSK (DPSK) is

$$p_b = \frac{1}{2}\exp(-E_b/N_0) \qquad (4.17)$$

and for binary FSK (frequency-shift keying) with non-coherent detection the bit error rate is

$$p_b = \frac{1}{2}\exp\left(\frac{-E_b}{2N_0}\right). \qquad (4.18)$$

Figure 4.11 shows the bit error rate of various modulation schemes in the Gaussian channel.

4.2.4 Bit Error Rate in the Ricean and Rayleigh Fading Channel

The Fading Model. Multipath fading occurs if – due to reflections of the signal – a multitude of signal paths exists between transmitter and receiver. We consider the fading to be frequency non-selective, i.e. the frequency response of the channel does not vary over the signal bandwidth, cf. Sect. 3.3. To a good approximation, this holds if the delay differences between the signal paths are much smaller than a symbol duration. In this case, the influence of fading can be represented by a multiplication of the transmitted signal with

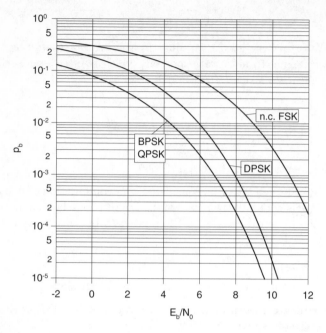

Fig. 4.11. Bit error rate p_b for BPSK, QPSK, DPSK, and non-coherent FSK in the Gaussian channel

a complex time-varying fading process $\underline{a}(t)$ characterizing the fading channel (multiplicative fading).

In the following, we consider the model of Ricean fading, which can be applied to radio channels if a direct line-of-sight signal path plus a number of reflected paths are present. In particular, the Ricean model is often used for describing mobile satellite channels [HDL+87, LCD+91]. Limiting cases of the Ricean channel are the Gaussian channel, where no fading is present, and the Rayleigh channel, where no direct signal path is available.

The Ricean fading process is composed of a constant component μ, representing the directly received signal, plus two quadrature fading components $x(t)$ and $y(t)$, which are independent Gaussian processes with zero mean and with variance σ^2:

$$\underline{a}(t) = \mu + x(t) + \mathrm{i}y(t). \tag{4.19}$$

We presume that

$$E\{|a(t)|^2\} = 1, \tag{4.20}$$

i.e. the mean received signal-to-noise ratio is not affected by the fading.

The severity of fading is described by the Rice factor c, which is defined as the power ratio between the direct signal and the reflected signal:

$$c = \frac{\mu^2}{2\sigma^2}, \tag{4.21}$$

and because of the requirement stated in Eq. (4.20), we have

$$\mu^2 = \frac{c}{1+c} \tag{4.22}$$

and

$$\sigma^2 = \frac{1/2}{1+c}. \tag{4.23}$$

The fading process is complex, i.e. it affects the amplitude and the phase of the transmitted signal. However, for non-coherent symbol-by-symbol detection the phase variations are irrelevant, and for coherent detection it can be assumed that the carrier recovery module in the receiver tracks the phase variations perfectly, as long as the fading is slow compared to the bit rate. In both cases, only the amplitude of the fading process

$$a(t) = \sqrt{(x(t) + \mu)^2 + y^2(t)} \tag{4.24}$$

is relevant for the bit detection. The probability density function of the Ricean fading amplitude can be derived as

$$f_a(a) = 2a(1+c)\, e^{-a^2(1+c)-c}\, I_0(2a\sqrt{c(1+c)}) \tag{4.25}$$

with I_0 designating the modified Bessel function of order zero. For the limiting case of Rayleigh fading ($c \to 0$) the probability density function of the fading amplitude is

$$f_a(a) = 2a\, e^{-a^2}. \tag{4.26}$$

In the limit $c \to \infty$, the channel amplitude becomes constant (Gaussian channel): $a(t) \equiv 1$.

Bit Error Rate. For the computation of the bit error rate, the fading process is assumed to be stationary and slowly varying compared to the symbol duration T_s, such that it is approximately constant during one symbol (slow fading). The fading process can change, however, over the time duration of several symbols.

According to the time-varying fading amplitude $a(t)$, the received bit energy $a^2(t)E_b$ and the bit error rate vary with time. For a given fading amplitude a the BER of BPSK is

$$p_b(a) = \frac{1}{2}\operatorname{erfc}\sqrt{a^2 E_s/N_0}. \tag{4.27}$$

Corresponding equations hold for other modulation schemes.

During deep fades ($a(t) \to 0$), $p_b(a)$ increases up to 0.5. This means that bit errors are no longer uniformly distributed over time, but mainly occur in bursts, coinciding with deep fades.

In order to disperse the error bursts, the transmit symbols can be interleaved, with appropriate de-interleaving at the receiver. Interleaving, however, causes additional effort and delay.

For sufficient interleaving, the bit errors can be considered statistically independent, and the fading channel can be characterized by the average BER $\overline{p_b}$:

$$\overline{p_b} = \int_0^\infty p_b(a)f_a(a)da. \tag{4.28}$$

For the Rayleigh channel, the integral in Eq. (4.28) can be solved analytically, and for BPSK and QPSK we get [Hag80]

$$\overline{p_b} = \frac{1}{2}\left(1 - \frac{1}{\sqrt{1 + N_0/E_b}}\right). \tag{4.29}$$

Here, E_b/N_0 represents the mean signal-to-noise ratio of the fading channel, which because of Eq. (4.20) is identical to the signal-to-noise ratio without fading. For DPSK the mean bit error rate is

$$\overline{p_b} = \frac{1/2}{1 + E_b/N_0} \tag{4.30}$$

and for non-coherent binary FSK we get [Hag82]

$$\overline{p_b} = \frac{1}{2 + E_b/N_0}. \tag{4.31}$$

Figure 4.12 shows the bit error rate of a BPSK- or QPSK-modulated signal in the Ricean fading channel.

Let E_{LOS}/N_0 be the signal-to-noise ratio resulting from the line-of-sight satellite link budget. With Rice factor c, the total mean signal-to-noise ratio including the multipath power results in

$$\overline{E_b}/N_0 = E_{\mathrm{LOS}}/N_0 \cdot \left(1 + \frac{1}{c}\right). \tag{4.32}$$

This value of the signal-to-noise ratio must be used to compute the bit error rate in the Ricean satellite channel.

In order to decrease the bit error rate, diversity techniques can be applied (e.g. antenna diversity) [Hag82]. In the following, however, only error control by means of channel coding will be considered.

4.3 Channel Coding (Forward Error Correction, FEC)

With channel coding, the bit error probability can be substantially reduced. Accordingly, a specified transmission quality can be obtained with a smaller transmit power.

The principle of channel coding is to add some redundancy to the information to be transmitted. The redundancy can be used by the receiver to detect

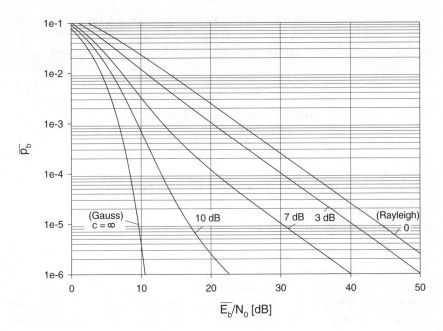

Fig. 4.12. Average bit error rate for BPSK and QPSK in the Ricean channel for different values of the Rice factor c

and correct transmission errors. This kind of channel coding works without feedback from the receiver and is therefore called forward error correction (FEC).

For systematic channel coding, redundant parity bits are added to the unmodified information bits. For non-systematic coding, the information bits are transformed into a larger number of code bits including redundancy.

For convolutional coding, the code bits are continuously generated by convolving the information bits with the bit response of a tapped shift register. For block coding, the information bits are divided into consecutive blocks, and the redundancy bits are computed for and added to each separate data block.

In general, with channel coding, n code bits are generated from k information bits. The information content of the coded sequence is represented by the code rate

$$r = k/n \leq 1. \tag{4.33}$$

Relating the energy nE_s included in n code bits to the k information bits results in the energy per information bit:

$$E_b = \frac{E_s}{r \log_2 M}. \tag{4.34}$$

Since $n \geq k$ code bits must be transmitted instead of k uncoded bits, it is clear that channel coding increases the required bandwidth (if the number of modulation levels is maintained). With roll-off factor β the required channel bandwidth is

$$B_{ch} = \frac{1 + \beta}{r \log_2 M} R_b \qquad (4.35)$$

and the bandwidth efficiency of the coded signal is

$$\eta_b = \frac{R_b}{B_{ch}} = \frac{r \log_2 M}{1 + \beta} \quad \text{in} \quad \frac{\text{b/s}}{\text{Hz}}. \qquad (4.36)$$

It can be seen that channel coding decreases bandwidth efficiency.

4.3.1 Convolutional Coding

Coding. Figure 4.13 shows a convolutional coder for a simple rate-1/2 convolutional code. The information bits are fed into a shift register. The code bits are generated by tapping certain cells of the register and modulo-2 adding their content. The pattern of the taps is characterized by the two binary generators $G_1 = 101$ and $G_2 = 111$. For each information bit i read into the register, two code bits c_1 and c_2 are generated and transmitted ($k = 1$, $n = 2$). In general, rate-k/n convolutional codes are determined by n generators [CC81], and for each k information bits read in, n code bits are written out and transmitted.

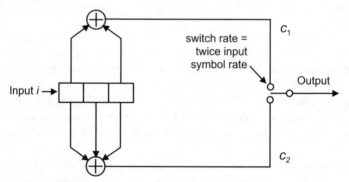

Fig. 4.13. Non-systematic rate-1/2 convolutional coder.
$K = 3 =$ number of shift register cells = constraint length.
$\nu = K - 1 = 2 =$ memory.
$\oplus =$ modulo-2 sum

The trellis diagram in Fig. 4.14 describes all possible code sequences (paths) for the coder shown in Fig. 4.13. The state of the coder corresponds to the current content of the last two shift register cells, representing the memory of the coder. Depending on whether 0 or 1 is read in as the information bit, the coder generates the code bits corresponding to the upper or lower branch, respectively.

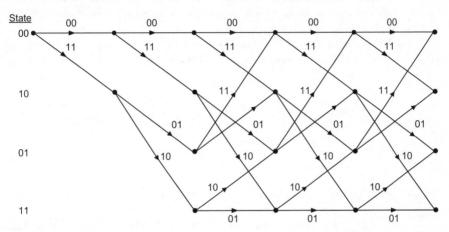

Fig. 4.14. Trellis diagram for the rate-1/2 convolutional code produced by the coder shown in Fig. 4.13 ($K = 3, \nu = 2$)

Decoding. Convolutional codes are usually decoded using the Viterbi algorithm. The Viterbi decoder compares the received signal with all possible code sequences, represented by the paths in the trellis diagram. The most likely code sequence is the path which is "nearest" to the received sequence. The "distance" of sequences can be characterized by a suitable metric. When the received sequence is represented by hard bit decisions, the Hamming distance (number of differing bits) can be used as metric, i.e., the most likely code sequence is the trellis path having the highest number of agreements with the received sequence.

Alternatively, the Viterbi decoder can use analog values y_i of the received bits (soft decisions). Then, for a Gaussian channel, the Euclidean distance

$$d = \sum_{\substack{\text{branches} \\ \text{of a path}}} (y_i - x_i)^2 \tag{4.37}$$

can be used as path metric [Bla83] with x_i denoting the branch labels. Using soft decisions, the performance on a Gaussian channel is improved by approx. 2 dB.

At every merging point of the trellis, the Viterbi decoder decides for the most probable path entering into the node. After some delay, the Viterbi decoder can output the most likely code sequence, thus approaching maximum likelihood decoding.

Performance of Convolutional Codes. The error correction capability of a convolutional code is primarily determined by its free distance d. This is the smallest number of differing code bits occurring in any two different paths (in the example shown in Fig. 4.14, $d = 5$). Besides, the bit error probability for a convolutional code is influenced by the number w_j (the weight) of the paths having distances $j = d, d+1, d+2, \ldots$.

For BPSK modulation in a Gaussian channel and soft-decision decoding the error probability of two sequences differing in j positions is given by [CC81]

$$P_j = \frac{1}{2} \operatorname{erfc} \sqrt{j E_s/N_0}. \tag{4.38}$$

With this, an upper bound for the bit error probability results in

$$p_b < \frac{1}{k} \sum_{j=d}^{\infty} w_j P_j. \tag{4.39}$$

Usually, the contribution of the summation terms decreases with increasing j.

Table 4.3 lists some parameters for rate-1/2 convolutional codes. Figure 4.15 shows the performance of several convolutional codes in a Gaussian channel, compared to uncoded BPSK transmission. Figure 4.16 shows the performance of some rate-1/2 convolutional codes in a Rayleigh fading channel, compared to uncoded FSK transmission. Table 4.4 lists the coding gain that can be obtained by a rate-1/2 convolutional code in the Gaussian channel, in a typical Ricean channel, and in the Rayleigh channel. It can be seen that an especially high coding gain can be achieved in the severely fading Rayleigh channel. In fact, uncoded transmission would require prohibitive transmit power levels for such channels.

Table 4.3. Generator polynomials (octal), free distance, and weight structure of the best rate-1/2 convolutional codes [CC81]

K	Generators	d	w_d	w_{d+1}	w_{d+2}	w_{d+3}
3	7, 5	5	1	4	12	32
4	17, 15	6	2	7	18	49
5	35, 23	7	4	12	20	72
6	75, 53	8	2	36	32	62
7	171, 133	10	36	0	211	0
8	371, 247	10	2	22	60	148

4.3.2 Block Coding

A binary block encoder generates from a block I with k information bits a block C with n code bits. This can be achieved e.g. by multiplying the

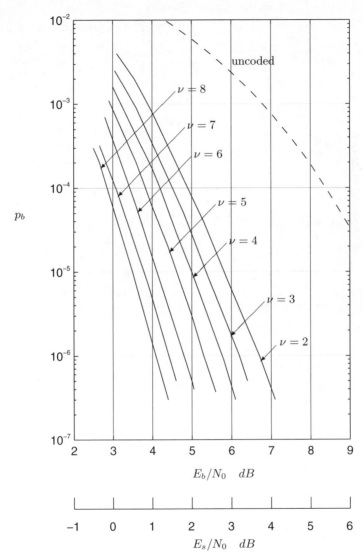

Fig. 4.15. Bit error rate for rate-1/2 convolutional codes in the Gaussian channel with BPSK modulation and soft-decision decoding [CC81].
$\nu = K - 1 =$ code memory.
uncoded: $E_s = E_b$; coded: $E_s = E_b/2$

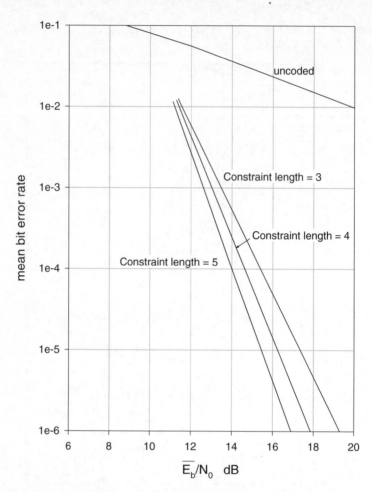

Fig. 4.16. Bit error rate for rate-1/2 convolutional codes in the Rayleigh channel. Non-coherent binary FSK modulation, soft-decision decoding, and perfect interleaving [Pro89]

Table 4.4. Required values of the signal-to-noise ratio $10 \log E_b/N_0$ for achieving a bit error probability of 10^{-6}

Transmission scheme	Gaussian channel	Ricean channel $c = 10$ dB	Rayleigh channel
Uncoded BPSK and QPSK	10.5 dB	23 dB	55 dB
Convolutional code, rate-1/2, K=7, soft decision	4.7 dB	6.5 dB	15 dB

information polynomial $I(x)$ by a generator polynomial $G(x)$:

$$C(x) = I(x)G(x). \tag{4.40}$$

Here, the information bits are modified through the coding process (non-systematic coding).

The coding procedure according to Eq. (4.40) maps the 2^k possible information patterns into 2^k points of the vector space $GF(2)^n$ with volume $2^n \gg 2^k$. The underlying code is denoted as (n,k) block code.

The performance of block codes is determined by the mutual "distance" of the codewords in the vector space. The larger the distance, the more reliably the received data blocks can be related to a codeword. Here, "distance" is measured as Hamming distance, which is equal to the number of differing bits in two n-tupels.

The minimum distance $d_{\min}$ of a block code is the smallest Hamming distance between any two codewords. Using a bounded-distance decoder, a code with minimum distance $d_{\min}$ allows to correct up to t bit errors and to detect e more errors if [CC81, ML85]:

$$2t + e < d_{\min}. \tag{4.41}$$

4.3.3 Error Protection with Cyclic Redundancy Check (CRC)

A check for transmission errors can be performed by transmitting a check sequence derived from the data. The receiver derives the check sequence from the received data too, and compares it with the received check sequence. If both sequences are not identical, a transmission error definitely occurred. On the other hand, from an equivalence of the check sequences an error-free transmission can be assumed with high probability. Pure error detection with cyclic redundancy check sequences can be considered a special kind of block coding.

Generally, it is assumed that in a channel with a bit error rate $\leq 10^{-4}$ a check sequence of 16 bits guarantees sufficient protection against undetected transmission errors. In fading channels a worst-case estimation can be obtained by assuming a bit error rate of $1/2$ during the transmission of a packet. Then, the probability p_u for undetected errors equals the probability of randomly receiving a check sequence matching an erroneous information block:

$$p_u = 2^{-(n-k)}. \tag{4.42}$$

4.3.4 RS Codes

The BCH codes (after Bose, Chaudhuri, and Hocquenghem) form a family of binary block codes. The single-error-correcting BCH codes are identical to the **Hamming codes**.

The family of **RS codes** (after Reed and Solomon) are multilevel codes. Their codewords consist of n code symbols which are represented by m bits each. The basic parameters of RS codes are

symbol size: m bits
block length: $n = 2^m - 1$ symbols $= m(2^m - 1)$ bits
information: k symbols $= mk$ bits
minimum distance: $d_{min} = n - k + 1$ symbols.

Table 4.5. Block length of RS codes

Number of bits per code symbol m	Block length in symbols $2^m - 1$	Block length in bits $m \cdot (2^m - 1)$
4	15	60
5	31	155
6	63	378
8	255	2040

Table 4.5 lists some possible block lengths of RS codes. The minimum distance of RS codes is related to symbols, because RS codes correct symbol errors. RS codes can correct up to t symbol errors and reconstruct up to e additional erasures, if

$$2t + e \leq n - k. \tag{4.43}$$

Shortening of Codes. In order to adapt a block code to a given data packet format, the block code can be shortened. For this purpose, a number l of information bits or symbols are not used (and not transmitted), resulting in an $(n - l, k - l)$ code. For decoding, the receiver inserts l zeroes into the received packet. The minimum distance of the code is maintained.

4.3.5 Performance of Block Codes

Let us consider a (not necessarily binary) block code for which a bounded-distance decoder can correct up to t errors [Bla83]. A packet error occurs if the received data block (including the redundancy) contains more than t errors. The packet error probability can be derived from the packet error distribution (PED) $P(i, n)$ defined as the probability that a data block consisting of n symbols contains i erroneous symbols. With this, the packet error probability is

$$p_p = \sum_{i=t+1}^{n} P(i, n) = 1 - \sum_{i=0}^{t} P(i, n). \tag{4.44}$$

For a Gaussian channel or a binary symmetric channel (BSC) with bit error probability p_b the bit errors occur independently of each other, and the error probability of a code symbol is given by

$$p_s = 1 - (1 - p_b)^m$$

with m = number of bits per code symbol.

Due to the independently occurring symbol errors, the PED has a binomial form:

$$P(i, n) = \binom{n}{i} p_s^i (1 - p_s)^{n-i}. \tag{4.45}$$

Accordingly, the packet error rate is

$$p_p = \sum_{i=t+1}^{n} \binom{n}{i} p_s^i (1 - p_s)^{n-i}. \tag{4.46}$$

In a reasonably dimensioned block code only the first terms of the sum are relevant. Figure 4.17 shows the packet error rate of RS codes in a Gaussian channel.

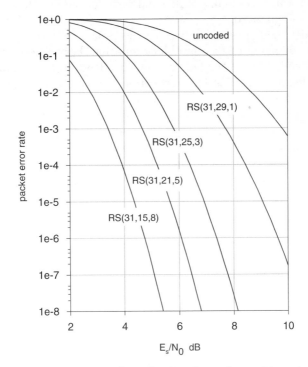

Fig. 4.17. Packet error rate for (n, k, t) RS codes with $n = 31$ symbols = 155 bits and different code rates. Gaussian channel, BPSK

Normally, long codes are more powerful than short codes with the same code rate k/n. However, long codes imply long data packets and require more decoding effort than short codes. Low-rate codes (i.e. codes with much redundancy) exhibit lower packet error rates than high-rate codes. However, low-rate codewords contain much overhead, leading to less efficient transmission. There is an optimum code rate achieving maximum data throughput, depending on n and p_s.

The bit error rate is often displayed versus E_b/N_0. However, in packet transmission schemes, usually a given transmission rate is assumed, determining E_s. Therefore, it is more convenient to plot the packet error rate against E_s/N_0, as shown in Fig. 4.17. By appending redundancy, E_b increases according to Eq. (4.34).

4.3.6 Performance of Block Codes in Fading Channels

For fading channels the symbol errors no longer occur independently of each other, but tend to occur in bursts. As shown in [BL95], the bandwidth of the fading process in relation to the symbol rate $1/T_s$ has a strong influence on the packet error rate of block-coded transmission.

For the limiting cases of fast fading and very slow fading, the packet error rate can be easily derived.

We consider the fading to be "fast" if the variations of the fading amplitude are fast compared to the data block duration. For this case we assume that for each code symbol the fading is independent. To a good approximation, fast fading pertains if several correlation times of the fading process are contained in a data block duration. Furthermore, the condition of independent fading for each code symbol is fulfilled if perfect interleaving is applied. This is the case most frequently considered in the literature.

For fast fading or perfect interleaving of code symbols, the bit errors occur independently of each other, as in the Gaussian or BSC channel, and the PED has a binomial form, corresponding to Eq. (4.45):

$$P(i,n) = \binom{n}{i} \overline{p_s}^i (1 - \overline{p_s})^{n-i}. \tag{4.47}$$

On the other hand, we consider the fading to be "very slow" if the fading amplitude is approximately constant during a data block duration. In this case the PED is determined by

$$P(i,n) = \binom{n}{i} \int_0^\infty p_s^i(a)(1 - p_s(a))^{n-i} f_a(a)\, da. \tag{4.48}$$

The approximation of very slow fading can be used as an upper bound for the packet error rate when code blocks are transmitted without interleaving. This is the case if messages have to be transmitted in single packets, perhaps confined to definite time slots. In particular, this is true for all kinds of slotted multiple access schemes where the user has to transmit a packet within a given time slot, according to the multiple access protocol (see Chap. 5). The same constraint is present in the case of TDMA-based multiple access of real-time multimedia services where stringent delay constraints must be obeyed (see Chap. 11).

In [BL95] the packet error rate is computed based on a multistate Markov model for the non-interleaved fading channel. Figure 4.18 compares the performance of a (43,21) Reed-Solomon code over GF(64), a binary (255,131) BCH code, and a rate-1/2 convolutional code (soft decision, zero tail) for the non-interleaved flat Rayleigh channel. All codes have approximately the same block length and code rate. For the convolutional code the bits were interleaved within the packets.

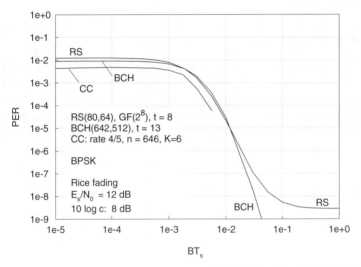

Fig. 4.18. Packet error rate versus normalized fading bandwidth BT_s for three rate-1/2 codes [BL99]

For small fading bandwidths, the block codes have approximately equal performance, whereas for large fading bandwidths marked differences occur. The lowest packet error rate was obtained with the convolutional code, but the comparison with the block codes is not quite fair, because we assumed hard decision for the block codes but soft-decision decoding for the convolutional code. The convolutional code with intra-packet interleaving is least sensitive to slow fading. For fast fading the BCH code is much better than the symbol error-correcting RS code, because the bit errors are randomly distributed over the whole packet.

4.4 Automatic Repeat Request (ARQ)

With forward error correction (FEC) the coded data is transmitted without any feedback from the receiver concerning the success of decoding. Thus, this kind of channel coding can work without a return channel. Another advantage of FEC is that the transmission delay caused by encoding and decoding is constant, which is most suitable for real-time services such as speech or video. On the other hand, if a code block cannot be correctly decoded by the receiver some loss of information occurs.

Automatic repeat request (ARQ), being based on retransmissions of erroneous data blocks, represents a different approach to error control. With ARQ, any loss of information can be avoided, which is most important for any kind of data transmission, such as file transfer or signaling. The time-varying transmission delay caused by block retransmissions is usually not a problem for such services.

4.4.1 Stop-and-Wait ARQ

Stop-and-wait ARQ is the simplest kind of automatic repeat request scheme. After the transmission of a data block the transmitter waits until it gets an acknowledgement from the receiver. Then the next data block is transmitted. In the case of a negative acknowledgement due to an erroneous data block at the receiver, the sender retransmits the last data block, Fig. 4.19. If a data block got lost, the sender time-outs and retransmits the missing block.

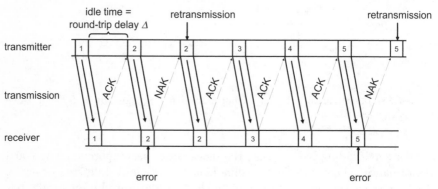

Fig. 4.19. Stop-and-Wait ARQ

The transmitter only has to store the currently transmitted data block, which is easy to realize. For satellite transmission, the efficiency of this technique is very low, however, because of the waiting times between the data blocks.

The throughput of an ARQ scheme is defined as the transmitted amount of information per time unit, related to the number of bits which could be transmitted during this period of time with continuous transmission. Let Δ be the transmission delay in both directions (round-trip delay) and let $R_s = 1/T_s$ be the transmission rate. During the time interval $\Delta + T_p$ (with T_p denoting the packet duration), $\Delta \cdot R_s + n$ bits of information could have been transmitted. The number of actually transmitted information bits is $k \cdot P$, where P is the probability of a data block being acknowledged by the receiver. Thus, the throughput of stop-and-wait ARQ is given by

$$\eta_{\mathrm{SW}} = \frac{k \cdot P}{\Delta \cdot R_s + n} = \frac{k}{n} \cdot \frac{P}{1 + \Delta/T_p}. \tag{4.49}$$

The code rate k/n can also take into account the packet overhead.

4.4.2 Go-Back-N ARQ

With this technique, the data blocks are being continuously transmitted. The acknowledgement for a data block arrives at the transmitter after a round-trip time Δ. During this period of time Δ/T_p additional data blocks have been transmitted. If the transmitter receives a negative acknowledgement, it cranks back to the negatively acknowledged data block and starts again with the continuous transmission of data blocks, Fig. 4.20. This means that $N = \lceil 1 + \Delta/T_p \rceil$ data blocks are repeated if a single block is in error.

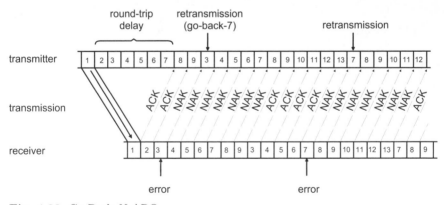

Fig. 4.20. Go-Back-N ARQ

The transmitter has to store N data blocks. Go-back-N ARQ is more efficient than stop-and-wait ARQ and requires only a small additional effort. The throughput of go-back-N ARQ is given by [LC83]

$$\eta_{\mathrm{GBN}} = \frac{k}{n} \cdot \frac{P}{P + (1 - P)N}. \tag{4.50}$$

4.4.3 Selective-Repeat ARQ

The data blocks are transmitted continuously. The transmitter repeats only those data blocks which are negatively acknowledged, Fig. 4.21.

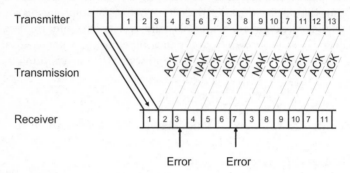

Fig. 4.21. Selective-Repeat ARQ

In order to maintain the correct order of the data blocks, sufficient memory must be available in the transmitter and in the receiver. Selective-repeat ARQ is the most efficient, but also the most complex repeat strategy. The throughput of selective-repeat ARQ is

$$\eta_{SR} = \frac{k}{n} \cdot P \tag{4.51}$$

and the mean transmission delay (until reception of ACK) is

$$D = \frac{\Delta + T_p}{P}. \tag{4.52}$$

The Inmarsat-C system uses selective-repeat ARQ for data transmission.

Hybrid ARQ. ARQ techniques are called hybrid if they use forward error correction as well as block retransmissions. Type-1 hybrid ARQ uses a fixed block code which is designed for error detection as well as for error correction. If a received block is detected as erroneous, the receiver first tries to correct the errors. If it does not succeed, it requests a block retransmission.

4.5 Typical Error Control Schemes in Mobile Satellite Communications

In mobile satellite communications the following error control schemes are typically used for continuous transmission:

– convolutional codes

- unequal error protection with convolutional codes
- concatenation of convolutional and block codes (typically used for data transmission)
- interleaving
- soft-decision decoding
- usage of channel state information
- trellis-coded modulation (TCM), e.g. trellis-coded 8-PSK.

Examples of channel coding schemes used for speech transmission in mobile satellite communication systems are:

- Ellipso: rate-1/3 convolutional code, $K = 9$, soft decision
- Globalstar: rate-1/2 convolutional code, $K = 9$
- Iridium: rate-3/4 convolutional code, $K = 7$
- Teledesic: trellis-coded 8-PSK modulation with rate 2/3.

For burst and packet transmission, usually a combination of block coding and automatic repeat request (hybrid ARQ) is applied.

Part II

Satellite Systems for Mobile/Personal
Communications

5. Multiple Access

5.1 Duplexing

A radio communication terminal, satellite, or ground station cannot simultaneously receive and transmit at the same frequency; the transmit signal would heavily jam the receive signal. Therefore, some scheme for separating the forward and return link transmission must be applied.

5.1.1 Frequency-Division Duplexing (FDD)

Frequency-division duplexing uses different frequency bands for forward and return link transmission.

Usually, the forward and return link frequency bands have equal bandwidth. Within these bands, corresponding forward and return channels are paired, i.e. they have constant frequency separation. This is taken into account in the allocation of frequency bands for radio communications.

FDD is the conventional duplexing scheme in satellite communications. As an example, the Globalstar system uses the following frequency bands:

User links	uplink (return transmission)	1610–1626.5 MHz
	downlink (forward transmission)	2483.5–2500 MHz
Feeder links	uplink (forward transmission)	5091–5250 MHz
	downlink (return transmission)	6875–7055 MHz.

5.1.2 Time-Division Duplexing (TDD)

With time-division duplexing, the forward and return transmissions use the same frequency band but occur in separate time intervals. Compared to FDD, TDD has the following advantages:

- No paired frequency bands are required; therefore, frequency planning is easier.
- TDD is more suitable for dynamic channel allocation.
- No separate satellite antennas are required for transmit and receive.
- No diplexer is required for a combined transmit/receive terminal antenna.

Disadvantages of TDD are:

– A guard time is required between up- and downlink transmission, which is
proportional to the signal delay. Alternatively, a suitable frame structure
can be used.
– Compared to FDD, a two-fold transmission rate is required, entailing a
two-fold peak transmit power.
– TDD requires synchronization between up- and downlink.
– TDD causes delay due to the frame structure.

Figure 5.1 shows the separation in time of the uplink and downlink data
slots for the Iridium mobile user link.

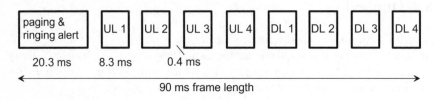

8 user slots per frame (4 transmit and 4 receive)
50 kb/s burst rate
4.6 kb/s channel gross rate
approximately half of the bits are available for user data

Fig. 5.1. Time frame for the Iridium mobile user link. UL = uplink, DL = downlink.
Burst duration = 8.28 ms $\hat{=}$ 414 bits

5.2 Multiplexing

Often, a multitude of signals is to be transmitted via the same medium. For
example, this is the case in a satellite feeder link, where many user signals
must be transmitted between the satellite and the ground station, in the
user downlink, where the signals belonging to a satellite or spot beam are
broadcast, or in the user uplink, where the users within a satellite spot beam
access the same satellite transponder. Here, the aspect of multiplexing arises:
How can the transmission medium be divided into "channels" such that each
signal can use a channel and the receiver can extract any signal out of the
signal multiplex?

In order to organize signals into a multiplex, the signals $s_i(t)$ should use
channels which are as orthogonal as possible, such that

$$\int s_i(t)s_j^*(t)\,dt \approx 0 \quad \text{for } i \neq j. \tag{5.1}$$

This can be achieved in different ways:

Frequency-Division Multiplexing, FDM. For FDM, the available frequency band is divided into non-overlapping frequency channels. In a single-transmitter scenario (point-to-point or broadcast), the transmitter uses a number of frequency channels to transmit the signal multiplex. In a multiple access scenario (e.g. for user uplinks to a satellite), the senders simultaneously transmit at different frequencies, and the receiver selects the signals with appropriate filters. This access method is called frequency-division multiple access, FDMA. In satellite networks, the FDMA receiver can be located either on board the satellite (e.g. if a different multiplexing scheme is used in the feeder link) or at the ground station.

Time-Division Multiplexing, TDM. For TDM, time is divided into non-overlapping intervals (time slots). One time slot is periodically allocated to each of the signals to be multiplexed. One period of slots constitutes a TDM time frame. The signals to be multiplexed are divided into data blocks, with each block fitting into a time slot.

In a single-transmitter scenario, the series of data blocks form a continuous signal (no guard times between the blocks are necessary). The signal is transmitted on a single radio carrier, and the users select their data blocks out of the received TDM frames by time filtering (gating).

In a time-division multiple access (TDMA) scenario, a specified time slot of the TDMA frames is allocated to each sender, within which it can transmit its data blocks. If the users are not bit synchronous, guard times must be included in the time slots.

Code-Division Multiplexing, CDM. Here, the multiplexed signals are transmitted at the same frequency and at the same time, but are separated by modulating them with orthogonal code signals (signatures). In a code-division multiple access (CDMA) scenario, the users modulate their signals with different signature code signals. On board the satellite or in the ground station the user signals are reconstructed by correlating the received signal mix with the respective code signals.

Satellites with Spot Beams, Space-Division Multiplex, SDM. Satellite spot beams can support FDM, TDM, or CDM by additionally separating users within different beams. However, spot beams do not achieve complete signal separation, especially if spot beams overlap. For frequency or time multiplexing, different frequency bands must be used in adjacent spot beams. For code multiplexing, adjacent spot beams must use different codes, but may use the same frequency band.

Space-division multiplexing (SDM) refers to the separation of single signals with the help of adaptive directional antennas. In this sense, satellites with spot beams represent a first step to the realization of SDM and space-division multiple access (SDMA).

Polarization Multiplexing. Polarization multiplexing is based on the separation of radio signals through different linear or circular electromagnetic polarizations. This method is normally used for satellite feeder links.

Figure 5.2 illustrates the principles of FDM, TDM, and CDM, and Fig. 5.3 extends them to the FDMA, TDMA, and CDMA multiple access scenarios. In addition to these basic multiple access schemes, combined schemes are possible, such as FDMA/TDMA or FDMA/CDMA (multi-frequency schemes).

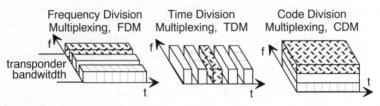

Fig. 5.2. Principle of FDM, TDM, and CDM

Exploitation of Speech Pauses. During a speech dialog, normally only one partner is talking, while the other one listens. By allocating the inactive channel to other connections, approximately 50% of the bandwidth can be saved. This method is called *digital speech interpolation* and can be applied to point-to-point connections (e.g. a satellite feeder link), for broadcasting (e.g. in a satellite user downlink), or in access networks with negligible delay.

5.3 Multiple Access

The task of multiple access arises when more than one sender wants to transmit to a single receiver. In particular, the users within a satellite spot beam (radio cell) are independent of each other, but should simultaneously use the corresponding satellite transponder. According to the underlying multiplexing scheme, the satellite uplink (the radio resource) is divided into a number of frequency channels, time slots, or code channels.

Satellites with spot beams are equipped with a separate input amplifier or transponder for each spot beam. Therefore, the issue of multiple access is limited to users within the same spot beam.

Multiple Access Protocol. If there are as many multiplexed channels available as there are users, a dedicated channel can be assigned to each user (fixed assignment multiple access). However, in the satellite networks considered here, there are much fewer channels than users. Therefore, communication channels are temporarily allocated to users who intend to set up a communication session (demand assignment multiple access). For this purpose, rules are required according to which the active users can access the frequency, time, or code channels. These rules are implemented through a multiple access protocol. This protocol must be adapted to the characteristics of the services to be provided, e.g. real-time/non-real-time services, continuous/packet transmission, etc.

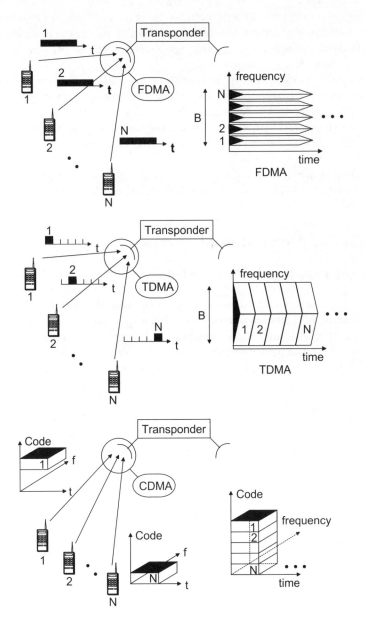

Fig. 5.3. Principle of FDMA, TDMA, and CDMA (after [MB93]). B is the transponder bandwidth © John Wiley & Sons Limited, reproduced with permission

For the transmission of single data packets from a user to the satellite or ground station (return direction), random access protocols are suitable, requiring no previous coordination. In particular, random access protocols are applied for transmitting connection setup requests, other signalling messages, or short data messages. A frequently used random access protocol is slotted Aloha, cf. Sect. 5.4.

For speech communication and for the transmission of larger data files it is advantageous to request the assignment of a traffic channel before transmission. Typically, this is done using slotted Aloha in a known random access frequency channel. Then, the network allocates a channel which, according to the multiplexing scheme used, can be a frequency channel, a time slot, or a code.

In the following, we consider only a single-satellite coverage area or spot beam. Cellular satellite networks will be treated in Chap. 6.

5.4 Slotted Aloha Multiple Access

5.4.1 The Principle of Slotted Aloha

Slotted Aloha is a random access scheme in the time domain. The time axis is divided into time slots with duration T_p corresponding to a packet duration plus a guard time accounting for synchronization errors. A terminal transmits a packet within the next upcoming time slot after the transmission has been initiated by the user. All transmitters use the same carrier frequency. If a packet is successfully transmitted, an acknowledgement is returned by the receiver (i.e. the satellite or the ground station) to the transmitter.

If a time slot is used by more than one terminal, the packets "collide" at the receiving satellite, the packets cannot be correctly received, and no acknowledgement is returned. After timing out, the terminals retransmit their packets after having waited for an additional random delay τ, Fig. 5.4.

Packets can also get lost, if they are transmitted during a bad channel state. In [Lut92], data transmission with slotted Aloha is analyzed, taking into account forward error correction and packet retransmission. The slotted Aloha scheme can be improved if the users transmit their packets only during good channel states [BJL93].

5.4.2 Throughput of Slotted Aloha

In order to assess the throughput of slotted Aloha, we assume a very large number of users with comparable activity level. Then, the stream of the total transmitted packets can be modeled as a Poisson process with average packet transmission rate λ [packets/s] [Abr77], and the total offered traffic including packet retransmissions is given by

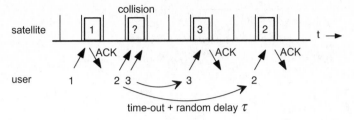

Fig. 5.4. Slotted Aloha

$$G = \lambda T_p \quad \text{in Erl.} \tag{5.2}$$

Let a packet be transmitted at time t_0. A collision will occur if further packets are generated in the interval

$$t_0 - T_p < t < t_0. \tag{5.3}$$

According to the underlying Poisson process [Pap65], the probability that no further packet is generated within this time interval is

$$e^{-\lambda T_p} = e^{-G} = \text{probability of success.} \tag{5.4}$$

The (average) throughput or the efficiency of a multiple access scheme can be defined as the transmitted amount of data per unit time, divided by the data rate which would be possible with a single continuous transmission using the same bandwidth and modulation scheme. According to this definition, the throughput of slotted Aloha is given by

$$\eta = G\,e^{-G} \tag{5.5}$$

(see Fig. 5.5). In steady-state, the throughput is equal to the total offered traffic generated by the users, excluding packet retransmissions. The maximum throughput can be achieved for $G = 1$ and amounts to

$$\eta_{\max} = 1/e \approx 0.368. \tag{5.6}$$

The average number of transmit attempts per packet is

$$N_t = G/\eta = e^G = \frac{1}{\text{probability of success}}. \tag{5.7}$$

At maximum throughput, $N_t = e \approx 2.72$.

Stability of Aloha. For an infinite number of users, Aloha is inherently instable: If the offered traffic G increases above the maximum of the throughput curve (e.g. due to traffic variations), the number of collisions will increase while the number of successful transmissions diminishes. Aloha can be stabilized by adapting the random delay τ to the estimated traffic volume, i.e., in overload situations retransmissions are more delayed. Alternatively, a collision resolution protocol (e.g. a tree algorithm) can be used.

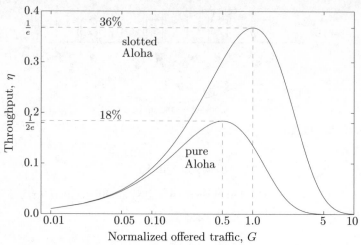

Fig. 5.5. Throughput of slotted Aloha and pure Aloha

5.4.3 Mean Transmission Delay for Slotted Aloha

The mean transmission delay is defined as the average time delay between the generation of a message and its successful reception at the receiver.

If no collision occurs, the average transmission delay is given by an average waiting time of $T_p/2$ before packet transmission, plus transmit duration T_p, plus the propagation delay of the satellite hop.

An acknowledgement returned by the receiver will be detected by the transmitter after the round-trip delay

$$\Delta = 2T_{\mathrm{prop}} + T_{\mathrm{ACK}} + T_{\mathrm{proc}} , \qquad (5.8)$$

where T_{prop} is the propagation delay between user terminal and satellite (we assume that the acknowledgement is produced by the satellite), T_{ACK} is the duration of the acknowledgement packet, and T_{proc} the processing time.

If after a time Δ no acknowledgement has been received (time-out), the packet will be retransmitted after a further random delay τ with mean value $\overline{\tau}$. The packet is retransmitted in the next upcoming time slot.

As long as the retransmission process is ongoing a newly generated packet has to wait. For a rough estimation of the delay we neglect this influence of a current retransmission process. Then, the mean transmission delay can be approximated by

$$D \approx 1.5T_p + \Delta + (N_t - 1)(\overline{\tau} + 1.5T_p + \Delta) \qquad (5.9)$$

(see Fig. 5.6). The delay increases with increasing N_t, i.e. with increasing throughput.

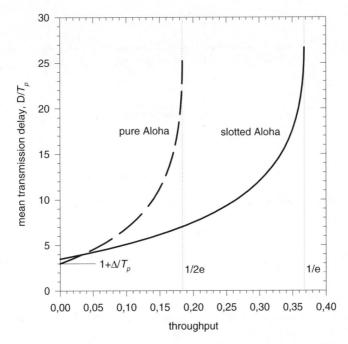

Fig. 5.6. Mean delay of slotted Aloha and pure Aloha as a function of throughput, for $\Delta = 2T_p$ and $\overline{\tau} = 10T_p$

Influence of the Channel. If for a (mobile) user the link to the satellite is shadowed, a transmitted packet will be heavily attenuated. Although the packet cannot be successfully transmitted, it causes no collision with a packet sent by another user in an unshadowed situation. Thus, the latter packet "captures" the channel (*capture effect*, see e.g. [Lut92]).

5.4.4 Pure Aloha Multiple Access

Similar to slotted Aloha, pure Aloha is a random access scheme in the time domain. However, in pure (asynchronous) Aloha no time slots are defined. Rather, a user transmits a packet whenever it is ready. If the packet is successfully transmitted, an acknowledgement is returned by the receiver.

If packets from different users overlap in time, the packets collide, and no acknowledgement is returned. After time-out the terminals retransmit their packets after having waited for an additional random delay, Fig. 5.7.

Throughput of Pure Aloha. Here we resort to the same assumptions as for slotted Aloha. Let a packet be transmitted at time t_0. A collision will occur if further packets are generated in the interval $t_0 - T_p < t < t_0 + T_p$, cf. Fig. 5.7. The probability that no further packet is generated within this time interval is

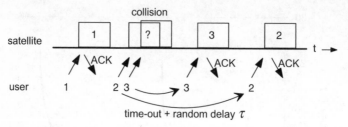

Fig. 5.7. Pure Aloha

$$e^{-2\lambda T_p} = e^{-2G} = \text{probability of success.} \tag{5.10}$$

With this, the throughput of pure Aloha is given by

$$\eta = G\,e^{-2G} \tag{5.11}$$

(see Fig. 5.5). The maximum throughput occurs for $G = 0.5$ and amounts to

$$\eta_{\max} = \frac{1}{2\,e} \approx 0.184. \tag{5.12}$$

Pure Aloha achieves a smaller throughput than slotted Aloha. The average number of transmit attempts per packet is

$$N_t = G/\eta = e^{2G} = \frac{1}{\text{probability of success}}. \tag{5.13}$$

At maximum throughput, $N_t = e \approx 2.72$, as for slotted Aloha.

Mean Transmission Delay of Pure Aloha. Under the same assumptions as for slotted Aloha, the mean transmission delay for pure Aloha can be approximated by

$$D = T_p + \Delta + (N_t - 1)(\overline{\tau} + T_p + \Delta), \tag{5.14}$$

cf. Fig. 5.6. The delay increases with increasing N_t, i.e. with increasing throughput.

5.5 Frequency-Division Multiple Access, FDMA

For frequency-division multiple access, the available frequency band is divided into subbands which are separated from each other by guard bands. Each subband can be accessed by modulating a radio carrier at the center frequency of the subband. At the receiver (regenerative satellite or ground station), the subbands are filtered out and the FDMA signals are reconstructed.

In mobile satellite networks with pure FDMA, each user modulates a carrier with a single speech signal. This scheme is called single channel per carrier, SCPC.

5.5.1 Adjacent Channel Interference

In spite of guard bands being included between frequency channels, adjacent channel interference (ACI) is caused by non-ideal channel filters and deviations of the carrier frequencies due to tolerances and Doppler shifts. The amount of ACI depends on the quality of the channel filters, the frequency inaccuracies, and the extent of the guard bands. For large guard bands, lower accuracy requirements can be accepted, or less ACI arises. For small guard bands, filters must be of higher quality and frequencies must be more exact, but better bandwidth efficiency can be achieved. Typically, guard bands are in the order of one-tenth of the wanted bands.

5.5.2 Required Bandwidth for FDMA

Considering a single radio cell, the required bandwidth (cell bandwidth) is

$$B_c = N_c \left(B_{\text{ch}} + B_g \right) = N_c \left[\frac{(1+\beta)R_b}{r \log_2 M} + B_g \right] \qquad (5.15)$$

with N_c denoting the number of FDMA channels in a radio cell (spot beam), and B_g being the guard band. The channel bandwidth B_{ch} has been inserted according to Eq. (4.35).

Solving the above equation for N_c, we can determine the possible number of active users for a bandwidth-limited transponder:

$$N_c = \frac{B_c}{B_{\text{ch}} + B_g} = \frac{B_c}{R_b} \cdot \frac{r \log_2 M}{1 + \beta + r \log_2 M \cdot B_g / R_b} . \qquad (5.16)$$

As already indicated in Sect. 5.4, the throughput or efficiency of multiple access schemes can in general be defined as

$$\eta_{\text{MA}} \quad = \quad \frac{\text{total effective bit rate with multiple access}}{\substack{\text{effective bit rate for a single continuous} \\ \text{transmission with bandwidth } B_c}}$$

$$= \quad \frac{N_c R_b}{R_{\text{cont}}} = \frac{N_c R_b (1+\beta)}{r \log_2 M \cdot B_c} . \qquad (5.17)$$

Here, it is assumed that the continuous transmission uses the same modulation scheme and channel coding rate as the multiple access transmission.

For FDMA, the efficiency of a bandwidth-limited transponder can also be expressed as

$$\eta_{\text{FDMA}} = \frac{\text{usable bandwidth}}{\text{radio cell bandwidth}} = \frac{N_c B_{\text{ch}}}{B_c} = 1 - N_c \frac{B_g}{B_c} = \frac{B_{\text{ch}}}{B_{\text{ch}} + B_g} . \qquad (5.18)$$

5.5.3 Intermodulation

Nonlinearities (in the satellite transponder) cause intermodulation products at the following frequencies:

$$f_{IM} = m_1 f_1 + m_2 f_2 + \cdots + m_{N_c} f_{N_c} \qquad (5.19)$$

with f_i denoting the FDMA carrier frequencies, and m_i being positive or negative integers. For modulated carriers, a continuous intermodulation spectrum arises.

For $N_c \gg 1$ the interference can be described by an approximately constant power spectral density L_{IM}. L_{IM} increases with increasing number of FDMA carriers and with increasing utilization of the transponder (causing increasing nonlinearity).

In order to reduce the nonlinearity, the transponder is not operated at saturation, resulting in output power P_{sat}, but backed off by a factor BO. With C_n designating the power of carrier n, the backoff factor is

$$BO = \frac{\sum\limits_{n=1}^{N_c} C_n}{P_{sat}} < 1. \qquad (5.20)$$

The reduction of the useful power by the intermodulation power $L_{IM} B_c$ is included in the backoff factor BO.

The spectral intermodulation power density L_{IM} depending on N_c and BO adds to the thermal noise power density N_0. The resulting noise power density is $L_{IM}(N_c, BO) + N_0$ and the total signal-to-noise ratio is reduced from E_b/N_0 by the factor

$$\frac{N_0}{L_{IM}(N_c, BO) + N_0} \cdot BO.$$

For a power-limited transponder the possible FDMA throughput is reduced accordingly.

There is a trade-off between BO and L_{IM}, resulting in an optimum value for BO. Because L_{IM} increases more than linearly with N_c, BO must be chosen lower with increasing N_c. Therefore, for a power-limited transponder the achievable FDMA throughput decreases strongly with increasing number N_c of FDMA carriers. Typical backoffs for satellite TWT transponders are in the range of $3 \ldots 10$ dB. The resulting efficiency of FDMA can be compiled as follows:

$$\eta_{FDMA} = \begin{cases} 1 - N_c \dfrac{B_g}{B_c} & \text{band-limited transponder} \\[2ex] \dfrac{N_0}{L_{IM}(N_c, BO) + N_0} \cdot BO & \text{power-limited transponder.} \end{cases} \qquad (5.21)$$

Voice Activation. Switching off the terminal transmit signal during speech pauses reduces the transponder load approximately by one-half, thus reducing intermodulation and enhancing efficiency.

5.5.4 Pros and Cons of FDMA

Advantages of FDMA are:

– unproblematic and mature technology
– low peak transmit power of terminals (continuous transmission)
– no equalizer required (low transmission rate, narrowband system)
– no user synchronization required.

Drawbacks of FDMA are:

– FDMA is inflexible and cannot adopt variable data rates.
– For a power-limited transponder, the capacity loss is strongly increasing with N_c.
– Uplink power control is required in order to limit adjacent channel interference (ACI) and the influence of nonlinearities.
– Exact frequency control is required in order to limit ACI.
– FDMA requires increased realization effort for diplexers (because of FDD) and channel filters.

For systems with analog modulation, FDMA is the only useful multiple access scheme.

5.6 Time-Division Multiple Access, TDMA

In time-division multiple access time is divided into frames, which are again divided into time slots. A basic traffic channel is formed by a certain time slot within each frame. The user signals for the return uplink are divided into segments corresponding to a frame duration. For digital speech, e.g., the signal segments may be speech packets containing a header and parity bits from channel coding, Fig. 5.8. These packets must be buffered until the time slot allocated to the user arrives. Then, the packet is transmitted as high-rate burst fitting into the time slot. In order to account for synchronization inaccuracies, bursts in consecutive time slots are separated by a guard time interval. All users transmit at the same carrier frequency. During a time slot, the corresponding user exploits the total available bandwidth.

In the forward direction (user downlink) usually the same frame structure is used as in the return link. In order to avoid simultaneous transmission and reception for a user, the corresponding time slots in the forward and return direction are separated in time.

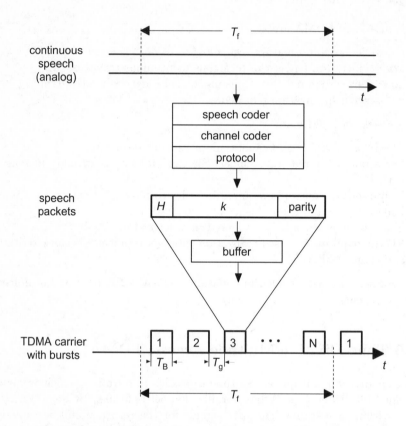

H = header (synchronisation preamble, signalling), e.g. 100 symbols
k = information bits from the speech coder
T_f = frame duration, e.g. 20 ms or 90 ms (Iridium)
T_B = burst duration
T_g = guard interval, e.g. 1 ms
N = number of channels, e.g. 4 or 8

Fig. 5.8. Principle of TDMA (burst generation)

5.6.1 Bandwidth Demand and Efficiency of TDMA

Here, we consider a single radio cell with a single TDMA carrier. We assume that a single time slot per frame is allocated to a user with information bit rate R_b. Let T_f denote the frame duration and

$$k = R_b T_f \qquad (5.22)$$

be the number of information bits per time slot (per speech packet). The information is assumed to be channel-coded with code rate r. Then, the number of code bits per time slot is $n = k/r$. For an M-ary modulation of the transmit signal (burst signal), the number of transmit symbols per time slot is

$$n_s = \frac{n}{\log_2 M} = \frac{R_b T_f}{r \log_2 M}. \qquad (5.23)$$

Each burst is assumed to begin with H additional header symbols for carrier and clock synchronization. Between each burst, a guard time T_g is foreseen to account for timing deviations. If the TDMA carrier should carry N channels, the required burst rate is

$$R_B = \frac{N(H + n_s)}{T_f - N T_g} \qquad \text{in symbols/s.} \qquad (5.24)$$

If the transmit signal is (square root) Nyquist-filtered with roll-off factor β, the bandwidth occupied by the TDMA carrier is

$$B_T = R_B(1 + \beta). \qquad (5.25)$$

By solving Eq. (5.24) for N, the number of TDMA channels (i.e. the possible number of active users) for a given bandwidth B_T can be derived as

$$N = \frac{B_T}{R_b} \cdot \frac{r \log_2 M}{(1 + \beta)\left(1 + \frac{H+G}{n_s}\right)}. \qquad (5.26)$$

Here, G is the guard interval in transmit symbol durations:

$$G = T_g R_B = \frac{T_g B_T}{1 + \beta}. \qquad (5.27)$$

In Eq. (5.26), the guard time G in transmit symbol durations and the burst header H are taken into account, relative to the n_s data transmit symbols (information plus parity) contained in a burst.

Efficiency of TDMA. In general, the efficiency or throughput of multiple access schemes is given by Eq. (5.17). For TDMA the following definition can be used:

$$\eta_{\text{TDMA}} = \frac{\text{usable time}}{\text{frame duration}} = 1 - N \frac{T_g + T_H}{T_f} \qquad (5.28)$$

with $T_H = H/R_B$ being the time duration of the header. With this relation and R_B according to Eq. (5.24) the TDMA efficiency can be expressed as

$$\eta_{\text{TDMA}} = \frac{1 - NT_g/T_f}{1 + H/n_s}. \tag{5.29}$$

The guard time T_g can be reduced by a more precise user synchronization. The synchronization preamble in the header H can be reduced by an improved receiver or by using differential or non-coherent modulation. If the user synchronization is sufficiently exact, the guard times can be avoided completely, and the multiplexed TDMA signal appears as a continuous bit stream. For this case of bit synchronous TDMA, also no preamble for bit timing recovery is necessary.

For TDMA, the decrease of throughput with increasing number N of channels is usually smaller than for FDMA.

The bandwidth efficiency of TDMA is given by the ratio of the total bit rate to the required bandwidth. With Eq. (5.26) we get

$$\eta_{b,\text{TDMA}} = \frac{NR_b}{B_T} = \frac{r \log_2 M}{(1 + \beta)\left(1 + \frac{H+G}{n_s}\right)} \quad \text{in } \frac{\text{b/s}}{\text{Hz}}. \tag{5.30}$$

This expression illustrates the various processes influencing TDMA bandwidth efficiency: channel coding, modulation, filtering, as well as TDMA header and guard time.

Example. Let us assume that we want to transmit a voice signal with bit rate 4.8 kb/s over a TDMA uplink. The FEC code rate is 1/2, the modulation is QPSK, and the roll-off factor is 0.25. The TDMA frames should have a duration of 50 ms (corresponding to the time duration of a voice packet) and contain 8 uplink channels. Each burst requires 100 symbols of header and a 1 ms guard time.

Applying the above equations, we get the following results: $k = 240, n_s = 240, R_B = 64.8$ ksymbols/s, $B_T = 81$ kHz. The efficiency of this TDMA scheme is $\eta_{\text{TDMA}} = 0.59$ and the bandwidth efficiency is $\eta_{b,\text{TDMA}} = 0.47$ b/s/Hz.

Compared to continuous transmission, TDMA transmission requires the terminals to transmit at a higher rate (the burst rate): $R_B \approx N \cdot R_{\text{cont}}$. In order to maintain the same signal-to-noise ratio, also the peak transmit power must be higher: $P_{\text{TDMA}} \approx N \cdot P_{\text{cont}}$. The average transmit power remains the same, however.

Voice Activation. By switching off the transmit signal during speech pauses, the mean transmit power is reduced approximately by one-half. The bandwidth demand of the TDMA uplink is not reduced, however.

Slotted Aloha. Similar to TDMA, a guard time T_g and a burst header H should also be taken into account for slotted Aloha. Then, we see from Eq. (5.29) that the throughput of slotted Aloha is reduced below the value given in Eq. (5.5) and results in $\eta_{\text{TDMA}} \cdot G\, e^{-G}$.

Multi-Frequency TDMA, MF-TDMA. In order to limit the peak transmit power for the terminals, a TDMA scheme with more than one carrier can be used, representing a combination of FDMA and TDMA (MF-TDMA or FDMA/TDMA). This is especially useful if a large number of terminals want to access a satellite or a base station. In the frequency domain, the TDMA carriers are separated by additional guard bands B_g. As an example, Fig. 5.9 shows the frequency plan for the Iridium mobile user link. Each 50-kb/s TDMA carrier is organized as shown in Fig. 5.1.

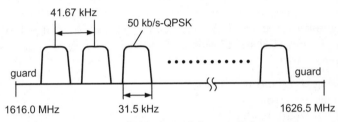

Fig. 5.9. Frequency plan for the Iridium mobile user link

If T carriers are used in a radio cell (spot beam), the MF-TDMA bandwidth required for the cell amounts to

$$B_c = T(B_T + B_g). \tag{5.31}$$

For a given cell bandwidth B_c, the number of channels within the cell becomes $N_c = TN$. Assuming a certain TDMA burst rate R_B, roll-off factor β, and guard band B_g, the number T of TDMA carriers per cell can be evaluated from Eqs. (5.25) and (5.31). The number N of channels per TDMA carrier is given by (5.26), and the number of channels or active users per radio cell can be derived as

$$N_c = TN = \frac{B_c}{R_b} \cdot \frac{r \log_2 M}{(1 + \beta + B_g/R_B)\left(1 + \frac{H+G}{n_s}\right)}. \tag{5.32}$$

A comparison with Eq. (5.26) shows that MF-TDMA is less bandwidth efficient than single-carrier TDMA.

5.6.2 Burst Synchronization in the Receiving Satellite

If the users do not transmit their TDMA bursts bit synchronously, the receiver has to synchronize to each single user burst. This is accomplished by means of the preamble within the burst header:

– For carrier synchronization (necessary for coherent detection), the preamble contains a series of 0's or 1's.

- Bit synchronization is achieved by means of a 1010... series.
- Further, the preamble contains a unique word (UW) indicating the start of the packet information (packet synchronization).
- The UW can also be used to resolve the phase ambiguity in the received signal (for coherent detection).

Packet Synchronization. The start of the packet information can be detected by comparing the received bit sequence with the UW pattern. The UW can be assumed as being detected if at least n_{th} bits within a window of received bits coincide with the UW, where n_{th} designates a threshold which must be appropriately chosen. The main criteria for the choice of n_{th} are the probabilities that the UW may not be detected or that a UW may erroneously be detected in an arbitrary bit stream. The probability to miss a UW increases with

- increasing threshold n_{th}
- increasing bit error rate
- increasing length of the UW.

The probability of erroneously detecting a UW (false alarm) increases with

- decreasing threshold n_{th}
- decreasing length of the UW.

The probability of a false alarm can be reduced by constraining the comparison of the bit stream with the UW to a time window in which the start of the packet can be expected.

The UW should be designed such that its autocorrelation function exhibits a single pronounced peak, see Sect. 5.8.1. Then, the begin of the data packet can be clearly detected.

5.6.3 Slot Synchronization in the Transmitting TDMA Terminals

Another synchronization requirement for TDMA is the synchronization of the bursts to the time slots of the TDMA frames. The bursts must be transmitted such that they fit into the TDMA time slots when they arrive at the satellite. Two situations can be identified:

1.) At the setup of a connection or during a satellite handover, initial slot synchronization is required.
2.) The slot synchronization must be maintained during a connection (tracking).

Two facts complicate TDMA slot synchronization in the satellite scenario:

- Depending on the user position within the satellite footprint, large delay differences occur.
- Due to the motion of non-geostationary satellites, the signal delay varies with time.

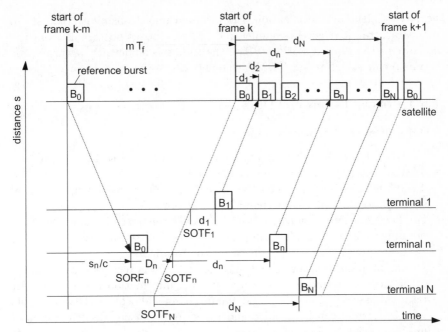

Fig. 5.10. Burst time plan and definition of the transmit frame of the terminals.
SORF = start of receive frame.
SOTF = start of transmit frame

Figure 5.10 shows how the terminals can synchronize the transmit time of their bursts according to a reference burst being broadcast by the satellite.

With T_f denoting the frame duration and s_n being the distance between the satellite and terminal n, the relation between SORF and SOTF is given by

$$\text{SOTF}_n - \text{SORF}_n = D_n = mT_f - 2s_n/c. \qquad (5.33)$$

Here, m is a positive integer, large enough that $D_n > 0$, i.e. $mT_f > 2s_n/c$. Terminal n transmits its burst after time delay d_n with regard to SOTF_n, i.e. the burst is transmitted at time

$$t = \text{SORF}_n + D_n + d_n. \qquad (5.34)$$

Equation (5.33) shows that the transmit frame of a terminal should begin the earlier the larger the distance is from the satellite. Two methods of determining the timing advance $2s_n/c$ can be distinguished:

For *open-loop synchronization*, the timing advance is determined from information about the positions of the terminal and the satellite, e.g. by using a satellite motion model and positioning means (such as GPS) integrated in the terminal. Guard times of 0.1 ms can be achieved by this method.

Closed-loop synchronization is based on repeated timing measurements and correction signals from the satellite or a reference ground station. For achieving initial synchronization, the terminal may transmit test bursts in a separate random access channel and receive correction signals from the satellite or the ground station. When the timing is sufficiently accurate, the terminal can switch to the TDMA channel.

5.6.4 Pros and Cons of TDMA

TDMA offers the following advantages:

- TDMA achieves higher efficiency than power-limited FDMA.
- TDMA technology is mature and well understood.
- TDMA terminals are cost effective, since they can extensively apply digital signal processing in the baseband.
- For TDMA, transmission and reception can be separated in time, therefore, no diplexer is required. This improves the receiver sensitivity.
- No exact power control is required for the terminals.
- During time slots without own transmission, TDMA terminals can monitor other time slots and TDMA carriers. This information can be used for terminal-assisted handover and dynamic channel allocation (DCA). In this way, interferers can be avoided and the system can be adapted to a non-uniform traffic load of the radio cells.
- For pure TDMA, no intermodulation occurs; therefore, the transponder can be operated near saturation. For MF-TDMA, less intermodulation occurs than for FDMA because of the lower number of carriers.
- Because of the high burst rate, TDMA exhibits lower sensitivity with regard to Doppler shifts and synthesizer phase noise.
- Flexible bit rates can be realized by allocating more than one time slot to a user.

TDMA has the following drawbacks:

- Because of the high burst rate, TDMA terminals must provide high peak transmit power.
- TDMA bursts contain overhead for burst synchronization.
- TDMA users must be synchronized to time slots.
- Because of the high transmission rate, adaptive equalizers may be required in the terminals and on board the satellite or in the ground station.

5.7 Code-Division Multiple Access, CDMA

In code-division multiple access the user signals are separated by modulating them with (nearly) orthogonal periodic signature sequences. This modulation results in a signal bandwidth which is much larger than the bit rate. Therefore, CDMA is also designated as spread spectrum multiple access, SSMA. Two modulation schemes are typically used for CDMA:

– frequency modulation (leading to frequency hopping CDMA, FH-CDMA)
– phase modulation (leading to direct-sequence CDMA, DS-CDMA).

At the receiver, a certain user signal can be reconstructed by correlating
the mix of received spread spectrum signals with the signature sequence corresponding to the wanted signal. The benefit of the correlation process with
regard to the wanted signal is called processing gain. The residual correlation of not perfectly orthogonal CDMA signatures causes interference in the
correlation process (multiple access interference).

Frequency Hopping CDMA (FH-CDMA). With frequency hopping
CDMA the data signal is multilevel frequency modulated with FSK or
CPFSK, e.g. with frequency steps being multiples of the signal bandwidth.
If an M-ary frequency modulation is used, each transmit frequency may be
determined by $\log_2 M$ bits of a binary signature sequence.

For fast frequency hopping several frequency modulation symbols are used
for each data symbol. For slow frequency hopping several data symbols are
transmitted during a single frequency modulation symbol. Some essential
features of FH-CDMA are:

– FH-CDMA signals are difficult to intercept and are robust to jamming.
– In combination with channel coding and interleaving, FH-CDMA signals
  are robust to multipath fading.
– Channel coding can be applied to restore erasures caused by hop collisions.
– A disadvantage of FH-CDMA is its low bandwidth efficiency.

FH-CDMA typically is applied for military communications, whereas modern
civil systems use direct-sequence CDMA. In the following we will therefore
concentrate on DS-CDMA.

5.8 Direct-Sequence CDMA (DS-CDMA)

In direct-sequence CDMA the spreading modulation is binary or quaternary
phase-shift keying (BPSK or QPSK). Alternatively, this can be viewed as
multiplying the baseband signal with a fast code sequence (signature), Fig.
5.11. An element of the code sequence is called a chip. Each terminal uses
a different code sequence from a set of (nearly) orthogonal code sequences.
In this way, the user signals can be well distinguished at the receiver (i.e.
the satellite or the ground station) which time-correlates the CDMA receive
signal with the user code signals and decides on the user data.

The basic procedure is to modulate each information bit (or code symbol)
of a user with one period of a periodic sequence consisting of P chips. An
alternative is to superimpose a user-specific spreading code with a very long
period P_{long} on the user signal. In this case, the bits are spread by successive
parts of length $P \ll P_{\text{long}}$ of the user-specific sequence.

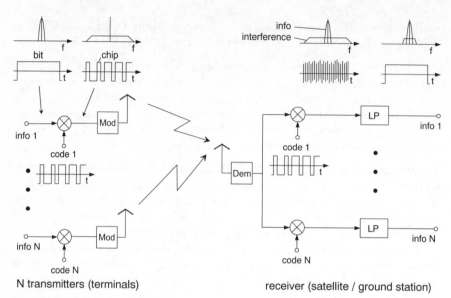

Fig. 5.11. Principle of DS-CDMA. LP = lowpass filter or integrate & dump

The users can transmit their signals asynchronously or synchronously (with regard to sequence periods and chips). For asynchronous CDMA no mutual synchronization of the user terminals is required, and hence it can be implemented more easily. On the other hand, for synchronous CDMA the code signals can be perfectly orthogonal, completely avoiding multiple access interference.

Coherent or incoherent detection can be applied in the receiver to detect the user signals. Incoherent detection is robust and therefore suitable for mobile communications; coherent signal detection requires a lower signal-to-noise ratio to achieve a given bit error rate.

5.8.1 Generation and Characteristics of Signature Sequences

Here we consider binary code sequences, which can be used as signature sequences for DS-CDMA. A periodic code sequence C is formed by stringing together periods of P elements, $c_1, c_2, \ldots, c_P$, with $c_i \in \{-1; 1\}$. Usually, each data bit is modulated by a period of the code sequence.

In order to assess the suitability of code sequences for CDMA the autocorrelation function (ACF) and the cross-correlation function (CCF) are considered. We take the autocorrelation function over one period of a code sequence C:

$$R_C(j) = \sum_{i=1}^{P} c_{i+j} c_i. \tag{5.35}$$

For $j = 0$ the ACF becomes

$$R_C(0) = \sum_{i=1}^{P} c_i^2 = P. \tag{5.36}$$

For $j \neq 0$ the ACF obeys

$$-P \leq R_C(j) \leq P. \tag{5.37}$$

Let C and D be two code sequences. Their cross-correlation function is defined as:

$$R_{CD}(j) = \sum_{i=1}^{P} c_{i+j} d_i \tag{5.38}$$

with

$$-P \leq R_{CD}(j) \leq P. \tag{5.39}$$

The code signals in Fig. 5.11 should have an autocorrelation function with a pronounced peak, enabling good signal detection. Moreover, the code signals should be well distinguishable, i.e. they should be as nearly orthogonal as possible; equivalently, they should exhibit small or zero cross-correlation.

These requirements can be well fulfilled by using independent stochastic sequences as code sequences (although these are not periodic). The problem with this approach is, of course, that the receiver cannot know the stochastic sequences used by the various terminals. Therefore, periodic deterministic sequences are used which approximate the features of stochastic sequences as well as possible. Such sequences are called pseudo-noise sequences (PN sequences). In the following, we will discuss the main types of PN sequences used for DS-CDMA systems.

Maximum Length Shift Register Sequences (m-Sequences). Maximum length shift register sequences can be generated by an m-stage binary shift register. The content of a register cell can be 0 or 1, and the shift register is fed back with the modulo-2 sum of a set of register outputs. This set is defined by taps of the shift register which can be described by coefficients $a_1 \ldots a_{m-1} \in \{0; 1\}$, Fig. 5.12.

The pattern of register taps determines the generated periodic sequence C. The initial content of the register determines the start phase of the period. If the taps represent a primitive polynomial[1] of degree m, the resulting sequence C exhibits the maximum possible period length

$$P = 2^m - 1 \tag{5.40}$$

and is called an m-sequence. A period of an m-sequence contains 2^{m-1} ones and $2^{m-1} - 1$ zeroes. At the output of the shift register the unipolar sequence is transformed into a bipolar m-sequence with $c_i \in \{-1; 1\}$ which is suitable for DS-CDMA, cf. Fig. 5.11. The ACF of a bipolar m-sequence C_μ is

---

[1] Primitive polynomial $p(x)$: irreducible and divisor of $x^m - 1$.

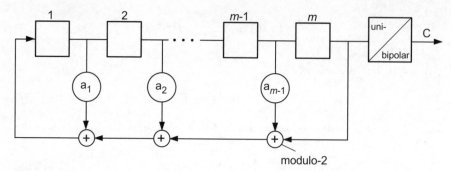

Fig. 5.12. Generating m-sequences by a maximum length shift register

$$R_\mu(j) = \begin{cases} P & j = 0 \\ -1 & j \neq 0 \end{cases} \qquad (5.41)$$

and thus approximates the ACF of a stochastic sequence which is

$$R_{\text{stoch}}(j) = \begin{cases} P & j = 0 \\ 0 & j \neq 0. \end{cases} \qquad (5.42)$$

Figure 5.13 shows the normalized ACF $R_\mu(\tau)/R_\mu(0)$ of a bipolar m-signal with rectangular pulses (chips) of duration T_c.

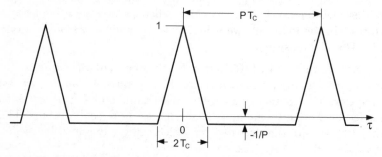

Fig. 5.13. Normalized ACF of a bipolar m-signal

Table 5.1 compiles the characteristics of m-sequences with different lengths of period. $R_{\mu\nu}(j)$ designates the cross-correlation function of two sequences C_μ and C_ν.

Figure 5.13 and Table 5.1 show that m-sequences have a very suitable ACF, but exhibit a very bad (large) CCF.

Gold Sequences. Gold sequences avoid the large CCF values of the m-sequences. A set of Gold sequences can be constructed by the following procedure:

1.) Choose an m-sequence A with period $P = 2^m - 1$.

Table 5.1. Characteristics of m-sequences

m	P	Number of different m-sequences with period P	$\max\lvert R_{\mu\nu}(j)\rvert/R_\mu(0)$
3	7	2	0.71
4	15	2	0.60
5	31	6	0.35
⋮	⋮	⋮	⋮
10	1023	60	0.37
11	2047	176	0.14
12	4095	144	0.34

2.) Find an m-sequence B with a small CCF $R_{AB}(j)$. If m is not divisible by 4, B can be taken to be the $(2^e + 1)$st decimation of A, i.e. $b_i = a_{i(2^e+i)}$, with $e = \lfloor m + 2 \rfloor/2$. The CCF of A and B will assume only three values:

$$R_{AB}(j) \in \{-1, -t(m), t(m) - 2\} \qquad j = 0 \ldots P - 1 \qquad (5.43)$$

with

$$t(m) = \begin{cases} 2^{(m+1)/2} + 1 & m \text{ odd} \\ 2^{(m+2)/2} + 1 & m \text{ even.} \end{cases} \qquad (5.44)$$

A and B are called a pair of preferred m-sequences [Pro89].

3.) Generate all P sequences $B \oplus \{P$ cyclic shifts of $A\}$, with $\oplus$ designating the modulo-2 sum.

4.) Together with the initial m-sequences A and B, a family of $P + 2$ Gold sequences has been obtained.

For different choices of A and B different families of Gold sequences arise. Gold sequences built from m-sequences with odd m exhibit especially useful characteristics. Gold [Gol68] has shown that the CCF for any pair of Gold sequences and the off-peak ACF of a Gold sequence (except for the initial m-sequences A and B) are three-valued according to Eqs. (5.43) and (5.44). Thus, the maximum value of the off-peak ACF and the maximum value of the CCF are

$$\max\lvert R_\mu(j \neq 0)\rvert = \max\lvert R_{\mu\nu}(j)\rvert = t(m). \qquad (5.45)$$

The maximum CCF value $t(m)$ approaches a lower bound (the Welch bound) until a factor of $\sqrt{2}$ for odd m and a factor of 2 for even m. The characteristics of Gold sequences are listed in Table 5.2.

Preferentially-Phased Gold Sequences. For constructing a set of preferentially-phased Gold sequences, the initial m-sequence A is omitted from the full Gold code set. In this way, a family of $P + 1$ preferentially-phased Gold sequences is obtained, with $R_{\mu\nu}(0) = -1$ for all sequences C_μ, C_ν of the family [MM91]. This is the optimal cross-correlation value at the origin that can

Table 5.2. Characteristics of Gold sequences [Pro89]

m	P	Number of different Gold sequences	$t(m)$	$t(m)/R_C(0)$
3	7	9	5	0.71
5	31	33	9	0.29
7	127	129	17	0.13
$\vdots$	$\vdots$	$\vdots$	$\vdots$	$\vdots$
11	2047	2049	65	0.03

be achieved, and the sequences are called quasi-orthogonal [GEV92]. Thus, preferentially-phased Gold sequences are optimal for synchronous CDMA. The sequence A has been omitted from the set because the above cross-correlation property would not hold for the full Gold sequence set.

Hadamard Sequences. Hadamard sequences are built on the basic Hadamard matrix

$$M_2 = \begin{bmatrix} 0 & 0 \\ 0 & 1 \end{bmatrix} . \tag{5.46}$$

Hadamard sequences can be constructed by expanding the basic Hadamard matrix according to

$$M_{2n} = \begin{bmatrix} M_n & M_n \\ M_n & \overline{M_n} \end{bmatrix}. \tag{5.47}$$

The rows of a Hadamard matrix M_n (with even n) represent n Hadamard sequences of length n. All these sequences mutually differ in $n/2$ positions, i.e. they are orthogonal. The cross-correlation of bipolar Hadamard sequences is

$$R_{\mu\nu}(0) = 0. \tag{5.48}$$

An example of a Hadamard matrix is

$$M_4 = \begin{bmatrix} 0 & 0 & 0 & 0 \\ 0 & 1 & 0 & 1 \\ 0 & 0 & 1 & 1 \\ 0 & 1 & 1 & 0 \end{bmatrix} . \tag{5.49}$$

Unfortunately, Hadamard sequences have unfavorable ACF characteristics. Without additional measures they can only be used for *synchronous* CDMA.

5.8.2 Investigation of Asynchronous DS-CDMA in the Time Domain

In this Section we will show how a DS-CDMA receiver processes the incoming signal in order to isolate a wanted user signal. This consideration will also reveal the importance of the correlation characteristics of the used code sequences. Let us consider a set of users n, with $n = 1 \ldots N$. $a_n(t)$ is the data signal of user n, with symbol duration T_s:

$$a_n(t) = \sum_{k=-\infty}^{\infty} a_{n,k} \, g_a(t - kT_s).$$ (5.50)

The data signal has a rectangular pulse shape with amplitude 1 and duration T_s:

$$g_a(t) = \begin{cases} 1 & 0 \le t < T_s \\ 0 & \text{otherwise.} \end{cases}$$ (5.51)

A full period of the code signal containing P chips is assumed to fit into one symbol duration T_s of the data signal. Then, the code signal $c_n(t)$ of user n has chip duration $T_c = T_s/P$ and can be expressed by

$$c_n(t) = \sum_{k} \sum_{i=0}^{P-1} c_{n,i} \, g_c(t - kT_s - iT_c).$$ (5.52)

Also the code signal is assumed to consist of rectangular pulses (chips):

$$g_c(t) = \begin{cases} 1 & 0 \le t < T_c \\ 0 & \text{otherwise.} \end{cases}$$ (5.53)

The transmit signal is formed by multiplying the data signal $a_n(t)$ with the code signal $c_n(t)$ and modulating it onto a radio carrier with frequency ω_0:

$$s_n(t) = \sqrt{2} a_n(t) c_n(t) \cos \omega_0 t.$$ (5.54)

The symbol energy is $E_s = T_s$. Figure 5.14 shows a simplified block diagram of the DS-CDMA transmitter and receiver.

Since we are considering asynchronous CDMA, the user signals arrive at the receiver with different delays τ_n and carrier phases φ_n. We relate the delay and the carrier phase to the (wanted) signal of user 1; hence $\tau_1 = 0$ and $\varphi_1 = 0$. The input signal of the receiver is the sum of the arriving user signals:

$$e(t) = \sum_{n=1}^{N} s_n(t - \tau_n).$$ (5.55)

Let us assume that the receiver coherently demodulates the received signal with regard to the signal of user 1. Then, the demodulated signal is

$$\begin{aligned} r(t) &= \sum_{n=1}^{N} s_n(t - \tau_n) \cdot \sqrt{2} \cos \omega_0 t \\ &= \sum_{n=1}^{N} a_n(t - \tau_n) c_n(t - \tau_n) \cos \varphi_n \; . \end{aligned}$$ (5.56)

transmitter:

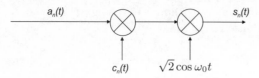

coherent receiver:

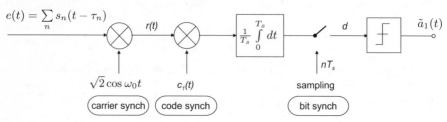

Fig. 5.14. Block diagram of the DS-CDMA transmitter and receiver

Here, the signal component at twice the carrier frequency has been filtered out. The demodulated signal $r(t)$ is correlated with the code signal of user 1, resulting in the decision variable for the data of user 1:

$$d = \frac{1}{T_s} \int_0^{T_s} r(t)c_1(t)dt = a_{1,0} + \sum_{n=2}^{N} \tilde{R}_{1n}(\tau_n) \cos \varphi_n. \tag{5.57}$$

Here, the summation term represents the interference caused by the other users ($n \neq 1$). The correlator (multiply by $c_1(t)$ plus integrate & dump) in Fig. 5.14 can also be viewed as a matched filter for the code signal $c_1(t)$. The function

$$\tilde{R}_{1n}(\tau_n) = \frac{1}{T_s} \int_0^{T_s} a_n(t - \tau_n)c_n(t - \tau_n)c_1(t)dt \tag{5.58}$$

is the normalized even or odd cross-correlation function of the code sequences C_1 and C_n, depending on the data signal of user n. In order to achieve a small CDMA interference, it is important that the CCF $\tilde{R}_{1n}(\tau_n)$ is small.

A special case is synchronous CDMA using orthogonal codes, for which $\forall n$: $\tau_n = 0$ and $R_{1n}(0) = 0$. Then, no interference due to other users arises and the despread signal is given by

$$d = a_{1,0}. \tag{5.59}$$

With the analysis in the time domain it can also be shown that CDMA can suppress multipath echoes. For this purpose we consider a user signal with an echo, which is characterized by echo amplitude E, echo phase φ_E, and echo delay τ_E. The demodulated receive signal is

$$r(t) = a_1(t)c_1(t) + E \cos \varphi_E \cdot a_1(t - \tau_E)c_1(t - \tau_E). \tag{5.60}$$

The despreading results in

$$d = \frac{1}{T_s} \int_0^{T_s} r(t)c_1(t)dt = a_{1,0} + E\cos\varphi_E \tilde{R}_1(\tau_E). \qquad (5.61)$$

The function

$$\tilde{R}_1(\tau_E) = \frac{1}{T_s} \int_0^{\tau_E} a_{1,-1}c_1(t-\tau_E)c_1(t)dt + \frac{1}{T_s} \int_{\tau_E}^{T_s} a_{1,0}c_1(t-\tau_E)c_1(t)dt$$

$$(5.62)$$

is the normalized even or odd autocorrelation function of the code sequence C_1, depending on the data signal of user 1. In order to achieve good echo suppression, it is important that the ACF $\tilde{R}_1(\tau)$ is small for $\tau \neq 0$.

As we will see in Sect. 5.9.2, a Rake receiver separately demodulates and despreads the echo component and adds the result to the decision variable d. In this way, also the energy contained in the echo signal is exploited for the data detection.

5.8.3 Investigation of Asynchronous DS-CDMA in the Frequency Domain

In order to derive the capacity and bandwidth efficiency of asynchronous DS-CDMA, we resort to a consideration in the frequency domain. To this end we make the following assumptions:

- There are N active users in the considered satellite spot beam.
- The user signals are channel-coded with rate r.
- Each transmit symbol is modulated by P chips of a signature sequence, with P usually representing a period of the sequence. This procedure corresponds to cascading channel coding and spreading.
- The spread user signals are transmitted asynchronously and with arbitrary carrier phases.
- Gaussian transmission channels are assumed (no signal shadowing, no multipath fading).
- The transmission power of the user terminals is perfectly controlled, i.e. the differences in distance and antenna gain are compensated. Thus, all user signals are received at the satellite with equal power. The analysis also holds for fading channels if power control can perfectly compensate for multipath fading and shadowing. Finally, power control can also compensate for deviations of the terminal antennas from a hemispherical characteristic.
- During speech pauses, the transmit signal is switched off. The voice activity state of user n is denoted $\alpha_n \in \{0;1\}$ with $E\{\alpha_n\} = \alpha \approx 0.5$. In real systems, the transmit signal is not switched off during speech pauses, but transmitted with a much lower bit rate and power. Then, α corresponds to the average transmit power divided by the transmit power during talk spurts.

With $T_s = T_b \cdot r \log_2 M$ denoting the duration of a transmitted symbol, the chip duration is given by

$$T_c = \frac{T_s}{P} = \frac{T_b \, r \log_2 M}{P} \tag{5.63}$$

and the chip rate is

$$R_c = \frac{1}{T_c} = P R_s = P \frac{R_b}{r \log_2 M} \; . \tag{5.64}$$

The processing gain is defined as the ratio of the chip rate to the bit rate:

$$G =: \frac{R_c}{R_b} = \frac{P}{r \log_2 M} \; . \tag{5.65}$$

This parameter describes the interference reduction potential of the asynchronous DS-CDMA scheme.

Differently from the previous section, the chip pulse shape is assumed to be ideally band-limited, according to a roll-off factor $\beta = 0$:

$$\begin{aligned} G_c(f) &= \begin{cases} T_c & |f| \leq 1/2T_c \\ 0 & \text{otherwise} \end{cases} \\ g_c(t) &= \frac{\sin(\pi t/T_c)}{\pi t/T_c} \; . \end{aligned} \tag{5.66}$$

As an example, Globalstar uses a CDMA pulse shape with $\beta \approx 0.02$.

The CDMA transmit signal has power $E_b/T_b = E_b R_b$, and the bandwidth of the CDMA carrier is

$$B_T = R_c = G R_b \gg R_s. \tag{5.67}$$

Thus, the processing gain can also be expressed as $G = B_T/R_b$. Finally, the power spectral density of the transmit signal is given by

$$E_b R_b/R_c = E_b/G \quad \text{for } |f - f_0| \leq 1/2T_c \tag{5.68}$$

and 0 otherwise.

Receive Signal and Interference Spectral Power Density. For asynchronous DS-CDMA, the user signals are transmitted with arbitrary time shifts and carrier phases. Moreover, they are modulated with independent data. Hence, for a considered (wanted) user all the other users' signals act as additive noise signals with power spectral density E_b/G, and on average add up in terms of power [Vit95]. Therefore, the power spectral density of the noise caused by other users (multiple access interference, MAI) results as the sum of the power spectral densities of $N - 1$ user signals:

$$L_i = \sum_{n=2}^{N} \alpha_n E_b/G. \tag{5.69}$$

If independent PN sequences are used for the in-phase and quadrature components of a quaternary transmit signal, the signal power is equally distributed among these components. Then, the user interference becomes independent of the carrier phase of the different user signals.

In the following, L_i will be approximated by its mean value

$$\overline{L_i} = \alpha(N-1)E_b/G. \tag{5.70}$$

The influence of the statistical variations of L_i will be neglected. Taking into account the thermal noise power spectral density N_0, the mean total noise power spectral density results in

$$L_{\text{tot}} = \overline{L_i} + N_0 = \alpha(N-1)\frac{E_b}{G} + N_0. \tag{5.71}$$

When correlating the received signal with the PN sequence of the wanted user, the corresponding data signal is despread (and thus reconstructed), but the power spectral density of the other-user interference remains unchanged. Therefore, the signal-to-noise ratio of the despread signal is given by

$$\frac{E_b}{L_{\text{tot}}} = \frac{E_b}{N_0 + \alpha(N-1)E_b/G}. \tag{5.72}$$

This equation shows that the interference from other users is reduced by the processing gain G with typical values ranging from 100 to 1000.

Bit Error Probability and User Capacity. For the determination of the bit error probability the total noise can be assumed to be Gaussian distributed, if $N \gg 1$. Under this assumption, the well-known formulas for the bit error probability in the Gaussian channel can be applied. For uncoded BPSK and QPSK, corresponding to Eq. (4.13), we get

$$p_b = \frac{1}{2}\text{erfc}\sqrt{\frac{E_b}{L_{\text{tot}}}} = \frac{1}{2}\text{erfc}\sqrt{\frac{E_b}{N_0 + \alpha(N-1)E_b/G}}. \tag{5.73}$$

Equation (5.73) shows that CDMA requires a higher transmit power than a single transmission being limited by thermal noise N_0 only.

In order to stay below a given maximum bit error probability, the system must accordingly provide a minimum signal-to-noise ratio E_b/L_{tot}. With this requirement, the maximum number of active users ("channels") follows from Eq. (5.72) as

$$\begin{aligned}
N &= 1 + \frac{G}{\alpha}\left[\left(\frac{E_b}{L_{\text{tot}}}\right)^{-1} - \left(\frac{E_b}{N_0}\right)^{-1}\right] \\
&= 1 + \frac{B_T}{R_b} \cdot \frac{1}{\alpha}\left[\left(\frac{E_b}{L_{\text{tot}}}\right)^{-1} - \left(\frac{E_b}{N_0}\right)^{-1}\right]. \tag{5.74}
\end{aligned}$$

N increases with processing gain G and available bandwidth B_T. The reduction of transmit power during speech pauses saves terminal power, reduces multiple access interference, and thus increases the capacity of CDMA.

Effect of Channel Coding. For a given bit rate R_b and a given CDMA signal bandwidth B_T, the number M of modulation levels as well as the channel code rate r do not appear in Eq. (5.74). However, the application of channel coding influences Eq. (5.74) because the required signal-to-noise ratio E_b/L_{tot} for coded transmission should be smaller than the signal-to-noise ratio required for uncoded transmission. Therefore, from Eq. (5.74) it can be seen that channel coding increases the possible number N of users without increasing the required bandwidth B_T of the CDMA carrier. However, with channel coding the ratio $T_s/T_c = P$ becomes smaller, resulting in a smaller number of possible spreading sequences.

Example. As an example, we consider a Gaussian channel and a BPSK-modulated signal which is channel-coded with a convolutional code with rate $r = 1/2$ and constraint length 7 ($\nu = 6$). A bit error probability of 10^{-6} shall be achieved. From Fig. 4.15 it can be seen that a signal-to-noise ratio of $E_b/L_{tot} = 5$ dB (corresponding to a factor of 3.16) is required. Assuming an average voice activity of $\alpha = 0.5$ and a processing gain of $G = B_T/R_b = 127$, the number of CDMA users, N, according to Eq. (5.74) is shown in Fig. 5.15 as a function of transmit power represented by E_b/N_0.

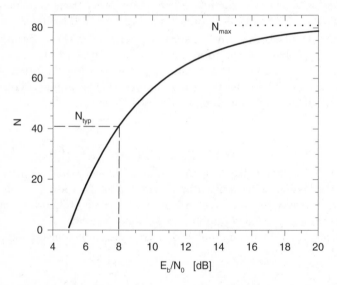

Fig. 5.15. Number of active users for asynchronous CDMA, single radio cell. $E_b/L_{tot} = 5$ dB, $G = 127$, $\alpha = 0.5$

In the limit of very high transmit power, $(E_b/N_0)^{-1}$ can be neglected, and the user capacity becomes

$$N_{max} = 1 + \frac{1}{\alpha} \cdot \frac{G}{E_b/L_{tot}} = 1 + \frac{1}{\alpha} \cdot \frac{B_T/R_b}{E_b/L_{tot}}. \tag{5.75}$$

For the above example we get $N_{\max} = 1 + 0.63 \cdot B_T/R_b = 81$.

In the limit of very high transmit power, however, the system is very sensitive to an imbalance in the received power (e.g. due to imperfect power control) and statistical variations of voice activity. Therefore, the transmit power is chosen at a lower value, e.g. such that the average interference power is equal to the thermal noise power [Gib96]:

$$\overline{L_i} = \alpha(N - 1)E_b/G = N_0. \tag{5.76}$$

With this, the total noise power spectral density becomes

$$L_{\text{tot}} = 2\alpha(N - 1)E_b/G. \tag{5.77}$$

Solving this equation for N, a typical number of CDMA users amounts to

$$N_{\text{typ}} \approx \frac{G}{2\alpha E_b/L_{\text{tot}}} \approx \frac{G}{E_b/L_{\text{tot}}} = \frac{B_T/R_b}{E_b/L_{\text{tot}}} \tag{5.78}$$

which is approximately one-half of the maximum number of users, $N_{\max}$. For the above example we get $N_{\text{typ}} \approx 0.32 \cdot B_T/R_b = 40$.

Alternatively to backing off the number of users by one-half, Eq. (5.78), less backoff from $N_{\max}$ may be applied, e.g. down to $0.7 \cdot N_{\max}$.

For a single radio cell as considered here, CDMA is substantially less bandwidth efficient than TDMA with quaternary modulation ($N \approx 0.6 \cdot B_T/R_b$). However, while the number of users is hard-limited for TDMA, it is soft-limited for CDMA. Increasing the number of users above the value given by Eq. (5.78) somewhat reduces the signal-to-noise ratio E_b/L_{tot} (graceful degradation), however, no blocking of calls occurs. Moreover, we will see that CDMA can develop its full potential in a cellular spot beam scenario, see Chap. 6.

Efficiency (Throughput) of CDMA. According to Eq. (5.17) the throughput of CDMA is

$$\eta_{\text{CDMA}} = \frac{NR_b(1 + \beta_{\text{cont}})}{r \log_2 M \cdot B_T}$$

$$\eta_{\text{CDMA,typ}} = \frac{1 + \beta_{\text{cont}}}{r \log_2 M \cdot E_b/L_{\text{tot}}} \tag{5.79}$$

with β_{cont} designating the roll-off factor for continuous transmission. The bandwidth efficiency of CDMA is given by

$$\eta_{\text{b,CDMA}} = \frac{\text{total bit rate}}{\text{CDMA bandwidth}} = \frac{NR_b}{B_T} \quad \text{in} \quad \frac{\text{b/s}}{\text{Hz}}$$

$$\eta_{\text{b,CDMA}} \approx \frac{1}{\alpha}\left[\left(\frac{E_b}{L_{\text{tot}}}\right)^{-1} - \left(\frac{E_b}{N_0}\right)^{-1}\right]$$

$$\eta_{\text{b,CDMA,typ}} \approx \frac{1}{E_b/L_{\text{tot}}}. \tag{5.80}$$

5.8.4 Multi-Frequency CDMA, MF-CDMA

In order to provide a larger number of channels within a radio cell, several CDMA carriers at different frequencies can be used in parallel. In the frequency domain, the carriers are separated by guard bands B_g. The guard bands required for MF-CDMA are usually smaller than those for MF-TDMA.

If T CDMA carriers are used for a cell, the total required bandwidth is

$$B_c = T(GR_b + B_g), \tag{5.81}$$

and, corresponding to Eq. (5.74), the number of active users per cell results in

$$N_c = TN = T + \frac{B_c - TB_g}{R_b} \cdot \frac{1}{\alpha} \left[\left(\frac{E_b}{L_{\text{tot}}} \right)^{-1} - \left(\frac{E_b}{N_0} \right)^{-1} \right]. \tag{5.82}$$

Globalstar is an example for an MF-CDMA system, see App. C.5. 13 CDMA carriers with 1.23 MHz bandwidth can be used in a spot beam, resulting in a cell bandwidth $B_c = 16.5$ MHz.

5.8.5 Qualcomm Return Link CDMA (Globalstar)

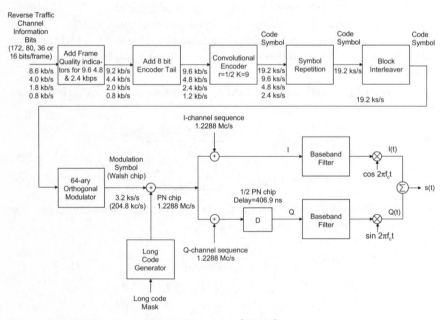

Fig. 5.16. CDMA return link of Globalstar [Gib96]

Globalstar uses asynchronous CDMA with non-coherent detection. A special feature of the Globalstar approach is the usage of an extremely long PN code with a period of $2^{42} - 1$, Fig. 5.16. Each terminal uses the same long PN code, however, the time offset of the code is determined by the user address [MC94]. Since every possible time offset is a valid address, a virtually unlimited number of signature codes exist. Hence, an individual signature code corresponding to an exclusive access channel can be allocated to each user. This allows soft handover and corresponds to perfect dynamic channel allocation, see Chap. 7.

For the transmission, modulation with orthogonal Walsh-Hadamard sequences and non-coherent maximum-likelihood envelope detection is used (fast inverse Hadamard transformation) [Vit95]. Non-coherent detection is robust and especially advantageous for low-rate mobile satellite systems, for which coherent detection is difficult because of signal shadowing, fading, and severe phase noise.

5.8.6 Synchronous Orthogonal DS-CDMA with Coherent Detection

For chip (and bit) synchronous CDMA, perfectly orthogonal spreading codes can be realized, e.g. by using Hadamard sequences. This approach eliminates the interference between the user signals (multiple access interference) and thus achieves the bandwidth efficiency of the underlying modulation scheme.

In the forward direction, chip synchronous CDM can easily be realized, because the signals are transmitted from a single entity (satellite or gateway). In order to aid coherent detection, an additional reference signal is broadcast by the satellite which can be used for carrier and chip timing synchronization in the terminals. This pilot signal is a known DS-CDMA signal which may be transmitted with enhanced power.

Synchronous CDMA is also proposed for the return link [GEV92]. For non-geostationary satellites and mobile users, however, it may be a formidable task to achieve chip synchronization of the users.

An attractive feature of synchronous CDMA is that it requires less precise power control than asynchronous CDMA.

On the other hand, synchronous CDMA can not be combined with satellite diversity, because for geometrical reasons it is in general not possible to synchronize separated users simultaneously to more than one satellite.

Bandwidth Efficiency of Synchronous DS-CDMA with Orthogonal Sequences. There are only P orthogonal spreading sequences with length P (e.g. Walsh-Hadamard sequences or preferentially-phased Gold codes). The efficiency of synchronous CDMA is therefore code book-limited. According to [DGL96], the bandwidth efficiency of synchronous DS-CDMA is given by

$$\eta_b = \min\left\{ G \log_2 M \left[\left(\frac{E_b}{L_{\mathrm{tot}}} \right)^{-1} - \left(\frac{E_b}{N_0} \right)^{-1} \right] ; r \log_2 M \right\}. \qquad (5.83)$$

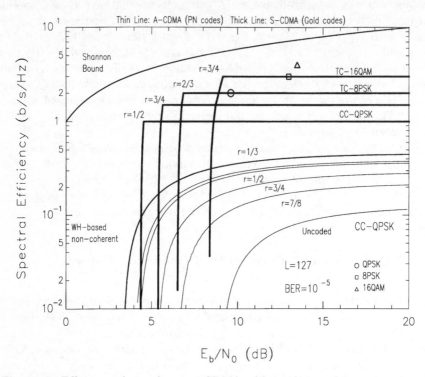

Fig. 5.17. Efficiency of asynchronous CDMA with incoherent detection and synchronous CDMA with coherent detection [DGL96]. $L \triangleq$ period length P. ©1996 IEEE

Figure 5.17 shows that the bandwidth efficiency is substantially improved compared to asynchronous CDMA. In the Gaussian channel, synchronous CDMA achieves the efficiency of ideal FDMA or TDMA.

Unfortunately, even with synchronous CDMA some interference can arise:

a) Other users in neighboring radio cells or in neighboring satellite footprints may not be synchronous to the users in the considered cell and therefore generate interference.

b) Multipath signal components with a delay of more than one chip are not orthogonal to the direct signal component.

c) A nonlinear satellite transponder modifies the correlation characteristics of the code signals and destroys the orthogonality of the user signals.

Examples for the application of synchronous CDMA are the Globalstar forward link (Qualcomm CDMA) and an earlier concept for the Mobile Satellite Business Network of ESA, which should have used quasi-orthogonal CDMA [GEV92].

5.9 CDMA Receivers

5.9.1 PN Code Synchronization in the CDMA Receiver

In order to despread a specific user signal, the code sequence used in the receiver for the correlation with the received signal must be synchronized to the code sequence contained in the received signal. At the start of a transmission the code phase of the received user signal must be acquired, and during the transmission the code phase has to be tracked.

Code Acquisition. Figure 5.18 shows how code phase acquisition may be realized.

In Fig. 5.18, BP designates a bandpass filter which is broadband with regard to the data signal $a(t)$, but narrowband with regard to the code signal $c(t)$. Thus, the output signal of the bandpass filter is

$$x(t, \Delta) = \sqrt{2}a_n(t - \tau_n)R_{nn}(\Delta) \cdot \cos\omega_0 t + \text{interference} \qquad (5.84)$$

ED designates an envelope detector plus a multiplication by $1/\sqrt{2}$. The output signal of ED is

$$y(\Delta) \approx |R_{nn}(\Delta)|. \qquad (5.85)$$

According to $y(\Delta)$, the code generator is controlled in the following way:

– If $y(\Delta)$ is smaller than a suitable threshold, the internal code phase of the receiver is shifted by half a chip duration, $T_c/2$, and $y(\Delta)$ is tested again.
– If $y(\Delta)$ is equal to or greater than the threshold, the code synchronization loop is switched to tracking.

Usually, prior to each of the above tests, $y(\Delta)$ is averaged over several periods of the code sequence.

As an alternative to the serial code acquisition described above, parallel synchronization can be realized by testing several possible code phases one at a time.

Because the code acquisition is performed incoherently, it can be performed before the carrier synchronization. The carrier phase is then extracted from the despread RF or IF signal [Vit95].

$$e(t) = \sqrt{2}a_n(t - \tau_n)c_n(t - \tau_n)\cos\omega_0 t + \text{interference}$$

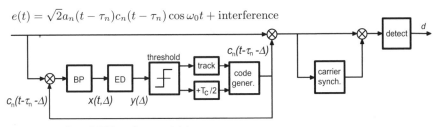

Fig. 5.18. Serial PN code acquisition

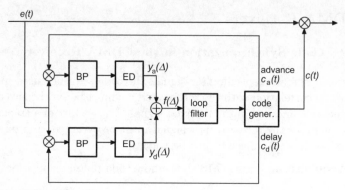

Fig. 5.19. PN code tracking

Tracking. The tracking loop adjusts the internal code phase of the receiver such that $y(\Delta)$ is maximized, i.e. $\Delta = 0$. A possible method of code tracking is shown in Fig. 5.19.

The code generator in the tracking loop generates the code signal

$$c(t) = c_n(t - \tau_n - \Delta), \tag{5.86}$$

which is shifted by Δ against the received code signal. Additionally, the code generator outputs an advanced code signal

$$c_a(t) = c_n(t - \tau_n - \Delta + T_c/2) \tag{5.87}$$

and a delayed code signal

$$c_d(t) = c_n(t - \tau_n - \Delta - T_c/2). \tag{5.88}$$

The output signals of the envelope detectors are

$$\begin{aligned}
y_a(\Delta) &= |R_{nn}(\Delta - T_c/2)| \\
y_d(\Delta) &= |R_{nn}(\Delta + T_c/2)|
\end{aligned} \tag{5.89}$$

and the error signal is

$$f(\Delta) = y_a(\Delta) - y_d(\Delta), \tag{5.90}$$

see Fig. 5.20. After being filtered, this error signal is used to control the receiver code phase. The range of the code phase error Δ for which this control loop works is given by

$$-\frac{3T_c}{2} < \Delta < \frac{3T_c}{2}. \tag{5.91}$$

As an alternative to the envelope detector a quadratic detector can be used.

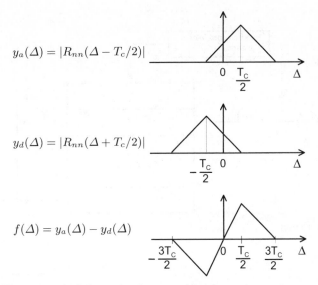

$$y_a(\Delta) = |R_{nn}(\Delta - T_c/2)|$$

$$y_d(\Delta) = |R_{nn}(\Delta + T_c/2)|$$

$$f(\Delta) = y_a(\Delta) - y_d(\Delta)$$

Fig. 5.20. Error signal of the code phase tracking loop

5.9.2 Rake Receiver

In radio communications, a CDMA signal can propagate to the receiver via more than one path. In such a situation, a Rake receiver can be used to improve the reception quality by detecting and evaluating several receive signal components. In satellite communications, more than one replica of a single transmitted signal can be received in the following scenarios:

– In an environment with frequency-selective multipath fading, echoes of a signal are received in addition to the direct signal. This situation occurs especially for mobile or personal satellite communications. Without a Rake receiver the echoes contribute to the CDMA interference.
– If satellite diversity is applied, a signal is transmitted via more than one satellite in parallel. This can be related to forward transmission, where the signal is transmitted from a ground station, as well as to return transmission, where the user terminal transmits via more than one satellite.
– During soft handover between spot beams or satellites, the situation is very similar to diversity transmission.

The principle of a Rake receiver is shown in Figs. 5.21 to 5.23. A Rake receiver consists of independent parallel spread spectrum demodulators (Rake fingers). Each Rake finger despreads a received signal component, using an accordingly delayed code sequence of the user signal to be detected. The despread signal components are then combined. In this way, a Rake receiver uses the energy of all received signal components for bit detection. As an example, Globalstar uses a Rake receiver with 2 fingers in the mobile terminals and a Rake receiver with up to 5 fingers in the fixed earth stations.

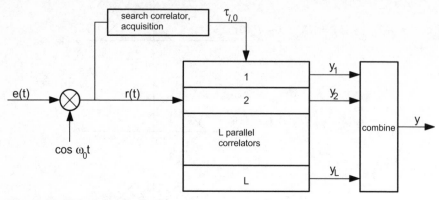

Fig. 5.21. Principle of a Rake receiver

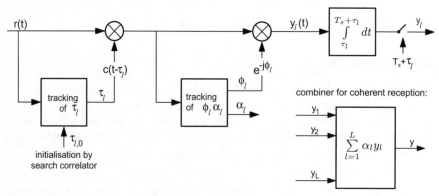

Fig. 5.22. Rake receiver for coherent reception

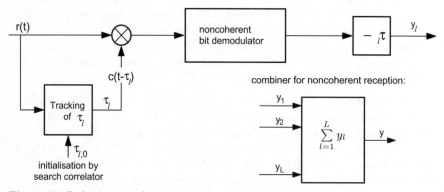

Fig. 5.23. Rake receiver for non-coherent reception

Each signal component (e.g. echo) $l = 1 \ldots L$ is determined by its delay τ_l, its amplitude α_l, and its phase ϕ_l. Accordingly, the total received signal is given by

$$e(t) = \sum_{l=1}^{L} \alpha_l a(t - \tau_l) c(t - \tau_l) cos(\omega_0 t + \phi_l). \tag{5.92}$$

Here, $a(t)$ denotes the data signal with symbol duration T_s, and $c(t)$ is the code signal with chip duration T_c.

The Rake receiver can distinguish signal paths for which the delay difference is at least in the order of a chip duration: $|\Delta\tau_l| \geq T_c$. However, for $|\Delta\tau_l| > T_s$, Rake reception is useless because in this case the echoes will be assigned to different information bits.

The task of the search finger is to scan the delay interval $0 \leq \tau \leq T_s$, to acquire the signal components (to determine $\tau_1 \ldots \tau_L$), and to initialize the Rake fingers.

A coherent Rake receiver tracks α_l and ϕ_l by using either a CDMA pilot or a channel estimator. This enables the coherent combination of multiple signal components. For this purpose, the correlator outputs are maximal ratio combined, which is the optimum kind of combining:

$$y = \sum_{l=1}^{L} \alpha_l y_l. \tag{5.93}$$

In a non-coherent Rake receiver, the values of α_l and ϕ_l are not available. Therefore, the correlator outputs can only be non-coherently combined, i.e. they are added up with equal weights:

$$y = \sum_{l=1}^{L} y_l. \tag{5.94}$$

Performance of a Coherent Rake Receiver. After despreading and phase adjustment, a signal component results in

$$y_l(t) = r(t) c(t - \tau_l) e^{-j\phi_l} = \alpha_l a(t - \tau_l) + \text{interference}, \tag{5.95}$$

and the signal at the output of the integrator is

$$
\begin{aligned}
y_l \;=\; &\pm \alpha_l T_s \\
&+ \text{interference from other signal components} \\
&+ \text{interference from other users.}
\end{aligned}
\tag{5.96}
$$

The maximal ratio combining of these components gives

$$y = \sum_{l=1}^{L} \alpha_l y_l = \pm T_s \sum_{l=1}^{L} \alpha_l^2. \tag{5.97}$$

a) Static Multipath Situation. Here, α_l, τ_l, and ϕ_l are constant, and the received bit energy is

$$E_b = T_s \sum_{l=1}^{L} \alpha_l^2 = \text{const.} \tag{5.98}$$

The Rake receiver achieves the same bit error probability as a conventional receiver with bit energy E_b. The Rake receiver acts as a matched-filter receiver for the transmit filter plus the multipath channel.

b) Mobile Radio Channel. Here, the multipath components (e.g. echoes) are Rayleigh processes with a bandwidth approximately corresponding to the Doppler frequency ν/λ. Hence, the amplitudes α_l are Rayleigh-distributed, and the echo phases are uniformly distributed between 0 and 2π.

The bit energy of an echo component is

$$E_l = T_s \alpha_l^2 \tag{5.99}$$

with mean value $\overline{E_l}$ and probability density function

$$p(E_l) = \frac{1}{\overline{E_l}} \exp(-E_l/\overline{E_l}). \tag{5.100}$$

Accordingly, also the bit energy resulting from the combination of the echoes is not constant. As an example, let us consider L Rayleigh components with equal mean bit energy $\overline{E_l}$. Then, the resulting bit energy

$$E_b = T_s \sum_{l=1}^{L} \alpha_l^2 \tag{5.101}$$

is chi-square distributed with order L:

$$p(E_b) = \frac{E_b^{L-1} \exp(-E_b/\overline{E_l})}{(L-1)!(\overline{E_l})^L} . \tag{5.102}$$

Compared to the Rayleigh distribution, Eq. (5.100), the probability of small bit energy values is substantially reduced. The Rake receiver acts like L-fold diversity with optimum maximal ratio combining, see Fig. 5.24.

For coherent detection of CDMA signals it is useful to despread the signal before coherent detection. Otherwise, the carrier has to be recovered from the spread signal with strong interference [MB93].

5.9.3 CDMA Multiuser Detection

In a conventional DS-CDMA receiver as discussed in Sect. 5.8.2, a particular user's signal is detected by correlating the total received signal $r(t)$ (the sum of all user signals) with the wanted user's signature code signal. For

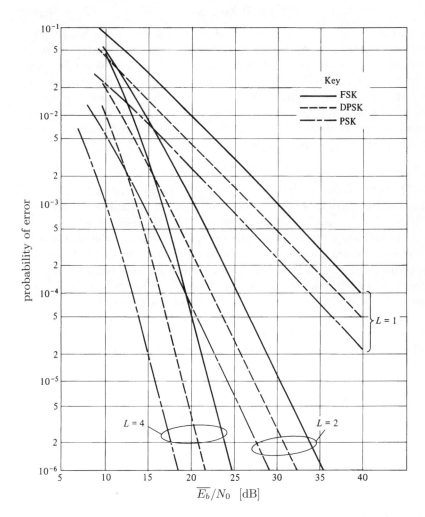

Fig. 5.24. Bit error probability for L-fold diversity with maximal ratio combining in the Rayleigh channel [Pro89]

non-orthogonal asynchronous CDMA the other users' signals cause multiple access interference, limiting the capacity of the system. User signals with high received power cause especially strong interference (near-far problem).

In matrix-vector notation, the output of the conventional CDMA detector can be described as [Mos96]

$$y = RAd + z. \tag{5.103}$$

For a synchronous system, the vectors d, z, and y are N-vectors containing the data, noise, and correlator (matched-filter) outputs of the N users, respectively. The matrix A is a diagonal matrix containing the received amplitudes; the matrix R is a symmetric $N \times N$ correlation matrix, whose entries contain the values of the correlations between every pair of codes. The matrix R can be broken up into two matrices, one representing the autocorrelations, the other the cross-correlations. With this, the conventional matched-filter detector output can be expressed as

$$y = Ad + QAd + z, \tag{5.104}$$

where Q contains the off-diagonal elements (cross-correlations) of R. The first term of Eq. (5.104), Ad, is the decoupled data weighted by the received amplitudes. The second term, QAd, represents the multiple access interference (MAI). For asynchronous CDMA systems, due to the overlapping time-shifted bits, the matrices are of a larger size than discussed above.

In multiuser detection (also called joint detection) information from several user signals with respect to their codes and timing (and possibly amplitude and phase) is exploited to better detect each user. In [Ver86] the optimal multiuser detector was introduced, yielding the most likely transmitted sequence d for a given received signal $r(t)$. Under the assumption that all possible transmitted sequences are equally probable, this detector is known as the maximum-likelihood sequence (MLS) detector [Pro89].

Due to the extremely high number of possible d-vectors, an exhaustive search is clearly impractical. For DS-CDMA, MLS detection can be implemented by a Viterbi algorithm following the conventional correlator bank [Ver86]. Unfortunately, the required Viterbi algorithm is not practical because its complexity still grows exponentially with the number of users.

Therefore, suboptimal multiuser detectors have been developed which are more feasible to implement. Most of the proposed detectors can be classified in one of two categories: linear multiuser detectors and subtractive interference cancellation detectors [Mos96].

In linear multiuser detection, a linear mapping (transformation) is applied to the soft outputs of the conventional detector to produce a new output with better performance. In subtractive interference cancellation detection, estimates of the interference are generated and subtracted out.

Linear Multiuser Detectors. Linear multiuser detectors apply a linear mapping, L, to the soft output of the conventional detector in order to reduce the MAI seen by each user.

Decorrelating Detector. The decorrelating detector applies the inverse of the correlation matrix, $\mathbf{L}_{dec} = \mathbf{R}^{-1}$, to the conventional detector output in order to decouple the data. The soft estimate of this detector is

$$\mathbf{d}_{dec} = \mathbf{R}^{-1}\mathbf{y} = \mathbf{Ad} + \mathbf{R}^{-1}\mathbf{z}. \qquad (5.105)$$

Thus, the decorrelating detector completely eliminates the MAI. An attractive property of this detector is that no estimate of the received amplitudes is required. A disadvantage is that it increases noise. Also, the computations needed to invert the matrix $\mathbf{R}$ are difficult to perform in real-time.

Minimum Mean-Square Error (MMSE) Detector. The MMSE detector takes into account the thermal noise and utilizes the knowledge of the received signal powers. It implements the linear mapping which minimizes $E\{|\mathbf{d} - \mathbf{Ly}|^2\}$, the mean-square error between the actual data and the soft output of the conventional detector [XSR90]. This results in [Mos96]

$$\mathbf{d}_{MMSE} = \left[\mathbf{R} + (N_0/2)\mathbf{A}^{-2}\right]^{-1}\mathbf{y}. \qquad (5.106)$$

The MMSE detector balances the desire to decouple the users with the desire not to enhance the thermal noise. Generally, it provides better bit error rate performance than the decorrelating detector.

A disadvantage of the MMSE detector is that it requires an estimation of the received amplitudes.

Interference Cancellation. In principle, subtractive interference cancellation detectors generate separate estimates of the MAI contributed by each user in order to subtract out some or all of the MAI seen by each user. Such detectors are often implemented with multiple stages which should provide increasingly reliable outputs. Typically, hard bit decisions are used to estimate the MAI (decision-feedback detection) requiring reliable estimates of the received amplitudes and carrier phases, however, a soft-decision approach is also possible.

Successive Interference Cancellation. In a successive interference cancellation detector, each one of the multiple stages of the detector decides, regenerates (re-spreads), and cancels out one additional user signal from the received signal, so that the remaining users experience less MAI in the next stage. Figure 5.25 shows a simplified diagram of the first stage of this detector. The first stage is preceded by a ranking of the signals in descending order of received powers, thus, the strongest interference is canceled out first.

The successive interference canceler requires only a minimal amount of additional hardware and can provide significant improvement over the conventional detector. However, there is a need to reorder the signals whenever the power profile changes. Of course, wrong bit decisions cause an increase of interference instead of its cancellation. Thus, a minimum performance level of the conventional detector is required for the interference canceler to yield improvements.

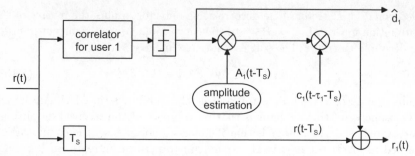

Fig. 5.25. First stage of a successive interference cancellation detector [Mos96]. User 1 = user with the highest received power

Parallel Interference Cancellation. The parallel interference cancellation detector estimates and subtracts out all the MAI for each user in parallel. This can be done in multiple stages [VA90].

The initial bit estimates $\hat{\mathbf{d}}(0)$ derived from a conventional detector are scaled by the amplitude estimates and re-spread by the user codes, producing a delayed estimate of the received signal for each user, $\hat{s}_n(t - T_s)$. Then, all respective interference estimates are subtracted in parallel from the delayed receive signal. In this way, stage m of the interference canceler yields [Mos96]

$$\hat{\mathbf{d}}(m+1) = \mathbf{y} - \mathbf{Q}\mathbf{A}\hat{\mathbf{d}}(m) = \mathbf{A}\mathbf{d} + \mathbf{Q}\mathbf{A}(\mathbf{d} - \hat{\mathbf{d}}(m)) + \mathbf{z} \qquad (5.107)$$

with the term $\mathbf{Q}\mathbf{A}\hat{\mathbf{d}}(m)$ representing an estimate of the MAI.

Zero-Forcing Decision-Feedback (ZF-DF) Detector. The zero-forcing decision-feedback detector [DH95, KKB96] performs a linear operation followed by successive interference cancellation. The linear operation *partially* decorrelates the users without increasing the noise, and the successive interference cancellation subtracts the interference from one additional user at a time, in descending order of signal strength. If all bit decisions are correct and the amplitude and phase estimates are reliable, the ZF-DF detector eliminates all MAI and maximizes the signal-to-noise ratio.

Limitations of Multiuser Detection. Multiuser detection has the following limitations:

- Multiuser detection is currently too complex to be implemented in mobile terminals. Hence, it is limited to the return link. Detection of multiple users in the satellite or ground station is required anyway.
- In CDMA, the same frequency band is used for every spot beam of the satellite (see Chap. 6). If other-cell interference is not included in the multiuser detection algorithm, the potential gain is reduced. Similarly, uncaptured multipath signals limit the potential gain.

In spite of these limitations, multiuser detection offers substantial capacity gain and improves spectrum utilization without the need for synchronous and orthogonal code sequences. Also, the transmit power of the mobile terminals and the precision requirements for power control may be reduced.

5.10 Characteristics of CDMA

Advantages of CDMA. CDMA multiple access offers a number of attractive features [Vit94]:

- *Soft capacity limit:* As shown in Sect. 5.8.3, CDMA exhibits no hard capacity limit, rather the signal-to-noise ratio degrades gradually with increasing number of active users (graceful degradation).
- *Universal frequency reuse:* Because the user signals are distinguished by nearly orthogonal PN codes, the same frequency band can be used in all radio cells (spot beams and satellites). This substantially improves the bandwidth efficiency of the system and eliminates the need for frequency planning. The latter advantage disappears, however, if different code families must be assigned to neighboring spot beams.
- *Band-sharing:* Several CDMA systems can jointly use an allocated frequency band with dynamic capacity distribution, while TDMA systems need exclusively allocated frequency bands and thus are less flexible. For example, in the USA, the band 1610–1621.35 MHz is allocated for multiple CDMA systems, whereas the band 1621.35–1626.5 MHz is exclusively assigned to Iridium.
- *Transmit signal with low spectral power density:* In certain frequency bands, given limits with regard to the power spectral density must be obeyed. In particular, this is the case when secondarily allocated frequency bands are used. CDMA systems can cope with such limitations. An example is the Omnitracs system using the Ku band on a secondary basis.
- *User synchronization:* Asynchronous CDMA requires no user synchronization; for TDMA, time slot synchronization of the users is required.
- *Constructive combination of multipath signals with a Rake receiver:* Multipath signals can be separately detected and combined by a Rake receiver, in this way exploiting the total received power. In satellite systems, the echo delays are usually smaller than 1 μsec, however. Therefore, CDMA signal bandwidths in the range of 1 MHz do not provide substantial gain.
- *Satellite diversity with a Rake receiver:* With a Rake receiver, the detection and combination of CDMA signals being simultaneously transmitted via more than one satellite can easily be accomplished. With maximum ratio combining, optimum satellite diversity can be achieved.
- *Soft handover for satellite networks:* For soft handover a connection via two satellites is maintained in the critical handover phase near a cell border, and the forward and return signals are transmitted over both satellites. The terminal and the ground station can receive both satellite signals and

combine them with a Rake receiver, thus improving transmission quality during handover.

- *Interference reduction and capacity enhancement by means of voice activation:* During speech pauses, the signal can be coded with fewer bits and transmitted with lower power. In this way, the interference to the other users is reduced and system capacity is improved.
- *Channel coding with CDMA:* With CDMA, channel coding requires no additional bandwidth. The processing gain is maintained. The number of possible code sequences is reduced, however.
- *Flexibility with regard to data rate:* Different and time-varying data rates (e.g. for the exploitation of speech pauses) can easily be realized without signaling overhead.
- *Polarization multiplex:* In CDMA satellite networks, both circular polarizations can be used in the mobile link, because the user signals in neighboring cells are additionally separated by different signature sequences [GJPW90]. This would substantially improve bandwidth efficiency but requires additional antenna effort on board the satellites. Also, for mobile terminals the cross-polarization attenuation may be very small. For mobile systems with TDMA or FDMA the usage of both polarizations is not useful.
- *User localization:* Due to the high chip rate, the spread CDMA signal can be used advantageously to determine the position of mobile users.
- *Privacy:* Signal spreading with user-specific codes improves privacy. Moreover, it is difficult to detect or intercept the spread signal.

Requirements for CDMA Networks. In contrast to the advantages achieved by using CDMA, CDMA also imposes some requirements on a satellite system:

- *Fast and exact power control* is required in order to achieve equal power levels of the user signals received by the satellite or ground station. Otherwise, strong user signals would suppress weaker signals (near-far problem), especially for DS-CDMA.
  - *Open-loop power control* is based on the received amplitude of a forward pilot. This power control method is used to compensate power variations due to changing path geometry and not too heavy signal shadowing. It is also used during connection establishment.
  - *Closed-loop power control* is based on correction signals received from the satellite or ground station and is used for fine correction. In satellite networks, the long propagation delay represents a problem for closed-loop power control, however. The fast power variations caused by fading (especially during signal shadowing) cannot be compensated. If sufficient link margin is available, the transmit power can be providently increased during signal shadowing.

For synchronous CDMA the requirements for power control are less critical than for asynchronous CDMA. For TDMA, only rough power control is required.

- *Backoff of the satellite transponder:* For CDMA, the satellite transponder has to be backed off in order to maintain linearity of transmission. TDMA allows the transponder to be driven up to a higher power level, because the user signals are transmitted one at a time.
- Rake receivers require *fast signal acquisition* in order to track the time-varying multipath components in the received signal.
- Due to high chip rates, complex signal processing, and multiple Rake signal demodulators (Rake fingers), CDMA requires *complex receivers*.

Planned S-PCN systems do not show a clear preference for TDMA or CDMA. Iridium and ICO will use TDMA, Globalstar and Ellipso will use CDMA. Each of these systems uses more than one carrier per spot beam (MF-TDMA or MF-CDMA, respectively).

5.11 CDMA for the Satellite UMTS Air Interface

UMTS is the next generation of mobile communication networks, which will include an integrated satellite component commonly designated as satellite UMTS (S-UMTS), cf. Sect. 1.1.3. A number of concepts for S-UMTS systems have already been proposed, which can be considered as the generation of satellite systems succeeding the generation of Iridium, Globalstar, and ICO. Most of these new proposals are based on proprietary CDMA multiple access schemes. In contrast to this, the European Space Agency has proposed a CDMA scheme which could become the basis of an open ETSI standard.

The ESA proposal comprises two options, a satellite wideband CDMA scheme (SW-CDMA) and a satellite wideband hybrid CDMA/TDMA scheme (SW-C/TDMA). Whereas the former scheme is more suited for global LEO/-MEO systems, the latter one is advantageous for regional GEO or HEO systems.

In order to favor the development of dual-mode UMTS terminals, the ESA S-UMTS proposal provides much commonality with the terrestrial UTRA (UMTS terrestrial radio access) approach.

5.11.1 The ESA Wideband CDMA Scheme

This scheme is based on wideband CDMA with frequency-division duplexing (FDD). The forward link uses synchronous CDM, and the return link employs asynchronous CDMA. The chip rate is 4.096 Mc/s with a 2.048-Mc/s half chip rate option which, due to its finer frequency granularity eases band sharing by different S-UMTS systems.

Figure 5.26 shows the frame structures of the forward and return links. The figures show the physical forward and return channels of a single user. The signals are structured in slots, with 16 slots building a frame. In the

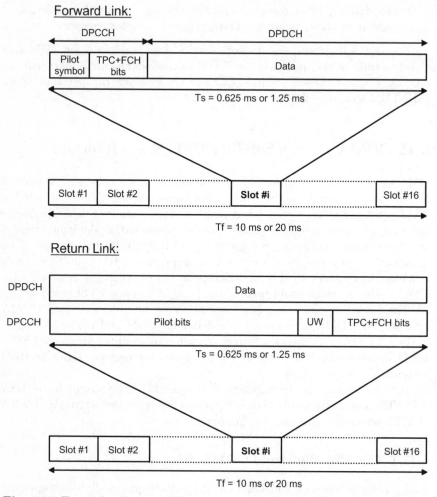

Fig. 5.26. Forward and return link frame structures of the ESA wideband CDMA scheme [ESA98a]

forward link the frames of different users are synchronous, in the return link the frames of different users are asynchronous.

The framing of the CDMA signal allows to map a dedicated physical data channel (DPDCH) together with a dedicated physical control channel (DPCCH) into the signal. In the forward link the DPDCH and DPCCH are time multiplexed, whereas in the return link they are transmitted with separate spreading (channelization) codes in phase quadrature. The DPCCH contains pilot symbols or bits, which support coherent detection both at the satellite/ground station and the mobile terminal. The transmit power control (TPC) bits serve for frame-by-frame closed-loop power control; the frame control header (FCH) defines the structure and the data rate of the frame, which may change frame by frame. Optionally, the return link may contain a UW for signal synchronization in satellite diversity operation. The five TPC/FCH bits are error protected by mapping them into a 16 bit bi-orthogonal Walsh code. Moreover, the pilot and control bits are dispersed over all slots of a frame to provide maximum robustness to fading. Multiple services belonging to the same connection are normally time-multiplexed.

Figure 5.27 shows the modulation and spreading applied in the forward and return links.

In the forward link, the data modulation is QPSK, except for very low bit rates (2.4 kb/s), for which BPSK is preferred, being less sensitive to phase errors. The same binary spreading and scrambling codes are used for the I and Q branches. The return link modulation is QPSK, and two different binary spreading codes are used for the I and Q branches.

As in UTRA, orthogonal variable spreading factor codes based on Hadamard codes are used in both links to allow for variable data rates between 2.4 and 64 kb/s. A binary scrambling code is used in the forward link; complex scrambling is used in the return link. The pseudo-noise scrambling codes are satellite-specific. Different spot beams of a satellite use cyclic shifts of the same scrambling code. Finally, the CDMA chips are filtered with a square root raised cosine filter with a roll-off factor of 0.22.

The following list summarizes the features of the ESA wideband CDMA scheme for S-UMTS:

- forward: synchronous DS-CDM, return: asynchronous DS-CDMA
- especially suited for LEO/MEO constellations with global coverage
- FDD
- full frequency reuse in each spot beam
- 4.096 Mc/s (2.048-Mc/s half chip rate option)
- forward link modulation: QPSK, return link modulation: BPSK with control channel on the quadrature carrier
- forward link spreading: BPSK, return link spreading: complex
- data rates: up to 32 kb/s for handheld, up to 144 kb/s for transportable terminals
- spreading factor: variable from 16 to 256, according to the data rate

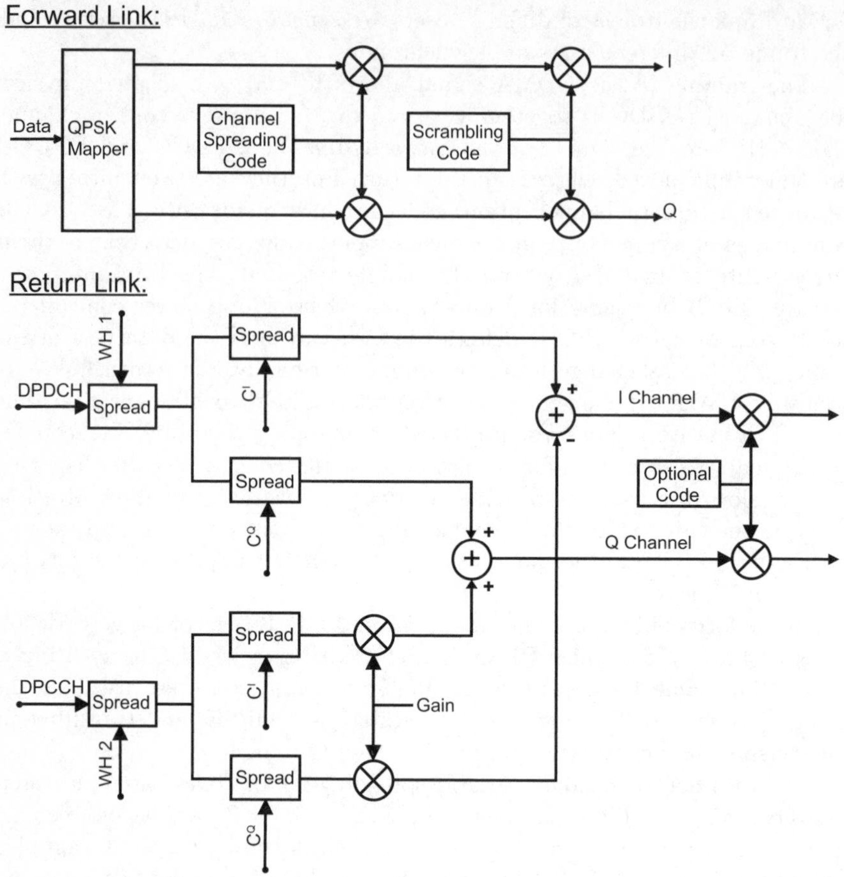

Forward Link:

Return Link:

Fig. 5.27. Forward and return link modulation and spreading of the ESA wideband CDMA scheme [ESA98a]

- multirate concept: orthogonal variable-rate coding (multicodes for data rates higher than 64 kb/s)
- coherent detection in up- and downlink
- satellite diversity is supported
- interference mitigation by linear MMSE is supported in the forward and return link.

The concept of the ESA wideband CDMA scheme offers much commonality with the ETSI UTRA and with the ARIB CDMA. By the use of pilot symbols it supports adaptive satellite antennas. Satellite diversity and soft handover are possible for the forward and return link. The scheme provides a high-penetration paging channel with increased transmit power and link margin. Finally, the CDMA signal provides the possibility of user localization.

5.11.2 The ESA Wideband Hybrid CDMA/TDMA Scheme

The proposed W-C/TDMA scheme is mainly characterized by a slotted frame structure and a quasi-synchronous operation of the uplink, resulting in a quasi-orthogonal multiple access to a single multibeam satellite [GEV92]. Frequency-division duplexing (FDD) and frequency/time-division duplexing (F/TDD) are foreseen. For the latter scheme, transmission and reception occur in different time slots and on different frequencies. This eliminates the requirement of a high-performing diplexer at the mobile terminal.

Figure 5.28 shows the frame structure of the forward and return link. A multiframe consists of 8 ordinary frames plus one extra frame for asynchronous initial access and forward link signaling.

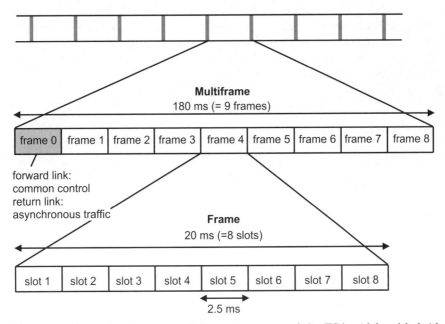

Fig. 5.28. Forward and return link frame structures of the ESA wideband hybrid CDMA/TDMA scheme [ESA98b]

The following list summarizes the features of the ESA wideband hybrid CDMA/TDMA scheme for S-UMTS:

− forward: orthogonal C/TDM
− return: quasi-synchronous DS-C/TDMA, asynchronous CDMA for random access
− especially suited for regional GEO/HEO systems
− FDD or F/TDD
− full frequency reuse in each spot beam

– 4.096 Mc/s (2.048-Mc/s half chip rate option)
– data modulation: QPSK or dual BPSK
– forward link spreading: QPSK, return link spreading: $\pi/4$-QPSK
– data rates: up to 183 kb/s
– spreading factor: variable from 16 to 256, according to the data rate
– multirate concept: orthogonal variable-rate coding
– pilot-aided quasi-coherent detection in up- and downlink
– satellite selection diversity or fast hard-handover.

$\pi/4$-QPSK carrier modulation causes less amplitude fluctuations of the transmit signal than QPSK. Hence, the terminal power amplifier can be operated in the nonlinear range producing a higher output power. This is especially beneficial in GEO/HEO systems exhibiting high signal attenuation. Similarly, the concept of quasi-synchronous CDMA is more effective for single-satellite systems. In LEO constellations, for example, it is not possible to synchronize the users with respect to all visible satellites. Satellite path diversity is not possible for C/TDMA, which, however, is no disadvantage for a single-satellite system.

6. Cellular Satellite Systems

6.1 Introduction

Virtually all satellite communication systems are cellular systems in the sense that the service area of the system is covered by more than one radio cell. For example, if the service area is covered by more than one satellite, radio cells are constituted by the illumination areas of the satellite antennas. If a satellite uses spot beams, the satellite footprint is subdivided into radio cells corresponding to the illumination areas of the spot beams, Fig. 6.1. In each cell a group of radio channels (characterized by their carrier frequency and bandwidth) is used which may be allocated to the cells in a fixed or in a dynamic manner.

As discussed in Chap. 3, the usage of spot beams is necessary if a high satellite antenna gain is required for the link budget. Spot beams focus the transmitted power onto a much smaller area than the total coverage area of the satellite. Another advantage of the spot beam approach is that a frequency band can be reused in different cells, depending on the multiple access scheme used. This improves the bandwidth efficiency of the system.

The following consequences arise for satellite systems with spot beams:

- The complexity of the satellite payload and of the satellite antenna is increased.
- Cell handovers are required for non-geostationary satellites.
- Co-channel interference (CCI) arises between cells using the same frequency.

This chapter will mainly deal with the following basic questions of a cellular satellite system:

- How can the available frequency band be efficiently used by the cells? Fixed channel allocation is assumed in this chapter; for dynamic channel allocation see Chap. 7.
- How much interference arises among the radio cells?
- How much bandwidth is required for the cellular satellite system?
- What is the traffic capacity of such a system?

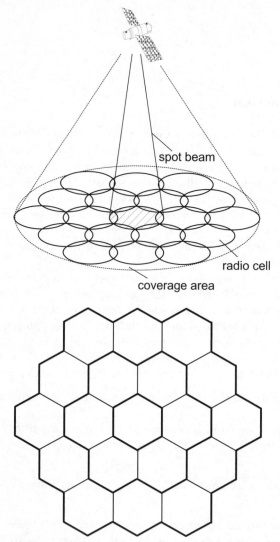

Fig. 6.1. Separation of a satellite footprint into radio cells (spot beams) and idealization of the cells by hexagons

6.1.1 Concept of the Hexagonal Radio Cell Pattern

The hexagon is used as an idealized radio cell shape. As shown in Fig. 6.1, it can be used to continuously fill out a coverage area without any overlapping of cells.

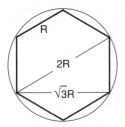

Fig. 6.2. Idealized radio cell. R = length of an edge = radius of circumference = radius of radio cell

According to Fig. 6.2, the area of the idealized radio cell is

$$A_c = \frac{3}{2}\sqrt{3}R^2. \tag{6.1}$$

Concerning the way in which to fill up a satellite footprint with cells, a typical approach is to surround a central cell (the nadir cell) with concentric rings of cells, Fig. 6.1. The number of cells in ring n is $6n$, and the total number of cells in a cell pattern with n rings is

$$Z = 1 + 6 + 2 \cdot 6 + \cdots + n \cdot 6 = 1 + 3n(n+1). \tag{6.2}$$

The following list gives some examples.

n:	0	1	2	3	4	5	6	7
Z:	1	7	19	37	61	91	127	169

For an example of a cell pattern with 7 rings, see Fig. 3.7. For a more detailed treatment of cell patterns, see App. A.2.

6.1.2 Cell Cluster and Frequency Reuse

For CDMA, the same carrier frequency can be used in all radio cells, since all signals are distinguished by nearly orthogonal code sequences.

For FDMA or TDMA, neighboring cells must use different frequency bands[1], since signals transmitted simultaneously in the same frequency band would strongly interfere with each other (co-channel interference, CCI). However, in radio cells, which are separated by a sufficiently large distance, the same carrier frequency can be used also for FDMA and TDMA (frequency reuse). The distance D required for this is called *reuse distance*. To guarantee a certain reuse distance D for fixed channel allocation, radio cells with different carrier frequencies are appropriately grouped into cell clusters.

---

[1] The frequency band used in a cell does not need to be continuous.

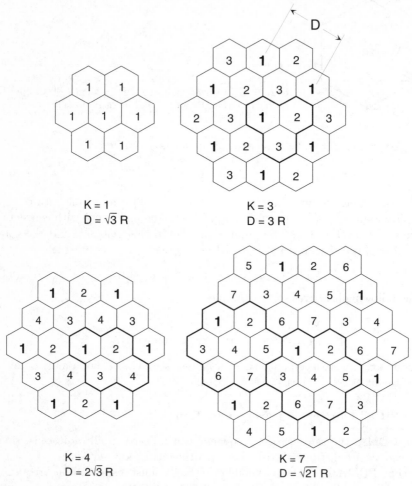

Fig. 6.3. Radio cell structures with cluster sizes $K = 1$, 3, 4, and 7. D is the frequency reuse distance. Numbers in cells denote frequency bands

Figure 6.3 shows examples of cell patterns which are constructed from cell clusters. For such regular cell patterns all frequencies have the same reuse distance D, which (from cell center to cell center) is given by

$$D = \sqrt{3K} R. \tag{6.3}$$

K designates the cluster size, which is the number of different frequency bands within the cluster. Regular cell patterns exist for cluster sizes

$$K = i^2 + ij + j^2 \tag{6.4}$$

with integer $i, j \geq 0$. Cells that use the same frequency bands are found by starting in the center of a cell, moving i cells across cell sides, turning $60°$, and moving j cells. Examples of the cluster size of regular cell patterns are $K = 1, 3, 4, 7, 9, 12, \ldots$.

For CDMA a cluster size of $K = 1$ can be chosen, which means that the same frequency is used in every cell. For FDMA and TDMA typical values of the cluster size in satellite systems are $K = 4$ or 7.

Frequency Reuse Factor. The frequency reuse factor[2] W describes how often a given carrier frequency is used (on average) in the service area of a satellite or a satellite constellation.

Let A_s be the service area. The number of cells (with area A_c) within the service area is

$$Z = A_s / A_c \tag{6.5}$$

and the frequency reuse factor is

$$W = \frac{Z}{K} = \frac{A_s}{K A_c}. \tag{6.6}$$

Here, $K A_c$ corresponds to the area of a cell cluster. W increases with increasing number of cells, Z, and decreasing cluster size K. As an example, the Iridium constellation contains $Z = 2150$ active cells, and the cluster size is $K = 12$. Thus, the global frequency reuse factor is $W = 179$.

For a given system bandwidth B_s the system capacity increases the more often the frequency band is reused. This means that the system capacity is proportional to W:

$$\text{system capacity} \sim W \sim \frac{1}{K A_c}. \tag{6.7}$$

The system capacity increases with decreasing cluster size K and decreasing area A_c of a radio cell.

---

[2] In the literature the term "frequency reuse factor" is often used for the cluster size.

6.2 Co-Channel Interference in the Uplink

In order to assess the capacity of cellular TDMA and CDMA satellite systems, it is necessary to investigate the co-channel interference (CCI) in the (mobile) user uplink. For TDMA systems, the CCI determines the useful cluster size; for CDMA systems, CCI reduces the system capacity.

The sources of co-channel interference in the uplink are other terminals transmitting in the same frequency band as the wanted user. Figure 6.4 shows the geometry of the satellite uplink with respect to the user and the characteristics of the satellite spot beam antenna.

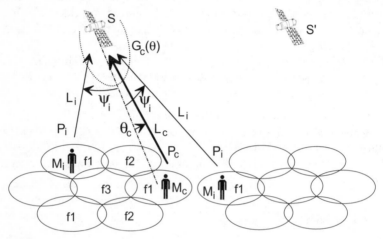

Fig. 6.4. Uplink interference scenario. Index c: user (carrier), index i: interference

In Fig. 6.4 the following designations have been used:

S: receiving satellite
S':   neighboring satellite
M_c: transmitting mobile terminal
M_i: interfering terminal
θ_c: receive angle of the wanted signal, relative to the center of the spot beam (boresight angle)
ψ_i: receive angle of the interfering signal, relative to the center of the spot beam of the wanted signal
$G_c(\theta)$:   antenna characteristic of the spot beam of the wanted signal
P_c, P_i: terminal transmit powers
L_c, L_i: propagation loss factors.

The antenna of the mobile terminal is assumed to be omnidirectional.

At the satellite, a signal with power C is received from the wanted user. In addition, the wanted user's satellite spot beam antenna receives interference with total power I, caused by other users transmitting at the same frequency as the wanted user. The signal-to-interference ratio is given by

$$\frac{C}{I} = \frac{P_c/L_c \cdot G_c(\theta_c)}{\sum_i P_i/L_i \cdot G_c(\psi_i)},\tag{6.8}$$

cf. Eq. (3.15). The sum in the denominator has to be taken over all other users transmitting simultaneously with and at the same frequency as the wanted user.

The spot beam antenna characteristic of the satellite is assumed to be rotationally symmetric, and typically is a tapered-aperture antenna [CCF+92], see Eq. (6.25). Alternatively, according to CCIR, the antenna main lobe can be approximated by a Gaussian characteristic, with a quadratic decrease of antenna gain (in dB), [CCI90]:

$$\begin{aligned} G_c(\theta) &= G_{\max} - 3(\theta/\theta_{3\mathrm{dB}})^2 & \text{in dBi} \\ G_c(\theta) &= G_{\max} \cdot 10^{-0.3(\theta/\theta_{3\mathrm{dB}})^2} & \text{in linear form.} \end{aligned}\tag{6.9}$$

The spot beam contour is usually defined by a 3-dB decrease of antenna gain, i.e. the cell pattern is designed such that the circumferences of the hexagons correspond to the -3-dB contour, Fig. 6.5.

Interference arises only if P_i and P_c are transmitted simultaneously. Interference bears fully for FDMA and for TDMA in the same time slot. For CDMA, interference is reduced through the signal correlation in the receiver, according to the processing gain G.

The amount of co-channel interference is influenced by further effects:

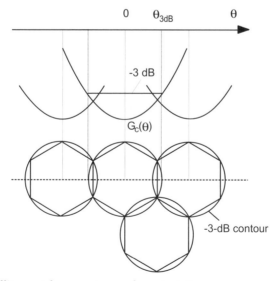

Fig. 6.5. Satellite spot beam antenna characteristic

– Signal shadowing and fading of the wanted and interfering signals modify the signal-to-interference ratio C/I.

– Power control in the wanted or interfering terminals via P_c and P_i influences C/I. For example, power control can compensate for variations in the slant range and in the signal shadowing of the wanted links. Here, different shadowing situations of the wanted and interfering signals must be taken into account.

6.2.1 Co-Channel Interference for FDMA and TDMA Uplinks

For FDMA and TDMA, different frequencies are used for each carrier within a cell cluster. A given frequency is used in the cluster only by one user at a time. Thus, according to Fig. 6.3, for each user channel there are six relevant co-channel interferers, which are located in the surrounding clusters. More distant interferers can be neglected.

In a worst-case scenario as shown in Fig. 6.6, the wanted user c is located at the edge of the corresponding spot beam. All interferers i are located at the edge of their corresponding spot beams, in the direction of the center of the cell of the wanted signal.

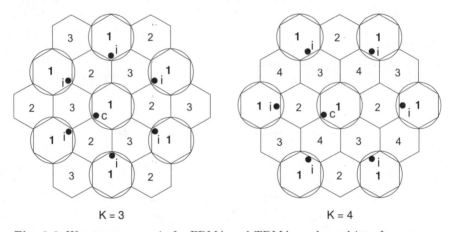

Fig. 6.6. Worst-case scenario for FDMA and TDMA co-channel interference

For the computation of the signal-to-interference ratio, the following additional assumptions are made:

– Differences in the slant ranges between users and satellite are neglected or are assumed to be compensated by transmit power control.
– Ideal channels are considered (no fading).
– Spot beam antennas with Gaussian characteristics are assumed.

– The distortion of the radio cells caused by the deviation from nadir and by the earth's curvature is neglected. This corresponds to an approximation for small nadir angles.
– Interference from users of other satellites is not considered.

Because the wanted user is located at the edge of the spot beam, the antenna gain for the wanted signal is

$$G_c(\theta_{3\mathrm{dB}}) = 0.5\,G_{\max}$$
$$= G_{\max} - 3\,\mathrm{dB} \quad \text{in dBi.} \tag{6.10}$$

The distance of the co-channel interferers from the center of the cell of the wanted signal (Figs. 6.6 and 6.3) is

$$
\begin{aligned}
\text{for } K = 3: &\quad d = 2R \\
\text{for } K = 4: &\quad d = \frac{3}{2}\sqrt{3}R \\
\text{for } K = 7: &\quad d = \sqrt{(2\sqrt{3}R)^2 + R^2} = \sqrt{13}R.
\end{aligned}
\tag{6.11}
$$

For the angle ψ_i we use the following approximation, which holds for small nadir angles:

$$\tan\psi_i = \tan\theta_{3\mathrm{dB}} \cdot d/R$$
$$\psi_i \approx \theta_{3\mathrm{dB}} \cdot d/R. \tag{6.12}$$

The antenna gain for the interfering signals is given by $G_c(\psi_i)$ according to Eq. (6.9). Considering six interferers, the signal-to-noise ratio is

$$C/I = \frac{0.5\,G_{\max}}{6\,G_c(\psi_i)}$$
$$C/I = G_{\max} - 3\,\mathrm{dB} - 10\log 6 - G_c(\psi_i)$$
$$= 3(\psi_i/\theta_{3\mathrm{dB}})^2 - 10.8\,\mathrm{dB} \quad \text{in dB.} \tag{6.13}$$

The following table compiles the results for different cluster sizes:

K	d	ψ_i	$G_c(\psi_i)$	C/I
3	$2R$	$2\theta_{3\mathrm{dB}}$	$G_{\max} - 12$ dB	1.2 dB
4	$1.5\sqrt{3}R$	$1.5\sqrt{3}\,\theta_{3\mathrm{dB}}$	$G_{\max} - 20.25$ dB	9.5 dB
7	$\sqrt{13}R$	$\sqrt{13}\,\theta_{3\mathrm{dB}}$	$G_{\max} - 39$ dB	28.2 dB

In practice, for $K = 7$, the side lobes of the spot beam antennas reduce the signal-to-noise ratio relative to the value given in the table.

Because the system capacity is proportional to $1/K$, see Eq. (6.7), K should be as small as possible. However, in order to limit co-channel interference, a minimum value for K must be maintained. In this sense, co-channel interference limits the capacity of cellular FDMA and TDMA networks. The following statements result from the above table:

- In analog FDMA systems with multipath fading a mean C/I of approximately 18 dB is required. To achieve this, a cluster size of $K = 7$ is necessary.
- For FDMA and TDMA with digital channel-coded transmission a C/I of a few dB is sufficient, and a cluster size of $K = 4$ can be chosen.

Resulting Signal-to-Noise Ratio. In order to relate the co-channel interference to the bit error rate, the signal-to-noise ratio must be derived in terms of symbol energy divided by interference power spectral density. The symbol energy is $E_s = CT_s$, with symbol duration T_s. The interference power spectral density is $L_i = I/B_{ch}$, with noise equivalent channel bandwidth $B_{ch} = 1/T_s$. With this, the signal-to-noise ratio follows as $E_s/L_i = C/I$, which is determined by Eq. (6.8). If we also take into account thermal noise, we have

$$
\begin{aligned}
L_{\text{tot}} &= L_i + N_0 \\
\left(\frac{E_s}{L_{\text{tot}}}\right)^{-1} &= \left(\frac{C}{I}\right)^{-1} + \left(\frac{E_s}{N_0}\right)^{-1}.
\end{aligned}
\tag{6.14}
$$

Example. Let $E_s/L_{\text{tot}} = 5$ dB be the signal-to-noise ratio required to achieve a certain bit error performance for a given transmission scheme. Let $C/I = 9.5$ dB be the signal-to-interference ratio achieved with a cell pattern with $K = 4$. Then, the required transmit symbol energy related to the thermal noise power spectral density is

$$
(E_s/N_0)^{-1} = (E_s/L_{\text{tot}})^{-1} - (C/I)^{-1} \quad \rightarrow \quad E_s/N_0 = 6.9\,\text{dB}. \tag{6.15}
$$

This example shows that in order to compensate for the co-channel interference, the transmit power must be somewhat higher than for a channel with pure thermal noise. In this example the difference is 1.9 dB.

Bandwidth Demand and Capacity of MF-TDMA. We assume that the satellite footprint contains multiple spot beams using different frequency bands according to a cluster size K. Further, we assume that each spot beam in the considered footprint uses the same number T of TDMA carriers, i.e. each cell occupies the same bandwidth, corresponding to a uniform traffic distribution over the coverage area of the satellite. Under these presumptions the bandwidth required for the satellite is identical to the bandwidth required for a cell cluster and is given by

$$
B_{\text{sat}} = KB_c. \tag{6.16}
$$

If the user traffic is not uniformly distributed, a worst-case cluster can be identified, consisting of cells with different frequencies, carrying the largest traffic. For LEO/MEO constellations, the worst-case cluster depends on the instantaneous spatial distribution of the traffic and on the time epoch of the satellite constellation.

Starting from the bandwidth B_{sat} available for one satellite, the cell bandwidth is $B_c = B_{\mathrm{sat}}/K$, and, corresponding to Eq. (5.32), the number of channels or active users per radio cell results in

$$N_c = TN = \frac{1}{K} \cdot \frac{B_{\mathrm{sat}}}{R_b} \cdot \frac{r \log_2 M}{(1 + \beta + B_g/R_B)\left(1 + \frac{H+G}{n_s}\right)}. \tag{6.17}$$

The number of channels per satellite is $N_{\mathrm{sat}} = ZN_c$.

For the limiting case of pure TDMA (one carrier per cell), N_c can be derived from Eq. (6.17) by letting $B_g = 0$. For FDMA, N_c results by letting $H = G = 0$.

Example. Let us assume that we want to transmit a voice signal with a bit rate of 4.8 kb/s over an MF-TDMA uplink. The FEC code rate is 1/2, the modulation is QPSK, and the roll-off factor is 0.25. The TDMA frame should have a duration of 50 ms and contain 8 user channels. Each burst requires 100 symbols of header and a 1 ms guard time. According to the example in Sect. 5.6.1, we get $n_s = 240$ and $R_B = 64.8$ ksymbols/s.

Now, we assume that the TDMA carriers are separated by a 15 kHz guard band and that the cells use the carriers according to a cluster size of 4. Then, the number of channels per cell is $N_c = 0.10 \cdot B_{\mathrm{sat}}/R_b$.

6.2.2 Co-Channel Interference for an Asynchronous DS-CDMA Uplink

In Chap. 5 we have shown how in a single radio cell the capacity of CDMA is limited by co-channel interference (multiple access interference) caused by the users within that cell. In a cellular network with CDMA, each radio cell uses the same frequency band (cluster size $K = 1$). Therefore, co-channel interference is produced by other users in the same cell as the wanted user as well as by the users in all other (mainly neighboring) cells. All users are transmitting simultaneously.

For synchronous CDMA, user signals within a given cell (own cell) may be orthogonal, producing virtually no interference. However, users in other cells normally use different code families and therefore are non-orthogonal with regard to the wanted user's own cell. Moreover, the users of different satellites are usually not synchronized, producing interference in the overlapping region of the satellite coverages. This is especially relevant for LEO constellations with multiple coverage. These considerations show that co-channel interference in CDMA satellite networks is an important issue.

Figure 6.7 shows the geometry used to analyze the uplink co-channel interference for asynchronous DS-CDMA. Also, we presume the following characteristics:

- In each radio cell we consider a single CDMA carrier within the same frequency band.

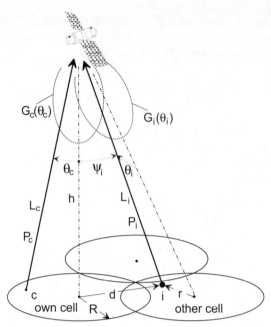

Fig. 6.7. Own- and other-cell interference scenario

- The cells are idealized by contiguous hexagons.
- We consider N users per cell, being uniformly distributed. The user density is

$$\rho = N/A_c = \frac{N}{\frac{3}{2}\sqrt{3}R^2}.$$
(6.18)

- The user signals are not synchronized, they add up with respect to their power.
- Gaussian transmission channels are assumed (no signal shadowing, no multipath fading).
- Within the cells the spot beam antenna characteristic and differences in the slant range attenuation L_i are perfectly compensated, such that the signals arrive at the corresponding spot beam antenna of the satellite with equal power, Fig. 6.8.
- In speech pauses the user signals are switched off or their power is reduced (average speech activity $= \alpha$).

Interference From the Own Cell. In the following, the own spot beam (the spot beam containing the wanted user) is considered the central spot beam in the nadir direction of the satellite.

For a user at nadir, the received uplink power is designated as $C = P_0/L_0 \cdot G_{\max}$ with $G_{\max} = G_c(0)$. For a user anywhere within the own spot beam the transmit power is assumed to be increased through power control by a

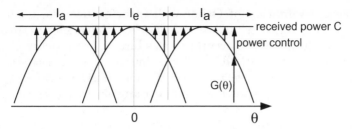

Fig. 6.8. Compensation of spot beam antenna characteristic by power control

factor of

$$\frac{L_c}{L_0} \cdot \frac{G_{\max}}{G_c(\theta_c)} \tag{6.19}$$

such that the differences in signal attenuation and antenna gain are compensated, and the received uplink power is maintained as C. With this, the interference power from the own cell is

$$I_e = \alpha(N-1)C \approx \alpha NC. \tag{6.20}$$

Interference From Other Cells. Similarly as for the users in the own cell, the transmit power of terminals in other cells is controlled according to

$$\frac{L_i}{L_0} \cdot \frac{G_{\max}}{G_i(\theta_i)} \tag{6.21}$$

such that the power received at their own spot beam antenna is C, too. $G_i(\theta)$ is the antenna characteristic of the interfering spot beam. The interference power caused by any terminal within another cell is therefore given by

$$C \cdot \frac{G_c(\psi_i)}{G_i(\theta_i)}, \tag{6.22}$$

with $G_c(\psi_i)$ representing the interference attenuation due to the spot beam antenna of the wanted user's beam.

Assuming uniformly distributed users, the interference from other cells is given by [Lut97]

$$I_a = \alpha C \rho \iint_{\substack{\text{other}\\\text{cells}}} \frac{G_c(\psi_i)}{G_i(\theta_i)}\, dA. \tag{6.23}$$

The index i designates the beam to which the interferer belongs.

The Other-Cell Interference Factor f. Corresponding to the approach in [VV93], we relate the other-cell interference power to the interference power of the own cell and define the other-cell interference factor f as

$$f = \frac{I_a}{I_e} = \frac{1}{A_c} \iint_{\substack{\text{other}\\\text{cells}}} \frac{G_c(\psi_i)}{G_i(\theta_i)}\, dA. \tag{6.24}$$

Numerical Evaluation of Other-Cell Interference. For the numerical evaluation of the other-cell interference factor f, the interferers are assumed to be uniformly distributed over the earth's surface, and are characterized by their longitude λ and latitude φ. Taking into account the curvature of the earth's surface, λ and φ define an interferer's boresight angle θ_i within its own beam, as well as the interference power attenuation $G(\psi_i)$ into the nadir beam.

The spot beam pattern of the satellite can be derived from a flat hexagonal cell pattern: The center of the cell at the origin is taken as satellite nadir. The distances of the cell centers from the origin are transformed into nadir angles of the spot beam centers. The azimuth angles of the cell centers are taken as the corresponding angles of the spherical coordinates. The antenna characteristics of all spot beams are assumed to be identical: $G_i(\theta) = G_c(\theta) = G(\theta)$. f is evaluated by numerically integrating the other-cell interference power over a satellite footprint, assuming that an interferer belongs to that spot beam from whose center its angular deviation θ_i is smallest.

The following scenarios are investigated:

- MEO system, $h = 10\,354$ km, minimum elevation $= 10°$, 169 spot beams, one-sided beamwidth $= 2.0°$. The integration extends over 36 most inner beams.
- LEO system, $h = 1414$ km, minimum elevation $= 10°$, 19 spot beams, one-sided beamwidth $= 14.9°$.

Two spot beam antenna patterns are considered:

i) A tapered-aperture antenna according to [CCF$^+$92, Col89], with $p = 2$ and $T = 0.9$:

$$G(\theta) = G_{\max} \cdot \left(\frac{J_1(u)}{2u} + 36\frac{J_3(u)}{u^3} \right)^2 \tag{6.25}$$

with

$$u = 2.07123 \cdot \frac{\sin\theta}{\sin\theta_{3\mathrm{dB}}}, \tag{6.26}$$

$G_{\max} =$ maximum antenna gain (in boresight direction), and
$\theta_{3\mathrm{dB}} =$ one-sided half-power (-3-dB) beamwidth.
$J_1(u)$ and $J_3(u)$ are the Bessel functions of the first kind and order 1 and 3, respectively.

ii) A Gaussian antenna characteristic with a quadratic decrease of antenna gain (in dB), according to Eq. (6.9).

Another parameter determining co-channel interference is the isolation of neighboring spot beams through their antenna characteristics. Assuming rotationally symmetric spot beam antenna patterns, the spot beam contour (the edge of the spot beam coverage) is defined by the boresight angle θ_s denoting the one-sided beamwidth of the spot beam. Usually, the spot beam contour θ_s is defined by a 3-dB decrease of antenna gain (i.e. $\theta_s = \theta_{3\mathrm{dB}}$), as shown in Fig. 6.5.

Higher beam isolation leading to less interference can be achieved by choosing a larger gain decrease at the spot beam contour. A convenient choice is a value of 4.3 dB, e.g., for which the absolute antenna gain at the spot beam contour is highest.

In order to investigate the influence of spot beam isolation, the gain decrease at the spot beam contour (identical to the gain decrease at the triple crossing point) is considered as variable s (in dB). Variation of s changes the shape of the antenna gain of a spot beam and therefore influences the amount of other-cell interference. This is different from the terrestrial case, where the signal attenuation is usually described by a power law of distance and, therefore, the shape of attenuation does not depend on the separation of the cells.

To account for a general spot beam isolation s, Eq. (6.26) is generalized to

$$u = G^{-1}(G_{\mathrm{max}} \cdot 10^{-s/10}) \cdot \frac{\sin \theta}{\sin \theta_s}. \tag{6.27}$$

Similarly, Eq. (6.9) generalizes to

$$
\begin{aligned}
G_c(\theta) &= G_{\mathrm{max}} - s(\theta/\theta_s)^2 && \text{in dBi} \\
G_c(\theta) &= G_{\mathrm{max}} \cdot 10^{-s/10 \cdot (\theta/\theta_s)^2} && \text{in linear form}
\end{aligned} \tag{6.28}
$$

with

$$\theta_s = \theta_{3\mathrm{dB}} \sqrt{\frac{s}{3.0103}}. \tag{6.29}$$

For these assumptions and scenarios, the results of the numerical evaluation of Eq. (6.24) are shown in Fig. 6.9. Some numerical values are compiled in Table 6.1.

Figure 6.9 indicates that other-cell interference can be reduced by increasing the spot beam isolation s. To achieve this, the satellite antenna diameter has to be increased, or the number of spot beams must be reduced. For geometrical reasons, f is larger for the LEO scenario than for the MEO scenario.

Lower Bound for Small Nadir Angles. For deriving a lower bound of f for small nadir angles, we apply the following:

− A Gaussian spot beam antenna characteristic is assumed.
− The curvature of the earth is ignored.
− Ideal channels are considered.
− Perfect power control of terminals is assumed.

For the geometry of the interference scenario we refer to Fig. 6.7. Neglecting the curvature of the earth, the receive angle of an interferer, relative to the center of the wanted (nadir) spot beam, is

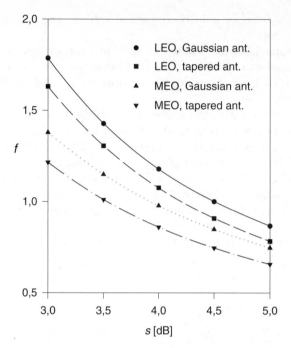

Fig. 6.9. Other-cell interference factor f versus spot beam isolation s

$$\psi_i = \arctan d/h. \tag{6.30}$$

With cell radius R, the contour of the spot beam is determined by the nadir angle

$$\theta_s = \arctan R/h. \tag{6.31}$$

Therefore, the spot beam antenna gain for an interferer (as a function of distance d of the interferer from the center of the cell of the wanted signal) is

$$G(d) = G_{\max} - s \cdot \left(\frac{\arctan d/h}{\arctan R/h} \right)^2 \quad \text{in dBi,} \tag{6.32}$$

with beam isolation $s = G_{\max} - G(R)$. For $h \gg d$ and $h \gg R$, Eq. (6.32) can be approximated by

$$G(d) \approx G_{\max} - s(d/R)^2 \quad \text{in dBi}$$
$$G(d) \approx G_{\max} \cdot 10^{-s/10 \cdot (d/R)^2} \quad \text{in linear form.} \tag{6.33}$$

This approximation holds for small nadir angles and is quite good for MEO systems with a large number of spot beams.

Equation (6.33) describes the attenuation of an interferer due to the antenna characteristic of the wanted (nadir) beam. Similarly, the power increase due to power control in the interferer's spot beam can be represented by

$$\frac{1}{G(r)} = G_{\max} \cdot 10^{s/10 \cdot (r/R)^2} \tag{6.34}$$

cf. Fig. 6.7, again taking into account the approximation for small nadir angles.

Assuming uniformly distributed users, the other-cell interference factor can be evaluated correspondingly to Eq. (6.24):

$$\begin{aligned} f &= \frac{1}{A_c} \underset{\substack{\text{other} \\ \text{cells}}}{\iint} \frac{G(d)}{G(r)} \, dx \, dy \\ &= \frac{1}{A_c} \underset{\substack{\text{other} \\ \text{cells}}}{\iint} 10^{s/10 \cdot (r^2 - d^2)/R^2} \, dx \, dy. \end{aligned} \tag{6.35}$$

Because the earth's curvature is neglected, the distance of an interferer from nadir can be expressed by the Cartesian coordinates x and y:

$$d = \sqrt{x^2 + y^2} \tag{6.36}$$

and the distance of an interferer from its own cell center is

$$r = \sqrt{(x - x_0)^2 + (y - y_0)^2} \tag{6.37}$$

with (x_0, y_0) designating the respective cell center.

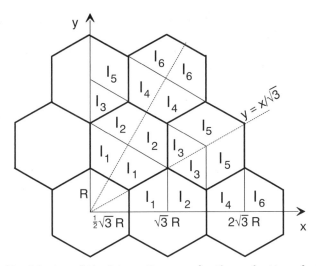

Fig. 6.10. Partitioning of the integration area for the evaluation of other-cell interference

Assuming a hexagonal cell pattern and partitioning the integration area according to Fig. 6.10, the integrals can easily be solved analytically. For s = 3 dB, the resulting integral values are

$$
\begin{aligned}
I_1 &= 0.2438R^2 & I_4 &= 0.0020R^2 \\
I_2 &= 0.0379R^2 & I_5 &= 0.0013R^2 \\
I_3 &= 0.0096R^2 & I_6 &\approx 0.
\end{aligned}
\tag{6.38}
$$

As expected, the most inner areas yield the main contributions, whereas third-tier cells can be neglected.

With cell area A_c according to Eq. (6.1) the normalized other-cell interference is given by

$$
f = \frac{12}{A_c} \cdot \sum_{\nu=1}^{6} I_\nu = \frac{8}{\sqrt{3}} \cdot 0.2946 = 1.36.
\tag{6.39}
$$

Due to the approximations involved, Eq. (6.35) is a lower bound for other-cell interference, which is independent of satellite altitude h and spot beam beamwidth θ_s.

For a tapered-aperture antenna, the evaluation of f according to Eq. (6.35) was performed numerically, resulting in the values given in Table 6.1.

Table 6.1 shows that for $s = 3$ dB interference is more than doubled due to the influence of other spot beams, leading to a decrease of more than 50% of system capacity. Introducing a 3-cell cluster FDM scheme would substantially reduce other-cell interference, but the capacity gain due to interference reduction would be more than outweighed by the capacity loss due to FDM. In principle, interference cancellation taking into account own- and other-cell interferers can reduce interference without decreasing system capacity.

For terrestrial CDMA, it is usually assumed that the signal power decreases with the fourth power of distance. Here, a typical value for the other-cell interference factor is $f = 0.44$ [VVZ94]. This shows that other-cell interference in satellite systems is much more critical than in terrestrial cellular systems. This should be taken into account when comparing CDMA and TDMA for (mobile) satellite systems.

Influence of Channel. The above results were derived assuming ideal propagation channels between (mobile) users and satellite. As mentioned before, the results also hold if multipath fading and shadowing are perfectly compensated by power control. Power control of an interferer impacts its wanted receive signal in the same way as the interference produced by him, because the interferer's signal is received in its own beam and in the victim beam via the same propagation path. Therefore, perfect power control can eliminate the effect of fading and shadowing on CDMA uplink interference. Power control can, however, increase interference to another satellite if it is in line-of-sight to a user, whose own satellite is shadowed. For the (less critical)

Table 6.1. Numerical values for the other-cell interference factor f

Scenario	Antenna	f		
		$s = 3$ dB	$s = 4$ dB	$s = 5$ dB
Lower bound	tapered-aperture	1.20	0.85	0.65
for small nadir angles	Gaussian	1.36	0.97	0.74
MEO system ($h = 10\,354$ km,	tapered-aperture	1.22	0.86	0.66
169 spot beams, $\varepsilon_{\min} = 10°$)	Gaussian	1.38	0.98	0.75
LEO system ($h = 1414$ km,	tapered-aperture	1.63	1.08	0.78
19 spot beams, $\varepsilon_{\min} = 10°$)	Gaussian	1.79	1.18	0.87

CDMA downlink, the situation is quite different, because the route of an interfering signal from the satellite to the victim terminal is different from its route to the wanted user.

As pointed out in [Mon95], power control may compensate for slow shadowing, whereas an additional power increase may compensate for degradation due to fading being too fast for power control. This additional power increase results in an increase of multiple access interference.

In mobile systems, strong signal shadowing may occur, which realistically cannot be compensated by power control. The resulting signal outages decrease the received interference power. However, due to the level variations, the wanted channel requires a higher mean signal-to-noise ratio for a given quality of service. Overall, it is expected that the performance degrades.

Usually, the time-share of shadowing increases with decreasing satellite elevation, cf. Table 3.3. This means that the nadir beam will receive lower other-cell interference than an outer beam. For a more detailed investigation of this matter an elevation-dependent channel model, e.g. [BWL96], should be used.

Number of Users per Radio Cell. From Eqs. (6.20) and (6.24), the total co-channel interference power is

$$I = I_e + I_a = \alpha(1 + f)NC. \tag{6.40}$$

With CDMA carrier bandwidth $B_T = GR_b$, Eq. (5.67), and $C = E_bR_b$, the interference power density becomes

$$L_i = \frac{I}{GR_b} = \frac{1}{G} \cdot \alpha(1 + f)NE_b. \tag{6.41}$$

The factor $(1 + f)$ represents the increase in interference caused by the users in neighboring cells.

If thermal noise is also taken into account, the overall noise power spectral density results in $L_{\text{tot}} = L_i + N_0$, and, according to Eq. (5.74), the number of active users for a single CDMA carrier is

$$N = 1 + \frac{G}{\alpha(1 + f)} \left[\left(\frac{E_b}{L_{\text{tot}}} \right)^{-1} - \left(\frac{E_b}{N_0} \right)^{-1} \right] \qquad (6.42)$$

with E_b/L_{tot} being the signal-to-noise ratio required to stay below a given bit error probability.

Let us assume that each spot beam in the satellite footprint uses the same frequency band $(K = 1)$ and the same number T of CDMA carriers (MF-CDMA). This means that each cell occupies the same bandwidth and offers the same capacity, which is suitable for a uniform traffic distribution over the footprint. Then, for a given satellite bandwidth B_{sat} (identical to cell bandwidth B_c), the number of active users per cell becomes

$$N_c = T + \frac{B_{\text{sat}} - TB_g}{R_b} \cdot \frac{1}{\alpha(1 + f)} \left[\left(\frac{E_b}{L_{\text{tot}}} \right)^{-1} - \left(\frac{E_b}{N_0} \right)^{-1} \right], \qquad (6.43)$$

according to Eq. (5.82). Corresponding to Eq. (5.78), a typical number of users results if the signal power is chosen such that the average interference power is equal to the thermal noise power. Neglecting guard bands B_g and taking $\alpha \approx 0.5$, a rule-of-thumb result is

$$N_{\text{typ}} = \frac{B_{\text{sat}}}{R_b} \cdot \frac{1}{(1 + f)E_b/L_{\text{tot}}} . \qquad (6.44)$$

Example. Let a total signal-to-noise ratio $E_b/L_{\text{tot}} = 5$ dB (corresponding to a factor of 3.16) be required to achieve a given bit error performance. Let us assume an other-cell interference factor $f = 1.36$. Then, a typical number of users results in $N_{\text{typ}} \approx 0.13 \cdot B_{\text{sat}}/R_b$.

A comparison with the example for Eq. (6.17) shows that for the assumptions made, CDMA is slightly more bandwidth efficient than TDMA.

CDMA can use the same frequency band in all cells. However, in neighboring cells the same code cannot be used for different users. Thus, several (e.g. three) code families with good cross-correlation properties are required. An alternative is to use a dedicated sequence for each user, as in the Globalstar system, for example.

6.3 Co-Channel Interference in the Downlink

In the downlink, co-channel interference is caused by the spot beam antennas of the own and other satellites simultaneously transmitting at the frequency of the wanted user's signal.

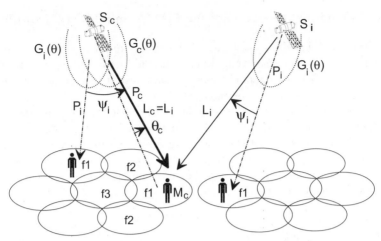

Fig. 6.11. Downlink interference scenario. Index c: user (carrier), index i: interference

Figure 6.11 shows the geometry of the satellite downlink with respect to the user and the satellite spot beams. The following designations have been used:

S_c: transmitting own satellite
S_i: neighboring satellite
M_c: receiving mobile terminal (wanted user)
θ_c: angle of the wanted signal, relative to the center of the wanted user's spot beam (boresight angle)
ψ_i: angle of the interfering signal, relative to the center of the interferer's spot beam
$G_c(\theta)$:   antenna characteristic of the spot beam of the wanted user
$G_i(\theta)$:   antenna characteristic of the spot beam containing the interferer
P_c, P_i: satellite transmit powers
L_c, L_i: propagation loss factors.

The antenna of the mobile terminal is assumed to be omnidirectional.

At the user terminal, a signal with power C is received from the own satellite spot beam. In addition, the terminal receives interference with total power I from other satellite spot beam antennas transmitting at the same frequency. The signal-to-interference ratio is given by

$$\frac{C}{I} = \frac{P_c/L_c \cdot G_c(\theta_c)}{\sum\limits_i P_i/L_i \cdot G_i(\psi_i)}, \tag{6.45}$$

cf. Eq. (3.15). The sum in the denominator has to be taken over all satellite spot beam antennas transmitting at the same frequency as the spot beam antenna of the wanted user.

6.3.1 Co-Channel Interference for FDMA and TDMA Downlinks

For FDMA and TDMA a satellite spot beam antenna transmits at a certain frequency only to one user at a time. Thus, as in the uplink, according to Fig. 6.3, for each radio channel there are six relevant co-channel interferers. More distant interferers can be neglected.

In a worst-case scenario, the considered user is located at the edge of the corresponding spot beam, Fig. 6.12. If there is no power control within the spot beams, the position of the interfering users within their spot beams is irrelevant. For perfect power control, the worst case appears if the interferers are located at the edge of their spot beams. The resulting worst-case C/I is the same for both scenarios.

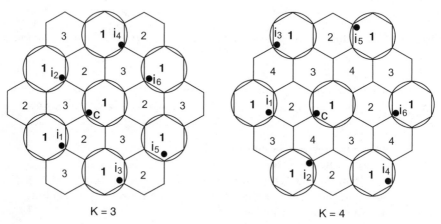

Fig. 6.12. Worst-case scenario for FDMA and TDMA downlink co-channel interference

For the analysis we make the following assumptions:

- The antenna characteristics of all spot beams are assumed to be Gaussian and identical: $G_i(\theta) = G_c(\theta) = G(\theta)$.
- Ideal channels are assumed (no fading).
- Only interference from the same satellite is considered. In this case, the wanted signal and the interfering signals propagate along the same path, therefore they are subject to the same signal attenuation: $L_i = L_c$.
- Differences in the transmit power of different spot beams (compensating for slant range differences with regard to the respective users) are neglected. This is a good approximation for small nadir angles.

The spot beam antenna gain for the wanted user is

$$G(\theta_c) = G(\theta_{3\mathrm{dB}}) = G_{\max} - 3\,\mathrm{dB}. \tag{6.46}$$

Neglecting the earth's curvature, $G(\psi_i)$ is determined by the distance of the wanted user from the cell center of the interfering spot beam, which is given in the following table for $K = 3$ and $K = 4$:

Interfering signal	Distance for $K = 3$	Distance for $K = 4$
i_1	$d_1 = 2R$	$d_1 = \sqrt{7}R$
i_2	$d_2 = \sqrt{7}R$	$d_2 = \sqrt{7}R$
i_3	$d_3 = \sqrt{7}R$	$d_3 = \sqrt{13}R$
i_4	$d_4 = \sqrt{13}R$	$d_4 = \sqrt{13}R$
i_5	$d_5 = \sqrt{13}R$	$d_5 = \sqrt{19}R$
i_6	$d_6 = 4R$	$d_6 = \sqrt{19}R$

When we approximate the angle of the interfering signals with respect to their own cell centers, ψ_i, according to Eq. (6.12), we get

Interfering signal	ψ_i for $K = 3$	ψ_i for $K = 4$
i_1	$2\,\theta_{3dB}$	$\sqrt{7}\,\theta_{3dB}$
i_2	$\sqrt{7}\,\theta_{3dB}$	$\sqrt{7}\,\theta_{3dB}$
i_3	$\sqrt{7}\,\theta_{3dB}$	$\sqrt{13}\,\theta_{3dB}$
i_4	$\sqrt{13}\,\theta_{3dB}$	$\sqrt{13}\,\theta_{3dB}$
etc.		

According to Eq. (6.9), the spot beam antenna gain $G(\psi_i)$ for the interfering signals results in

Interfering signal	$G(\psi_i)$ for $K = 3$	$G(\psi_i)$ for $K = 4$
i_1	$G_{\mathrm{max}} - 12\,\mathrm{dB}$	$G_{\mathrm{max}} - 21\,\mathrm{dB}$
i_2	$G_{\mathrm{max}} - 21\,\mathrm{dB}$	$G_{\mathrm{max}} - 21\,\mathrm{dB}$
i_3	$G_{\mathrm{max}} - 21\,\mathrm{dB}$	$G_{\mathrm{max}} - 39\,\mathrm{dB}$
i_4	$G_{\mathrm{max}} - 39\,\mathrm{dB}$	$G_{\mathrm{max}} - 39\,\mathrm{dB}$
etc.		

Concentrating on the strongest interferers, the resulting signal-to-noise ratio for $K = 3$ is

$$C/I = G_{\mathrm{max}} - 3\,\mathrm{dB} - (G_{\mathrm{max}} - 12\,\mathrm{dB}) = 9\,\mathrm{dB} \qquad (6.47)$$

and for $K = 4$:

$$C/I = G_{\mathrm{max}} - 3\,\mathrm{dB} - (G_{\mathrm{max}} - 21\,\mathrm{dB}) - 3\,\mathrm{dB} = 15\,\mathrm{dB}. \qquad (6.48)$$

This analysis shows that the TDMA or FDMA downlink is much less critical than the uplink. Thus, overall, $K = 4$ is sufficient for TDMA and $K = 7$ is sufficient for analog FDMA, cf. Sect. 6.2.1.

6.3.2 Co-Channel Interference for CDMA Downlinks

In the downlink chip-synchronous CDM with orthogonal spreading codes can easily be realized, because the signals are transmitted by a common source (satellite or gateway station). Therefore, no interference occurs between the synchronous user signals of a satellite (although neighboring satellites may cause interference).

Moreover, by means of a pilot tone transmitted by the satellite, coherent detection is possible at the terminal, allowing to work at a lower signal-to-noise ratio E_b/L_{tot}.

Thus, as for TDMA, also for CDMA the downlink is less critical than the uplink.

6.4 Bandwidth Demand and Traffic Capacity of Cellular Satellite Networks

6.4.1 Total System Bandwidth

Systems with Fixed Channel Allocation. Here, we derive the total system bandwidth required for systems with a fixed channel allocation to the satellite radio cells. Such systems can be FDMA or TDMA systems with a fixed allocation of carriers to cells, or CDMA systems using a fixed set of spreading codes per cell. The bandwidth required for a satellite is given by $B_{sat} = KB_c$, Eq. (6.16).

In a LEO/MEO satellite system, the coverage areas of the satellites within an orbital plane overlap. Thus, for systems with fixed channel allocation, neighboring satellites must be separated in frequency.

For an even number of satellites in the plane two different frequency bands can be allocated to the satellites in an alternating manner. For an odd number of satellites in the plane the coverage area of each satellite can be divided into two areas containing leading and trailing spot beams, respectively. Also in this case two different frequency bands are required in an orbit plane. An example of this approach is the ICO system [ICO97].

Also, different orbit planes of the satellite constellation must be separated in frequency. Let K_{orbit} designate the number of different frequency bands. For a LEO/MEO constellation with inclined orbits (e.g. a Walker constellation):

$$K_{orbit} = \text{number of orbits.} \qquad (6.49)$$

For polar LEO/MEO constellations, two different frequency bands are required:

$$K_{orbit} = 2. \qquad (6.50)$$

If the polar constellation contains an even number of planes, alternating frequency bands can be used in the planes. For an odd number of planes the coverage area of each satellite can be divided into a left and a right part, each of which can use one of the two frequency bands. During the satellite passage over the poles the frequency allocations must be exchanged.

With these considerations the total system bandwidth results in

$$B_s = 2K_{\text{orbit}}KB_c = K_sB_c \tag{6.51}$$

with $K_s = 2K_{\text{orbit}}K$ representing an effective cluster size for the total system. The overall frequency reuse factor of a satellite system with Z cells is

$$W_s = \frac{Z}{K_s} = \frac{A_s/A_c}{2K_{\text{orbit}}K} . \tag{6.52}$$

Systems with Dynamic Channel Allocation. In a TDMA satellite system with ideal dynamic channel allocation, each satellite is able to use all available channels in each radio cell, as long as the co-channel interference situation is acceptable. The bandwidth required by a satellite is determined by the traffic distribution within its coverage area. Ideal dynamic channel allocation in TDMA systems will, moreover, assign different radio carriers to overlapping cells of neighboring satellites. This will even be the case for time-varying overlapping in Walker constellations. Alternatively, fully overlapping cells can be switched off. No fixed frequency separation of satellites within the same orbit or in different orbits is required.

The channel allocation at a certain time can be characterized by an average frequency reuse factor W_s for the system, corresponding to an effective cluster size

$$K_s = Z/W_s. \tag{6.53}$$

W_s and K_s will vary with time.

In CDMA systems, ideal dynamic channel allocation can be achieved by assigning a specific signature code to every user in the system, as is done in the Globalstar system. Then, each satellite can use the same frequency band in every radio cell, and the satellite bandwidth is determined by the cell carrying the maximum traffic.

In CDMA systems, the same frequency band can be used in overlapping satellites. The CDMA capacity has just to be divided up among the satellites. Even different satellite systems can share a frequency band. In this case, the system operators have to negotiate the respective shares of the total capacity.

6.4.2 Traffic Capacity per Radio Cell

In order to derive the traffic capacity, we make the following assumptions:

- Let an infinitely large number of users per radio cell produce statistically independent telephone calls. The total arrival of calls constitutes a Poisson process with constant arrival rate $\lambda \; [s^{-1}]$.

– The durations of the calls are exponentially distributed with mean duration $1/\mu$ [s].

Thus, the average traffic per radio cell (the average number of ongoing calls) is

$$V_c = \lambda/\mu \quad \text{in Erl.} \tag{6.54}$$

Due to variations in the arrival and duration of calls, the actual traffic per cell varies according to a Poisson probability density. Thus, traffic peaks might occur which cannot be served by the system. For FDMA and TDMA systems with fixed channel allocation to radio cells, blocking of a new call will occur if all N_c channels are already in use (busy).

In CDMA systems, no hard blocking occurs if more than N_c users are active within a cell. Rather the signal-to-noise ratio decreases below the nominal value (graceful degradation). CDMA will be treated here like FDMA and TDMA, however. In [Vit95] the blocking (outage) for CDMA is described in more detail, taking into account the statistics of the speech activity.

For hard blocking and the assumption that blocked calls are lost, the blocking probability is given by the Erlang-B formula [Kle76]:

$$p_B = \frac{V_c^{N_c}/N_c!}{\sum\limits_{n=0}^{N_c} V_c^n/n!}. \tag{6.55}$$

For a given number N_c of channels and a maximum allowed blocking probability p_B the allowable average traffic V_c, i.e. the *traffic capacity of the radio cell*, can be obtained from Eq. (6.55).

For realistic values of the blocking probability, the traffic capacity V_c of a cell is smaller than the number of available channels, N_c. The ratio between these two numbers can be understood as the traffic efficiency of the cell:

$$\eta_{\text{Erl}} = \frac{V_c}{N_c}. \tag{6.56}$$

As an example, let us accept a blocking probability of $p_B = 1\%$. The table below shows the traffic capacity and the traffic efficiency of a cell for different numbers of channels per cell, N_c.

$p_B = 1\%$: $N_c = 10$ $\rightarrow$ $V_c = 4.5$ Erl $\eta_{\text{Erl}} = 0.45$
            $N_c = 100$ $\rightarrow$ $V_c = 84$ Erl $\eta_{\text{Erl}} = 0.84$
            $N_c = 1000$ $\rightarrow$ $V_c = 971$ Erl $\eta_{\text{Erl}} = 0.97$.

6.4.3 Traffic Capacity of the System

Assuming a system with Z radio cells with equal capacity V_c, the traffic capacity of the system (system capacity) is

$$C_s = ZV_c \quad \text{in Erl.} \tag{6.57}$$

Using Eqs. (6.56), (5.17), (6.51), and (6.52), the following relation can be derived:

$$C_s = \eta_{\mathrm{MA}} \eta_{\mathrm{Erl}} W_s \frac{B_s}{B_{\mathrm{ch}}}. \tag{6.58}$$

Here, η_{MA} denotes the efficiency of the multiple access scheme, and W_s is the overall frequency reuse factor of the system. Equation (6.58) shows that the system capacity is proportional to the overall frequency reuse factor W_s and, hence, proportional to $1/K_s A_c$, cf. Eq. (6.52).

Spectrum Efficiency. The spectrum efficiency η_s describes the efficiency of the system with regard to the "spectral" parameters bandwidth, area, and time:

$$\begin{aligned}
\eta_s &= \frac{\text{traffic capacity of the system}}{\text{system bandwidth} \cdot \text{service area}} \\
&= \frac{C_s}{B_s A_s} = \frac{V_c}{B_s A_c} \quad \text{in} \quad \frac{\mathrm{Erl}}{\mathrm{MHz \ km}^2}.
\end{aligned} \tag{6.59}$$

User capacity of the system. Only a small percentage of the subscribers of the system will be active at a given time. Thus, the system can adopt many more subscribers than indicated by its traffic capacity. The subscriber or user capacity of the system is

$$C_u = \frac{C_s}{\text{user activity}} \quad \text{users}. \tag{6.60}$$

6.4.4 Required User Link Capacity of a Satellite

In order to provide a rough estimation of the required capacity in the mobile user links, we make the following assumptions:

- The satellite system shall serve a total number N_u of users (subscribers) with an average activity of A_u in Erl. Then, the required total system capacity in terms of duplex channels can be estimated as $C_s = N_u A_u$.
- The subscribers are assumed to be uniformly distributed on the continents, covering 26% of the earth's surface.
- The global satellite constellation shall comprise S satellites, which are assumed to be uniformly distributed over the orbital sphere. S may be given from a system design or may be estimated as a function of orbit altitude and minimum elevation, cf. Sect. 2.3. Then, the capacity required for the mobile user link of a single satellite can be estimated as

$$C_{\mathrm{sat}} = \frac{1}{0.26} \cdot \frac{C_s}{S} = \frac{1}{0.26} \cdot \frac{N_u A_u}{S}. \tag{6.61}$$

Taking Iridium as an example and assuming 1 million subscribers with a 5 mErl busy hour activity results in 292 channels per satellite.

Capacity Estimation through Simulation. Equation (6.61) takes into account neither user accumulation in certain areas, nor the usage of more than one satellite channel per user to exploit satellite diversity. A more realistic capacity estimation can be achieved by assuming a realistic user distribution, and by simulating the satellite constellation, taking into account satellite elevation, diversity and/or handovers.

Figure 6.13 shows the simulation result for an Iridium satellite, exhibiting high traffic peaks when the satellite crosses areas of high user density. Assuming dynamic channel allocation to spot beams and taking into account the Erlang-B blocking probability (see Eq. (6.55)) for the total satellite, the required up/downlink channels for a satellite can be determined.

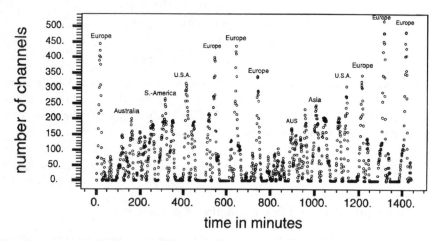

Fig. 6.13. Time-varying capacity requirements in the up/downlink of an Iridium satellite [Wer95]. $N_u = 1$ million users, $A_u = 5$ mErl user activity, no satellite diversity, handover always to the highest-elevation satellite

6.4.5 Overall Network Capacity Considerations

S-PCNs serve a globally distributed user group with a relatively high requirement for long-distance traffic. The capacity of the various communication links in the satellite network must be large enough to carry the traffic generated by fixed and mobile users. Moreover, upper limits for blocking probability and message delay have to be taken into account. The necessary capacity of the links depends on

- the number, geographic distribution, and traffic requirements (e.g. activity) of the users,
- the pattern of the satellite constellation including the ISL topology,
- the size and cell structure of the satellite coverage areas,

– the number and positions of gateway stations, and
– the routing strategy for long-distance traffic.

The global traffic requirements within and between suitably defined regions can be described by a traffic matrix. The traffic requirements can be matched to the topology (including the connection matrix) of the satellite network, and the required link capacities can be determined [WJLB95]. Based on the connection matrix, suitable routes are selected, where terrestrial long-distance links or intersatellite links may be preferred, or ISLs may be used for average traffic and terrestrial trunks may be used for peak traffic. By considering a series of time steps, it can be taken into account that the load of a link varies due to both satellite movement and traffic activity variation with the time of day. Maximum link loads as well as mean values can be evaluated. These investigations result in

– the number of mobile user channels per spot beam and per satellite,
– the number of feeder link channels per satellite,
– the number of intersatellite link channels (if ISLs are used), and
– the number and length distribution of terrestrial long-distance connections.

7. Network Aspects

7.1 Architecture of Satellite Systems for Mobile/Personal Communications

In this section we will discuss the main network elements of a satellite system for mobile/personal voice communications. Figure 7.1 shows the basic system architecture.

Space Segment. The space segment consists of an ensemble of satellites, which may be interconnected by intersatellite links (ISLs) allowing long-distance transmission within the space segment. Each satellite covers a circular area on the earth's surface, which increases with increasing orbit height and decreasing minimum elevation angle $\varepsilon_{\min}$. The choice of orbit planes and

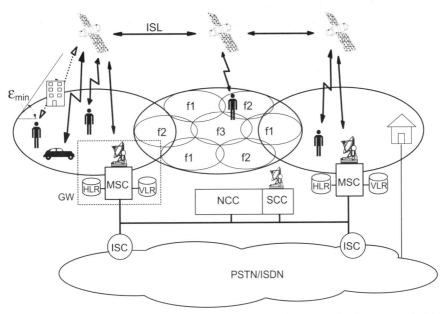

Fig. 7.1. Network architecture of a satellite personal communication network (S-PCN)

the satellite phasing within the orbits must guarantee continuous coverage of the service area (being the full surface of the earth for global systems).

Direct communication via satellite using a handheld terminal with low transmit power and an omnidirectional antenna requires a high antenna gain on board the satellite. Accordingly, the coverage area of an antenna beam (spot beam) is rather small, and the coverage area of the satellite is composed of several radio cells. The spot beam approach allows the reuse of frequency bands f_1, f_2, ... in separated cells, increasing the bandwidth efficiency of the system.

Apart from their communications task, the satellites can also perform decentralized tasks in the area of radio resource management, such as the allocation of radio channels to users.

Ground Segment. *Gateway stations (GWs)* are important elements of the ground segment. They comprise fixed earth stations, which provide communication links to the satellites and via the satellites to the mobile users within the GW service area. On the other hand, the GWs are connected via international switching centers (ISCs) to terrestrial fixed networks. Thus, GWs constitute interfaces between the satellite network and terrestrial networks.

With respect to a mobile user we distinguish two types of GWs. The *home gateway* is the GW which serves the home area of the mobile user. The *visited gateway* is a GW which serves a mobile user outside the service area of the home GW.

The GWs are equipped with databases allowing contact to be maintained with globally mobile users by means of GSM-like mobility management functions. The *home location register* (HLR) contains the data of subscribers residing within the service area of the relevant GW and being registered in that GW. If a user is currently located in a different GW service area, the HLR contains the corresponding VLR address (the visited GW). The *visitor location register* (VLR) contains information (e.g. the location area) of "visitors" currently staying in the service area of the visited GW. As an example, Fig. 7.2 shows the block diagram of an Iridium GW.

The *network control center (NCC)* controls access to the satellite network and is involved in RR. Among other tasks it allocates spot beam frequencies and distributes routing tables to the satellites. Also, it performs network management tasks, such as performance management and fault management. The *satellite control center (SCC)*, via telemetry, tracking, and command (TT&C) links, keeps the satellites in their correct orbital positions.

Gateways and control stations are interconnected by a fixed terrestrial network consisting of private or leased lines. Similarly, each satellite always has a connection to the NCC or to other control stations with corresponding functionalities.

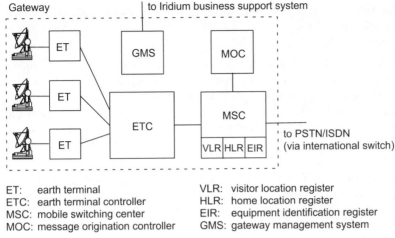

Fig. 7.2. Iridium gateway block diagram [HL95]

7.2 Network Control

7.2.1 Tasks of Network Control

Network control comprises all supportive tasks required for the communication of network users. In this respect, cellular mobile networks are characterized by specific features and requirements:

- The service area is divided into radio cells, and the users are mobile. Therefore, *mobility management* (MM) is required to stay in contact with the users. An important question related to mobility management is, in which radio cell the user is located at the moment. This information, which is required when an incoming call must be routed to the mobile user, is maintained by means of location update functions. Further tasks of mobility management are the authentication of mobile users and the provision of privacy.
- For TDMA systems, radio cells use different frequency channels, and frequencies are reused in sufficiently separated cells. For CDMA systems, it may be necessary to allocate spreading codes. Generally, the radio channels are non-ideal and are allocated only when requested by active users. This requires *radio resource management* (RR) concerning the establishment, maintenance, and release of radio channels. A basic functionality of RR is the selection of a proper radio cell for the mobile user (cell selection) which must be performed before a radio channel can be allocated. Due to user mobility and satellite movement, it may become necessary to change the radio channel, and thus a handover function is required. Also, informing mobile users about an incoming call (paging) is a function of RR.

– The establishment, maintenance, and release of end-to-end connections are essentially contained in the *call control* (CC) part of the *connection management* (CM).

The network control tasks listed above are performed in a cooperative effort by the network control center, the gateway stations with their home and visitor location registers, and the satellites.

Most S-PCN systems in operation or in development are based on the GSM network platform. Therefore, it is sufficient to discuss the above-mentioned network functions in the context of the GSM standard, e.g. [MP92, EV98].

Relevant Protocol Layers. Only the lower ISO/OSI layers are relevant for mobile communications. Table 7.1 lists the main tasks performed by the layers 1 to 3 of the GSM protocol architecture. The functions of MM, RR, and CC mainly reside in the GSM layer 3. Figure 7.3 shows how the functions of CM, MM, and RR can be implemented in the protocol stacks of the mobile terminal and the satellite GW. The signaling protocol architecture shown has been adopted from GSM.

Table 7.1. Relevant protocol layers for mobile communications

GSM protocol layer	Main tasks
Layer 3: network layer[a]	Transmission of messages, connection management, call control, mobility management (location update, call routing, etc.), radio resource management (paging, handover, random access, channel assignment, etc.)
Layer 2: data link layer	Segmentation and reassembly of messages, frame structuring, multiplexing, error detection and correction, transmission protocol, flow control
Layer 1: physical layer	Radio transmission

[a] The GSM layer 3 does not fully comply with the ISO/OSI layer 3.

7.2.2 Signaling Channels of the Air Interface

From the above considerations it is clear that network control requires dedicated communication, which is called signaling. For this purpose various signaling channels are provided, which for CM, MM, and RR extend across the mobile user link (radio interface), Fig. 7.3. Broadcast channels are used to distribute network information. Multiple access channels allow mobile users to get into contact with the network. Dedicated control channels are set up for the signaling between a specific user and the network.

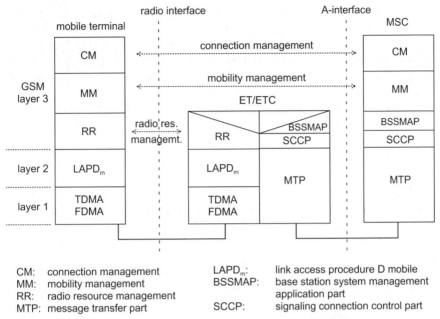

Fig. 7.3. Simplified GSM protocol architecture for the signaling between mobile user and satellite gateway station. After [Rah93]

Whereas carrier frequencies, time slots, or spreading codes constitute *physical channels* dedicated to the transmission of bits, *logical channels* define the mapping of the transmitted information to a specific signaling or traffic channel. Signaling channels and traffic channels are logical channels associated with functions of protocol layers above the physical layer.

Broadcast Channels. Broadcast channels are used by each GW or satellite to distribute network information to the user terminals. The broadcast control channel (BCCH) contains information concerning

- the identification of spot beam and location area,
- the BCCHs of neighboring spot beams,
- the radio channel configuration (mapping of logical onto physical channels), and
- network parameters (timers, etc.).

All non-active terminals (terminals in standby mode) monitor the BCCH of their current spot beam.

Further broadcast signaling channels are the frequency correction channel (FCCH) used for carrier synchronization and the synchronization channel (SCH) used for frame synchronization.

Common Control Channel (CCCH). The random access channel (RACH) is a multiple access CCCH signaling channel, which is typically used with a slotted Aloha protocol. The RACH is used by mobile users to contact the network (e.g. for requesting a dedicated control channel or for responding to paging).

The access grant channel (AGCH) is a broadcast CCCH which is used by the network for responding to RACH user requests (e.g. for the assignment of a signaling or traffic channel to a user).

Another broadcast CCCH is the paging channel (PCH), which is used by the network to contact a specific user (e.g. to inform the user about an incoming call).

Dedicated/Associated Control Channels. These are channels dedicated to the signaling between a specific user and the network.

A stand-alone dedicated control channel (SDCCH) is assigned to a terminal for point-to-point signaling with regard to MM and connection setup. This signaling channel can also be used for the transmission of short messages.

As opposed to stand-alone dedicated control channels, control channels associated to a traffic channel can also be assigned as follows. The fast associated control channel (FACCH) is a fast inband signaling channel, used, for example, for handover signaling. In TDMA systems the FACCH may use a full traffic burst ("steal" it from the traffic channel). The slow associated control channel (SACCH) is a slow inband signaling channel, used, for example, for power control and channel measurement reports.

7.3 Mobility Management

Mobility management is a functionality dedicated to the support of two kinds of mobility:

– Terminal mobility means to use a terminal at different places, in different radio cells, and possibly also in different networks (roaming).
– User mobility can be achieved by using an SIM card (subscriber identity module card) in different terminals, even in different networks.

For the purpose of MM, the system makes use of the HLR and the VLR of the GWs.

More specifically, the main MM procedures in GSM are:

– *IMSI (international mobile subscriber identity) attach* and *IMSI detach*: subscriber (re)activation and deactivation
– *registration*: registration to a network, e.g. for roaming
– *location update*: reporting a change of the location area
– *identification* and *authentication*: verification of the mobile user identity and prevention of unauthorized network access
– *TMSI (temporary mobile subscriber identity) (re)allocation*: use of an alias to protect the mobile user's identity
– *encryption*: use of ciphering to preserve the privacy of conversations.

In the following we will not try to describe all the functions of MM, rather we select a few functions and aspects of MM that are especially relevant for mobile satellite networks.

7.3.1 Service Area of a Gateway Station

An important concept with regard to user mobility in satellite networks is the service area of a GW. Here we must bear in mind that there is a great difference between satellite systems with and without ISLs.

For systems without ISLs we can define a *momentary service area* of a GW as the momentary area with simultaneous satellite visibility from the GW and from any user terminal within that area. Besides the location of the GW, this area is determined by the momentary satellite positions, Fig. 7.4(a).

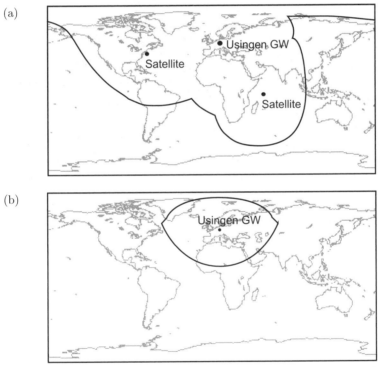

Fig. 7.4. (a) Momentary and (b) guaranteed service area of the Usingen GW of the ICO system. After [SIAK95]

The relevant service area of a GW is the *guaranteed service area*. This is defined as a fixed geographic area around the GW, which is served for all of the time, i.e. which provides continuous simultaneous satellite visibility from the GW and from any terminal within that area, Fig. 7.4(b).

The GWs of a satellite system without ISLs must be placed such that the intended service area of the system is fully covered by the guaranteed service areas of the gateways.

In a satellite system with ISLs, the service area of a GW can be deliberately defined, Fig. 7.5. Thus, a GW can serve an area of any extension.

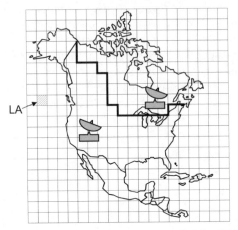

Fig. 7.5. Service area of a GW and location areas for the Iridium system; LA is the location area. After [HL95]

For Iridium, Fig. 7.5, the service area of a GW is defined by a set of corresponding location area codes (LACs). Some 2047 unique LACs are available for each GW. During network access, the home GW selects a serving GW based upon the current LAC of the mobile terminal. All GWs worldwide know which LACs belong to which GW [HL95].

7.3.2 Location Area

The location area (LA) is the smallest area used for the localization of a mobile user. It is characterized by a location area code.

In terrestrial cellular systems the location area usually consists of a group of neighboring radio cells. Due to the movement of non-GEO satellites, the cell-based location area concept used in the GSM system is not applicable for S-PCNs. Two alternative location updating methods have been developed in [SAI95]: the guaranteed service area approach and the terminal position-based approach.

In the guaranteed service area approach, the location area in systems without ISLs is defined as the guaranteed service area of a GW, cf. Fig. 7.4(b). Each GW broadcasts the LAC of its service area via the broadcast control channels of the spot beams (possibly of more than one satellite) overlapping the guaranteed service area. Neighboring location areas are served by other

GWs. A mobile will initiate a location update (see Sect. 7.3.3) whenever it receives a new location area identity through the selected satellite. In this way, the guaranteed service area approach excludes the effect of satellite dynamics in the location update procedure.

For systems with ISLs, the guaranteed service area of a GW can be defined freely.

For terminals with position determination capability, the guaranteed service area can be divided into several location areas. Figure 7.5 shows an example of this approach for a system with ISLs.

An alternative for terminals with position determination capability is to define the location area as a circular area around the terminal position. If a mobile's traveling distance exceeds a certain range, the mobile terminal initiates a location update procedure.

The position of a mobile terminal can for example be determined through a set of delay and Doppler measurements via satellites of the LEO/MEO system. These measurements can be performed by the mobile terminal or the GW. Alternatively, the mobile terminal position can be determined by a GPS receiver integrated into the terminal.

7.3.3 Registration and Location Update

Registration to the satellite network is required when a terminal is switched on outside the service area of terrestrial mobile radio networks or when the user leaves the service area of a terrestrial mobile radio network, but stays within the service area of the satellite network [CBTE94].

For registration, the terminal searches for a broadcast control channel at the known frequencies. According to the received information, the terminal may choose the network (terrestrial or satellite) and will identify the current radio cell. The broadcast control channel indicates the RACH which is subsequently used to inform the gateway currently responsible for the mobile user about the location area the user belongs to. If the user is situated within the service area of his or her home gateway the location area is registered in the corresponding HLR, cf. Fig. 7.1. If the user position belongs to another gateway, the location area is stored in the corresponding VLR, and the address of the currently visited gateway is reported to the HLR in the home gateway.

A location update is required when a non-active mobile user changes the location area. This is detected by monitoring the BCCH or the terminal position. Alternatively, the network can request a periodic location update [FPEH+95]. A location update procedure with change of VLR includes the following steps (see also Fig. 7.6):

− The terminal requests a signaling channel via the RACH.
− The satellite/GW assigns an SDCCH (or a TCH) via AGCH.
− The terminal transmits a location update request via SDCCH.

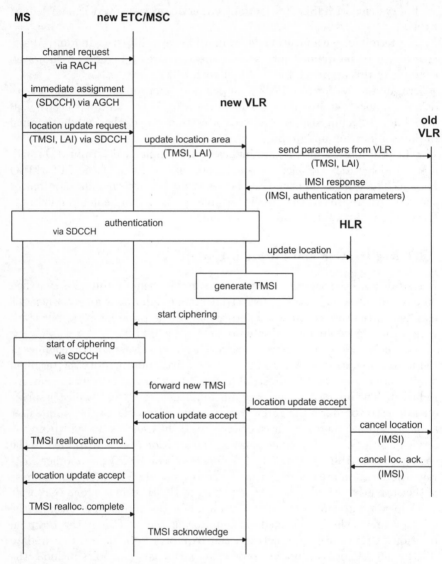

Fig. 7.6. Simplified GSM-like location update procedure with change of VLR

– The visited GW authenticates the mobile user and decides whether he or she is admitted to the area of the visited GW. For this purpose, authentication information from the old VLR is used.
– In the positive case, ciphering is started and a new TMSI is allocated.
– The address of the new visited GW is stored in the HLR of the home GW, and the location area of the user is registered in the VLR of the new visited GW.
– The previously visited GW is notified and the user is deleted from the corresponding VLR.

7.4 Paging

Although paging in GSM is a function of radio resource management, we prefer to describe it in terms of mobility management, since it serves to find and alert a mobile terminal in the location area.

When a mobile terminated call arrives, the mobile user is informed via the paging channel of the currently responsible gateway. The paging signal is broadcast within the user's current location area, which normally comprises a number of spot beams. Therefore, a trade-off arises between the signaling effort for paging and for location updating, respectively: large location areas (containing many spot beams) require more signaling for paging but infrequent location updates; small location areas require less paging effort but frequent location updates.

The user can be paged either in parallel via several satellites and spot beams, or successively in each beam of the location area. Parallel paging implies redundant signaling but causes less delay. Serial paging reduces the average signaling effort but delays the connection setup.

Using a one-step parallel paging scheme, the whole location area can be paged in a paging step. More than one paging step may be necessary because of signal shadowing. The required number of paging steps can be reduced if the mobile terminal continuously chooses the best visible satellite by means of a spot beam reselection procedure.

With the two-step paging scheme [He96], in the first step the mobile is only paged in the spot beam with the highest probability of user presence, thus minimizing the paging signaling overhead. If the first paging step fails, the paging signal is retransmitted in the whole location area. If satellite diversity is applied, the paging signal is broadcast via two satellites.

The most probable spot beam(s) can be determined from the spot beam and the time of the terminal's last contact with the GW (for location update, connection setup, etc.). For terminals with position determination capability, the most probable spot beams are the beams covering the registered terminal position.

7.5 Call Control

Call control (CC) is a subdomain of connection management (CM). GSM call control is responsible for setting up, maintaining, and releasing circuit-oriented services, including speech and circuit data. Here we will concentrate on the setting up of a call which can be initiated by the mobile user or from the fixed network side, respectively.

7.5.1 Setup of a Mobile Originating Call

For mobile originating calls, the terminal requests a connection via the RACH of the currently responsible gateway. For satellite systems without ISLs, the responsible gateway is also the access point to the fixed PSTN. For systems with ISLs, the responsible gateway can determine any suitable gateway to be the PSTN connecting gateway (e.g. the GW nearest to the called PSTN subscriber), based on the digits dialed by the mobile subscriber.

The connection setup of a mobile originating call is performed according to the following steps (see also Fig. 7.7):

- During standby, the mobile user monitors the BCCHs of the satellites and determines their quality.
- The mobile user dials.
- The terminal requests a signaling channel via the RACH of the selected satellite.
- The GW assigns an SDCCH (or a TCH) via AGCH.
- Authentication is performed, using appropriate information from the HLR.
- Ciphering is started.
- The visited gateway assigns a traffic channel including SACCH and FACCH to the user, and the call is activated.
- The visited gateway continues to control the call and the mobile radio link.
- For satellite systems with ISLs, the user data can be routed via ISLs directly to the PSTN-connecting GW, without going through the visited GW (cut-through routing [HL95]).

7.5.2 Setup of a Mobile Terminating Call

In mobile satellite systems a mobile terminating call from the PSTN normally arrives at the home gateway of the user. If the user is not in the service area of the home gateway, the currently visited gateway is read from the HLR and the mobile user is paged by the visited gateway within the current location area. After the user has responded to the paging (via the RACH) and a suitable satellite has been selected, the mobile connection is established and a fixed connection is extended from the home gateway to the visited gateway.

In S-PCNs with ISLs, calls from the PSTN can in principle be accepted at any gateway (e.g. at the GW nearest to the calling subscriber).

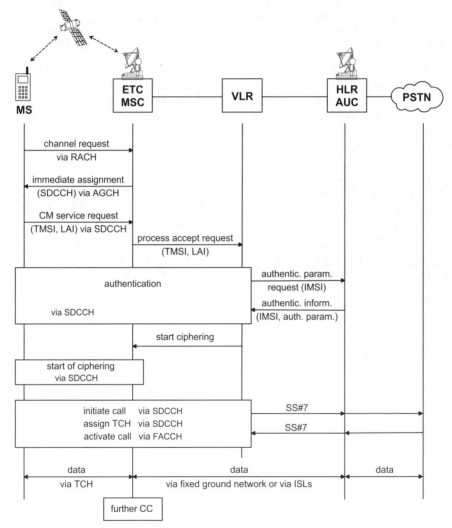

Fig. 7.7. GSM-like mobile originating call setup from a mobile user located outside the home area

The connection setup of a mobile terminating call is performed according to the following steps (see also Fig. 7.8):

- The call coming in from the PSTN is initially routed to the home gateway of the called mobile user.
- The home gateway checks the mobile user's authorization and service profile.

Let us assume that the mobile user is located outside the home area. Then:

- The address of the visited GW of the mobile user is read from the HLR.
- The call is routed to the visited GW.
- The LAI of the user is read from the VLR of the visited GW.
- The mobile user is paged within his location area.
- The called user responds via the currently available RACH and requests a signaling channel. The selected RACH determines the satellite to be used.
- The GW assigns an SDCCH (or TCH) via AGCH.
- Authentication is performed and ciphering is started.
- The visited GW assigns a traffic channel including SACCH and FACCH, and the call is activated.
- The visited GW continues to control the call and the mobile radio link.
- For satellite systems with ISLs, the user data can be routed via ISLs directly from the PSTN-connecting GW to the mobile user, without going through the visited GW (cut-through routing [HL95]).

7.6 Dynamic Channel Allocation

Since the frequency bands allocated to S-PCN are strictly limited, the resource of radio frequencies is precious and has to be used as efficiently as possible. Therefore, radio resource management is a very important and extensive group of tasks within mobile communication networks. Some aspects of RR (especially paging) have already been addressed in this chapter. Other aspects of RR are the selection of a satellite and a spot beam for a mobile user, the allocation of a traffic channel, and the determination of the transmit power for the satellite and the mobile terminal. These tasks arise when a new call is set up, but also during a handover procedure, i.e. when the channel, the spot beam, or the satellite must be changed during an ongoing call.

Here we will concentrate on the task of allocating traffic channels to the spot beams of a satellite and to mobile users. An important efficiency criterion related to channel allocation is frequency reuse: for a given system bandwidth, the system capacity increases with increasing frequency reuse, as in Eq. (6.7). Co-channel interference, however, sets a lower limit for the reuse distance.

When speaking about a (logical) traffic channel (TCH in GSM), we mean

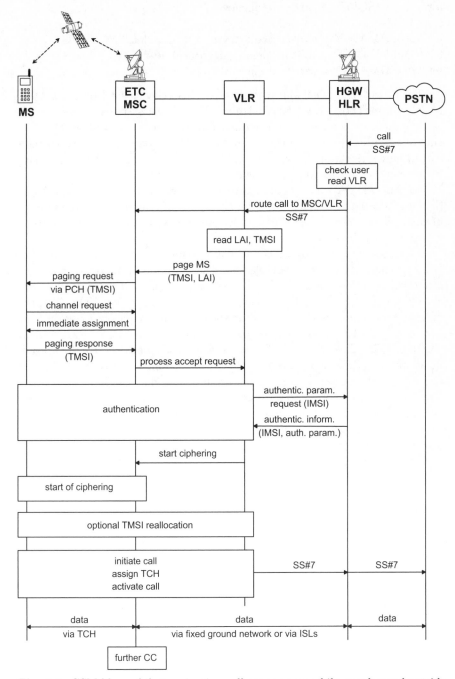

Fig. 7.8. GSM-like mobile terminating call setup to a mobile user located outside the home area

for FDMA (SCPC): a carrier frequency,
for TDMA: a time slot,
for MF-TDMA: a carrier frequency and a time slot,
for CDMA: a code sequence (signature), and
for MF-CDMA: a carrier frequency and a code sequence.

The aspect of channel allocation is especially relevant for FDMA/TDMA-based systems (MF-TDMA in particular). For CDMA systems, all spot beams can use the same frequency, and channel allocation reduces to assigning different code families to neighboring cells. For CDMA systems with user-specific spreading codes (Globalstar), the problem of channel allocation disappears, because each user has his or her own specific "channel".

For MF-TDMA we must take into account whether only full TDMA carriers can be allocated to a spot beam (we speak of granularity as the number of channels per carrier), or if carrier frequencies and time slots can be freely distributed to the spot beams (as in the terrestrial DECT system).

A straightforward way of channel allocation is *fixed channel allocation* (FCA), where the set of available channels is divided into K (equally big) groups of channels. Each radio cell uses one of the K channel groups, and clusters of K cells using different channel groups are formed. The footprint of a satellite is filled with cell clusters, such that the frequency reuse distance is maximized. This approach to channel allocation has been dealt with in Chap. 6. With fixed channel allocation, a new or handed-over call can only be adopted if a free channel is available in the channel group allocated to the respective spot beam. If all channels are in use, a new call is blocked and a handed-over call is forced to terminate, cf. Sect. 6.4. In the light of this, fixed channel allocation is rather sensitive to geographic traffic variations.

Generally, channel allocation at call setup should aim for a low blocking probability (without deteriorating ongoing calls too much). Channel allocation during a handover should minimize the probability of forced termination. In principle, these goals contradict each other; however, blocking of a new call can be more acceptable than the forced termination of a call.

Here, we will concentrate on *dynamic channel allocation* (DCA), which can cope better with these requirements than FCA.

For DCA there is no fixed correspondence between channels and spot beams. All channels available to a satellite are kept in a common pool, from which any channel can be allocated to any cell, as long as a certain frequency reuse distance or a certain signal quality is maintained. A channel is allocated to a user only for the duration of a call. After this, the channel is returned to the common pool of available channels. DCA requires an algorithm for selecting a channel from the pool.

There is a large number of different DCA schemes. In centralized schemes, a channel from the central pool is assigned to a call by a centralized controller, which may reside in the satellites. In the "first available" scheme, the first available channel satisfying the reuse distance encountered during a channel search is assigned to the call. With "locally optimized dynamic assignment", channel selection is based on the blocking probability in the vicinity of the cell in which a call is initiated.

Interference-based DCA [BN96] relies on downlink interference measurements of a mobile requesting a channel, and allocates the channel with lowest interference I_{down}, provided I_{down} is smaller than a given maximum $I_{\mathrm{down,max}}$, and I_{up} (measured by the satellite) is smaller than $I_{\mathrm{up,max}}$. A DCA scheme based on the carrier-to-interference ratio C/I uses estimated values of $(C/I)_{\mathrm{up}}$ and $(C/I)_{\mathrm{down}}$ for channel allocation.

Channel reuse optimization schemes try to maximize the utilization of every channel in the system, e.g. by defining a cost function for the allocation of a channel, which is low, if the channel resembles an FCA scheme. The 1-clique scheme uses a global channel reuse optimization approach based on graph theory.

For DCA, the satellites must be able to change frequency between the time slots. The terminals must be able to select a determined frequency and a determined time slot. Moreover, they must be able to change frequency between the time slots, in order to monitor the satellite beacons, as information for a possible handover.

7.6.1 C/I-Based DCA

The basic principle of C/I-based DCA is to apply the MaxMin scheme [GGV93], which is a way of achieving a capacity bound. From the set of channels available for the spot beam, CH, the channel ch is allocated, which maximizes the minimum carrier-to-interference ratio C/I in the set MT of mobile terminals using channel ch and therefore being affected by its allocation:

$$ch = \max_{ch \in CH} \; \min_{mt \in MT(ch)} \{C/I(mt, ch)\}. \tag{7.1}$$

Additionally, it must be guaranteed that

$$\min_{mt \in MT(ch)} \{C/I(mt, ch)\} \geq (C/I)_{\mathrm{min}} \tag{7.2}$$

with $(C/I)_{\mathrm{min}}$ denoting the minimum required C/I. Criterion (7.1) is usually applied for the uplink. For FDD with paired channels, condition (7.2) must be checked as well. The channel request is rejected if (i) all channels are already in use by the satellite, or (ii) an allocation of the channel would reduce the minimum C/I below a required value.

An efficient method for interference estimation is to use inactive time slots for interference measurements and to store the results for a future channel request. Interference measurements can also be based on BER measurements, measurements of the receive power of the signal and interference, or on information from the demodulation process.

According to [GYG97] DCA can advantageously be combined with power control. The transmit power for an uplink or downlink connection should be increased if the respective C/I is too small; the transmit power can also be reduced if the C/I is greater than required.

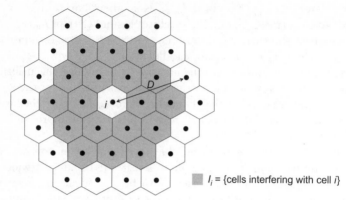

I_i = {cells interfering with cell i}

Fig. 7.9. Cell i and interfering cells for $D = \sqrt{21}R$ ($K = 7$) [DFG95]

7.6.2 DCA Using a Cost Function

Let us consider a certain cell i. Let I_i be the set of cells interfering with cell i, i.e. being closer to cell i than the required frequency reuse distance D, Fig. 7.9.

For a new connection or for a handover only these channels in cell i can be used which are used neither in i nor in I_i.

Let Λ_i be the set of channels available for cell i. For serving a connection request in cell i, the channel $ch \in \Lambda_i$ to be allocated is determined by using a cost function $C_i(ch)$. The channel ch^* is allocated to cell i, for which the cost is minimum:

$$C_i(ch^*) = \min_{ch \in \Lambda_i} C_i(ch). \tag{7.3}$$

For the cost function $C_i(ch)$, $ch \in \Lambda_i$, there are several alternatives. One example is to allocate channels in such a way that an FCA with a minimum reuse distance D is approximated as far as possible:

Let F_i be the set of channels in cell i corresponding to an FCA with reuse distance D. To these channels low costs are assigned. Accordingly, high costs are assigned to channels from the sets F_k of the neighboring cells $k \in I_i$.

Channel Rearrangement. To improve the performance of DCA, it is necessary to rearrange the channels in a cell each time a channel becomes free (at connection tear-down or outbound handover). This is done in such a way that the channel with the highest cost is returned to the pool of free channels. Without such a rearrangement the performance of the DCA would be considerably lower.

Hybrid Channel Allocation. For hybrid channel allocation, the channels are divided into two groups. One group is assigned according to FCA, the other group constitutes the DCA pool.

Channel Borrowing. Channel borrowing is a simple step in the direction of DCA. Initially, channels are allocated according to FCA. If all channels within a spot beam are occupied, and a new call arrives, a channel is borrowed from a neighboring spot beam, if this channel provides the necessary reuse distance to other cells currently using that channel. After any call within the own spot beam is finished, the channel allocation is rearranged, and the borrowed channel is returned.

CDMA with User-Specific Codes (Globalstar-CDMA). This approach corresponds to a completely dynamic channel allocation, up to the capacity of a radio cell. If the cell capacity is exceeded, graceful degradation occurs, rather than hard blocking.

7.7 Handover

In cellular radio networks for mobile communications different kinds of handover can occur. The radio channel of an ongoing connection can be changed (intra-cell handover), e.g. because the propagation or interference situation changes. In terrestrial networks, user mobility can require a change of the radio cell during an ongoing connection (intercell handover). For this kind of handover, channel allocation in the new cell is required (except for CDMA with user-specific codes).

In non-geostationary satellite systems, the *satellite movement* causes spot beam and satellite handovers to occur in a deterministic sequence, Fig. 7.10. MEO satellites can be visible for approximately 2 hours, a period much longer than a typical telephone call. LEO satellites may be visible for less than 10 minutes; therefore, handover of a call to another satellite will most probably be necessary. In LEO systems most of the satellite handovers take place within an orbit plane ("street of coverage"). The overlap of radio cells and satellite footprints provides a certain time window for executing the handover.

A satellite handover can also be caused by the *movement of a mobile user*, depending on the propagation situation in the user's environment. With regard to spot beam handovers, user movement can be neglected compared to the satellite's velocity.

High traffic load in a satellite or spot beam may be alleviated by satellite or spot beam handovers. Finally, bad co-channel interference situations may be resolved by changes of carrier frequency (intra-spot beam handover).

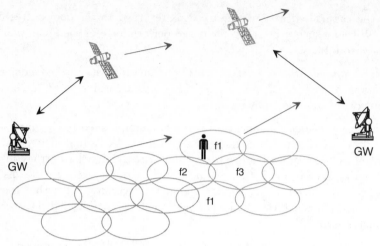

Fig. 7.10. Handover in non-geostationary S-PCNs

A satellite handover requires more signaling than a spot beam handover; also, it can in principle entail a GW handover. This is usually avoided, however, because it would have an impact on the routing within the terrestrial fixed network, and, especially, it would presuppose widely deployed intelligent network functionality.

In *satellite-fixed cell systems*, the cells move over the earth according to the satellite's motion, and handovers between spot beams of a satellite arise, occurring more frequently than satellite handovers.

In *earth-fixed cell systems*, cells are fixed geographic regions on earth [RM96b]. During the time interval a beam is assigned to a cell, the beam is continuously steered in order to compensate for the satellite's motion. Within a satellite footprint, no beam handovers occur. If the satellite footprint leaves an earth-fixed cell, the respective beam is redirected to a new incoming cell. This eliminates uncertainty as to when a handover occurs, resulting in a near-to-zero probability of handover failure. However, earth-fixed cell systems require a larger number of satellites compared to satellite-fixed cell systems [RM96b]. Moreover, the satellites must be capable of performing beam steering and cell switching. An example of an earth-fixed cell system is the present Teledesic concept.

If the satellite network is integrated with a terrestrial mobile network, inter-network handovers may occur, e.g. if the mobile user leaves the service area of the terrestrial network.

The following aspects are related to a handover:

– measurement of connection quality
– handover decision (determination of the time of a handover)
– handover procedure (signaling)

– channel allocation in the new spot beam (except for CDMA with user-specific codes).

For a satellite handover in non-GEO satellite networks, the following additional tasks arise:

– compensation of Doppler jumps
– compensation of different signal propagation times by buffering
– change or "extension" of the ISL route.

In the following, an example of the procedure for a terminal-initiated handover is given:

– The terminal monitors the BCCHs of neighboring spot beams or satellites, e.g. for TDMA in time slots not used by the terminal. If the signal quality of a neighboring beam is better than that of the current one, the terminal sends a message to the GW and requests a handover.
– In the GW a second transceiver is assigned, which will take over the connection.
– The satellite/GW assigns a channel.
– The connection is rerouted (in the satellite or in the GW).

7.7.1 Handover Decision

Under ideal line-of-sight conditions, the criteria for initiating a satellite handover could be based on the deterministic geometric properties. Since the satellite footprints overlap, different handover strategies exist:

1. The satellite with maximum elevation is always used.
2. A satellite is used until its elevation falls below a minimum value (ε_{min}).

Strategy 1 yields better channel characteristics but needs more handovers compared to strategy 2 [BW94].

In the mobile/personal environment, the handover process should also consider the impact of the satellite channel. Optimizing the quality of service implies that the best mobile link should always be chosen in terms of received signal power or signal quality. This would produce a large number of handovers, however, corresponding to a high signaling overhead in the network. Applying a power hysteresis threshold or a time-out period to handover initiation reduces the frequency of handovers and leads to a trade-off between service quality and signaling load [CB95]. In [ZTE96] a handover algorithm is proposed which is based on user terminal position and signal strength measurements.

7.7.2 Handover Procedure

Two fundamental versions of the satellite handover procedure are known: (i) backward handover and (ii) forward handover. Whereas the first method is very similar to handover in GSM, the second method is especially suited for satellite networks.

Figure 7.11 illustrates the different procedures. For a *backward handover*, the mobile station (MS) submits the handover request via the old satellite, and after a handover command is received from the network, it tunes to the new satellite to acquire a new traffic channel. Between the reception of the handover command from the old satellite and the handover complete message sent to the new satellite a handover break occurs with duration

$$T_{\text{break,b}} \approx 3\,T_{\text{prop}} + T_{\text{proc,b}}. \tag{7.4}$$

Here, T_{prop} is the propagation delay between MS and GW, and $T_{\text{proc,b}}$ represents the accumulated processing time (approx. 150 ms for GSM), including the time required to tune and synchronize to the new satellite. The break for a backward handover may last 200 ms for LEO satellites and up to 500 ms for MEO satellites.

Another disadvantage of the backward handover scheme is that, if the signal quality on the channel to the old satellite substantially decreases between the handover request and the handover command, the connection will be dropped.

This problem can be avoided by using a *forward handover* procedure. Here, the handover request is submitted to the new satellite, and the old traffic channel is maintained until the new channel has been acquired. Then, switching to the new satellite, MS sends an assign complete message. After receiving and processing this message, GW switches to the new satellite as well. Thus, a handover break occurs only between the sending and receiving of assign complete including some additional (but compared to backward handover significantly reduced) processing time:

$$T_{\text{break,f}} \approx T_{\text{prop}} + T_{\text{proc,f}}. \tag{7.5}$$

Typical breaks for forward handover in LEO systems last around 60 ms [EVW96].

7.7.3 Channel Allocation at Handover

Channel allocation is required for handovers between spot beams as well as for handovers between satellites. It is, however, not required for CDMA with user-specific codes (Globalstar).

The channels available in a satellite cell must be used for setting up new connections as well as for serving handover requests. To this end, different strategies can be adopted [LMN94]:

Method without Priorities. A handover request is treated like a connection setup request. The handover is blocked (i.e. the connection is terminated) if no free channel is available.

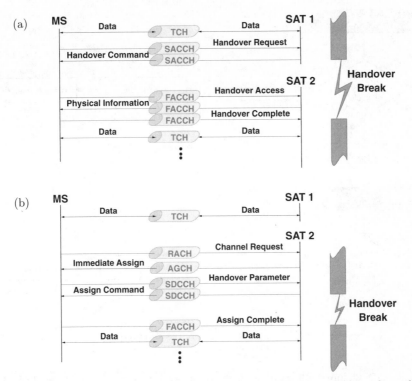

Fig. 7.11. Simplified signaling flow, corresponding GSM-like signaling channels, and handover break for (a) backward handover and (b) forward handover [EVW96]

Method with Reserved Channels. A number of channels are reserved for handover requests. However, handover requests can also be served with normal channels, see Fig. 7.12. Thus, handovers are prioritized against call setups.

Methods with Handover-Queueing. These methods exploit the overlapping of the radio cells: if no channel is available, the handover request is stored in a queue. The old channel is used further until a new channel is available and the handover can be executed. If in the meantime the terminal leaves the area of overlapping, the connection is dropped.

Dynamic Channel Allocation (DCA). With DCA, the channels available to the whole satellite are assigned to the spot beams on demand. A channel can migrate from beam to beam and remain associated with a given call during the whole satellite passage [RM96c]. Hence, the call dropping probability is reduced. If no free channel is directly available, a rearrangement of channels in the neighboring cells may alleviate this situation [Del95]. Finally, in [DFG95] a handover strategy is investigated which combines DCA with queueing of handovers.

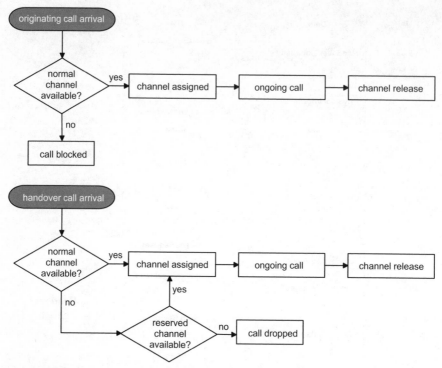

Fig. 7.12. Channel allocation with channels being reserved for handover [LMN94]: top, setup of a new connection; bottom, handover

7.8 Call Completion Probability

Similar to voice quality, connection setup delay, and messaging delay, the probability to successfully set up, maintain, and tear down a connection is a basic measure of service quality [Dos95].

Paging. Paging must be performed for call setup *to* a mobile user. Performance parameters are the probability of successful paging, P_{paging}, and the mean time duration until the user has been contacted.

Call Setup. Performance parameters for call setup are:

- probability of satellite visibility (or waiting time until a satellite becomes visible),
- probability and delay of a successful random access (via the RACH), and
- probability of a free channel in the satellite and in the GW. The blocking probability is

$$P_{\text{B}} \approx P_{\text{B,sat}} + P_{\text{B,GW}}. \tag{7.6}$$

Maintenance of the Connection, Handover. Let

T be the connection duration with probability density function $f_T(T)$
$n_{\mathrm{HO}}(T)$ the number of required handovers during time duration T
P_{HO} the probability of successful handover
$P_{S>T}$ the probability that the duration of the satellite visibility (good channel state) is larger than the duration of the connection.

Then, the probability of maintaining the connection is

$$P_M(T) = P_{\mathrm{HO}}^{n_{\mathrm{HO}}(T)} \cdot P_{S>T}. \tag{7.7}$$

Probability of Successful Connection.

$$
\begin{aligned}
P_{\mathrm{suc}}(T) &= P_{\mathrm{paging}}(1 - P_B)P_M(T) \\
\overline{P_{\mathrm{suc}}} &= \int_0^\infty P_{\mathrm{suc}}(T)f_T(T)\,dT.
\end{aligned}
\tag{7.8}
$$

7.9 Routing

For every connection between two end users, or every information entity (for instance, a packet) transferred between them, a route through the communication network must be chosen. Routing is a central task in large networks and influences network performance as well as quality of service (e.g. message delay).

A good reference for routing in circuit-switched networks are the books by Ash [Ash97] and Girard [Gir90], and routing in packet-switched networks is extensively discussed in the textbooks by Tanenbaum [Tan96], Schwartz [Sch87], and Bertsekas and Gallager [BG87]. The following discussion of routing in LEO/MEO satellite networks is largely based on [Wer97] and [WM97].

7.9.1 Routing in LEO/MEO Satellite Networks

Routing is especially important in LEO and MEO satellite constellations employing intersatellite links (ISLs). Consequently, in the following only ISL-based systems will be considered.

ISL Topology Dynamics. Throughout this section, Iridium will be used as an example, as it is the first operational LEO satellite communication system to employ ISLs, and allows some major challenges to be highlighted for connection-oriented routing.[1] Recalling Chap. 2, polar or near-polar constellations exhibit a *seam* between two counter-rotating orbits, effectively dividing the constellation into two hemispheres of co-rotating orbits.

---

[1] It should be explicitly noted in this context that the routing concept presented here is indeed developed on top of Iridium-like dynamic ISL constellations, but it is *not* the Iridium proprietary concept.

At a given instant, the ISL topology of the system comprises a unique set of *intra-orbit* and *inter-orbit* ISLs as illustrated in Fig. 2.22. Intra-orbit ISLs have fixed distance and antenna pointing, whereas inter-orbit ones are subject to continuous variations of both [Wer95]. For counter-rotating orbits, moreover, a frequent switching of ISLs would be necessary, a fact that has led to the avoidance of links across the seam for Iridium, see Fig. 2.22. Besides the *continuous* distance changes on co-rotating inter-orbit ISLs there is also a *discrete-time* contribution to the ISL topology dynamics: inter-orbit ISLs are deactivated in polar regions, meaning on/off switching of links. As a result, tailor-made routing strategies are required to handle the rerouting of connections.

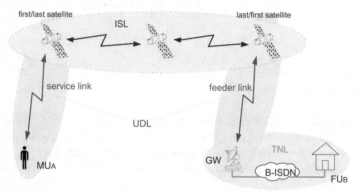

Fig. 7.13. Segments of a typical end-to-end connection in ISL-based LEO satellite systems

Network Segmentation. The end-to-end connection between a mobile user MU_A and a remote fixed partner FU_B comprises three segments, as illustrated in Fig. 7.13. The ISL segment forms a dynamically meshed sub-network in space. The up-/downlink (UDL) segment incorporates the service link between mobile users and satellites as well as the feeder link between satellites and fixed earth stations. These gateway stations (GWs) act as an interface between the satellite system and the terrestrial network link (TNL) segment. In the following, the satellites serving the service link and feeder link sub-segments will be referred to as the *first* and *last satellites* of a connection between two specified end users. Whenever the ISL subnetwork is considered separately without keeping track of specific end user connections, a pair of satellites are denoted as an *OD (origin–destination) pair* forming the end points of an ISL route.

For systems providing mobile telephony, the operation mode will naturally be connection-oriented. In the following presentation, we use ATM (asynchronous transfer mode) terminology and concepts which are common for connection-oriented multiservice networks. A short introduction to ATM principles is given in Sect. 10.3.1 in the context of broadband satellite networks.

In the considered LEO scenario we face the challenge of a multiple dynamic network topology, applying to both UDL and ISL segments. Inherently, this feature requires switching between subsequent different paths during the lifetime of a user connection. In the following, this is referred to as a virtual connection *handover (HO)*, extending the basic set of ATM concepts.

UDL and TNL Segments. The TNL segment is separable from the satellite network by assuming that the gateway either acts as a central switch between the satellite and the TNL part or is even implemented as an intermediate node between two different ATM networks. The networking of the TNL part itself is not considered further here.

For the UDL routing part, an extension of the *virtual connection tree (VCT)* concept – originally proposed by Acampora and Naghshineh [AN94] for terrestrial cellular ATM networks – is considered. In the original concept one root switch manages a fixed tree serving a certain mobile service area, Fig. 7.14(a). The VCT takes care of connection setup and VCI (virtual channel identifier) reservation and ensures fast and transparent handovers of roaming users in the mobile service area. To apply the VCT architecture and methodology to the LEO satellite environment, a dynamic satellite cluster headed by a *master satellite* is defined, Fig. 7.14(b). This satellite cluster builds a time-dependent VCT with spot beams, ensuring that fast and transparent spot beam and satellite handovers can be performed [Lut95]. The leaves of this tree may for instance contain the spot beams of all satellites that are above a minimum elevation angle for a certain user. In this way, a handover of virtual connections can be easily accomplished without dedicated signaling. For each user (group), or for each area covered by a VCT, the relevant VCT must be updated according to the satellite's movement.

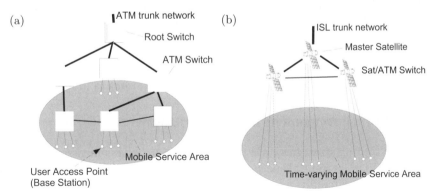

Fig. 7.14. Virtual connection tree (VCT) concept applied to handover/rerouting in (a) terrestrial cellular and (b) multibeam satellite ATM networks

ISL Segment. The time-varying network topology imposes specific challenges on the ISL routing:

- A route between a pair of serving satellites may not be optimum or even possible during the whole connection. Rather, route changes may be necessary, resulting in problems connected with flow control and critical delay variations during path switching.
- Due to time-limited satellite visibility, the first and last satellites may change during a connection (satellite handover). This requires additional route changes within the ISL network.

Virtual Topology. A possible route between a first and a last satellite can be modeled as a virtual path connection (VPC). Every section of this route, i.e. every ISL between adjacent satellites, in general contains a number of virtual paths (VPs), all sections using the complete set of VP identifiers (VPIs) independently. The routing function is incorporated in the respective dynamic VPI translation tables. By assigning VPs to the ISLs and setting up the corresponding VPI translation tables in the satellite switches, a *virtual topology* is defined on top of the physical one.

The basic idea behind this approach is that end-to-end virtual channel connections (VCCs) sharing the same first and last satellites at arbitrary time can be aggregated into one common VPC across the ISL subnetwork. Every transit satellite provides – on the basis of locally available switching tables – pure VP switching functionality between every pair of ISL ports, and thus the whole space segment becomes a pure and fast switching cross-connect network. Avoiding any switching on virtual channel (VC) level turns out to be especially favorable in the case of many simultaneous low bit rate connections on a trunk line, as in systems providing voice services. Moreover, the on-board storage for a pure VPI translation table can be kept small.

Figure 7.15 illustrates the overall concept comprising the integration of UDL and ISL segments. It specifically addresses the problem of guaranteeing a continuous connection in the case of an ISL subnetwork internal path handover. It has been shown [Wer97] that the operation of both basic types of handover – UDL satellite handover and ISL VPC/path handover – can be managed with the proposed concept but requires sophisticated optimization algorithms and techniques to keep the possible impairments to a minimum.

7.9.2 Off-Line Dynamic ISL Routing Concept

Classical routing strategies have traditionally more or less focussed on some kind of shortest or multiple path search in networks with a fixed topology. Dynamic routing capabilities are then only required for traffic adaptive routing or in reaction to unpredictable link or node failures.

In contrast to this, in ISL subnetworks it is generally required that the routing is dynamic already in order to cope with the permanent topological

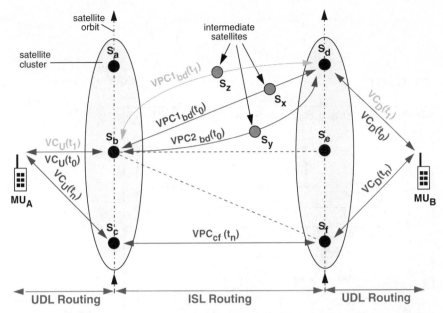

Fig. 7.15. Virtual-connection-based routing concept in satellite systems with ISLs. The continuous operation of an existing end user connection is illustrated for the case of VPC handover between the same first/last satellite pair from steps t_0 to t_1.

changes. *Dynamic virtual topology routing (DVTR)* [Wer97] has been proposed for use in such environments. Figure 7.16 illustrates the approach at a glance. The basic idea is to exploit the fact that all topology dynamics is periodically deterministic. This allows the setting up off-line of a time-dependent virtual topology providing continuous operation of end-to-end connections. This essentially builds the framework for later on-line traffic adaptive routing strategies as indicated in the figure, i.e. traffic adaptive routing can be performed within the limits of the given virtual topology.

The relevant parameters of the discrete-time network model and routing procedure are the following:

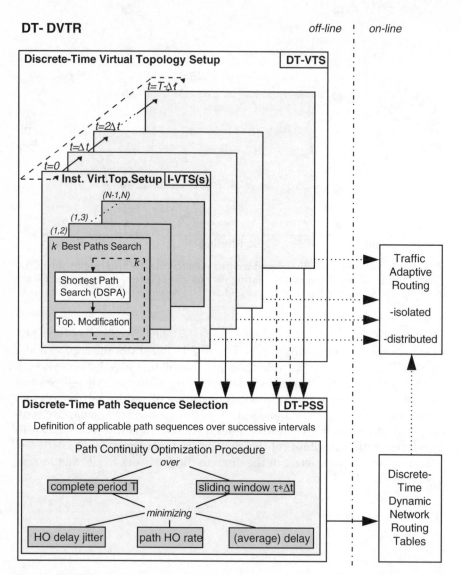

Fig. 7.16. Discrete-time dynamic virtual topology routing (DT-DVTR): concept and implementation

N	constant number of nodes (satellites)
T	constellation period

discretization:

S	number of constellation/topology snapshots
$\Delta t = T/S$	discretization step width
$t = s\Delta t,$	snapshot times
$s = \{0, \dots, S-1\}$	

network model:

$G(s) = (V, E(s))$	network graph at $t = s\Delta t$ with:
$V = \{1, \dots, N\}$	constant set of nodes
$E(s) = \{(i,j)_s\}$	time-varying set of undirected links between neighboring nodes i and j
$c_{ij}(s)$	costs associated with link $(i,j)_s$

routing:

W	constant set of $N(N-1)/2$ unordered origin–destination (OD) pairs
$w \in W$	single OD pair
K	maximum number of alternative OD paths (routes)
$P_w(s)$	set of up to K distinct loopless paths for w
$p(s) \in P_w(s)$	single path assigned to w
$\delta^{p(s)}_{(i,j)_s} \in \{0,1\}$	link occupation identifier: link $(i,j)_s$ used by path $p(s)$ or not

Assigning VPCs to OD pairs essentially determines an *instantaneous virtual topology* $VT(s)$ upon $G(s)$, covered by module I-VTS(s) in Fig. 7.16. This task is performed by a k-best path search algorithm for every OD pair, $k \in \{1, \dots, K\}$. The single shortest path search task can be formulated as finding the least-cost path $p(s)$, i.e. the path with minimum path cost

$$\min_{\forall p(s)} C_{p(s)} = \min_{\forall p(s)} \sum_{i,j} c_{ij}(s)\, \delta^{p(s)}_{(i,j)_s}. \tag{7.9}$$

An iterative approach based on successive calls of the Dijkstra shortest-path algorithm (DSPA) [Dij59] is indicated, which allows the introduction of topology modifications in between. For instance, by "eliminating" already occupied links one can force a set of disjoint paths, thus providing a base for simple and robust fault recovery mechanisms and building a sound framework for shaping network traffic flow at operation time. The path cost metric can for instance consist of the sum of all single link propagation and node processing/switching delays encountered on the path.

Performing I-VTS(s) for all $s = \{0, \dots, S-1\}$ constitutes the *discrete-time virtual topology setup DT-VTS*.

DT-DVTR is finally completed with another off-line module, namely *discrete-time path sequence selection DT-PSS*. This module solves the problem of selecting for all OD pairs sequences of VPCs over successive time intervals from the virtual topology provided by DT-VTS; here the routing becomes really dynamic and potential problems due to path switching can be avoided or minimized. Promising variants for such a *path continuity optimization procedure* are (i) minimizing HO delay jitter (i.e. instantaneous delay offsets during path handover) and (ii) minimizing path HO rate, with some restrictions also (iii) minimizing (average) delay. Concerning the duration of the optimization interval, two main approaches are incorporated:

1. The optimization is performed over the complete period T, then achieving an optimal network solution. The result is either one unique first-choice path sequence out of k^S possible ones, or a set of ordered (i.e. prioritized: first-choice, second-choice, etc.) sequences of such kind.
2. The optimization is performed in a time-distributed manner within a *sliding window* of discrete-time duration τ, i.e. extending over the interval $[s\Delta t, (s + \tau)\Delta t[$ (modulo T), $s = \{0, \ldots, S - 1\}$, $\tau \in \{2, \ldots, S - 1\}$. The results are s-unique first-choice path sequences, respectively s-unique sets of ordered sequences.

Figure 7.17 shows examples of instantaneous VPC sets over subsequent time intervals as found by an iterative Dijkstra link-disjoint SPA module of the ISLSIM[2] software. The effect of forced path switching becomes obvious as some ISLs are switched off when approaching the polar region.

Considering only one first-choice path sequence, both optimization methods are illustrated for the *minimize VPC HO rate* target function in Figs. 7.18 and 7.19, respectively.

For the whole system period, one unique *track* of VPCs is defined as the first-choice VPC sequence for each OD pair in the system period approach, Fig. 7.18. The criterion for defining this track is a minimum number of VPC handovers (VPC-HOs) over the whole system period. All end user connections that are served by the respective satellite pair will use the same unique VPC in a given interval. The rationale for this approach is to use a global minimization of the VPC handover rate, considering (i) that the overall signaling complexity decreases and (ii) that the critical HO delay jitter affecting single connections is reduced.

The main objective of the more sophisticated sliding window approach is to extensively exploit the optimization potential with respect to single short-term connections, paying the price of course of increased operational complexity. This is illustrated in Fig. 7.19, again considering one terminating satellite pair: a sliding time window (with window size typically exceeding the mean call holding time) is used as the discrete-time optimization interval, where the evaluation of the corresponding mathematical target function is performed. The latter is in essence a weighted sum of maximum and average delay jitter. In this way, optimal tracks through the "landscape" of alternative VPCs are selected. These VPC tracks finally determine the routes, including

---

[2] Proprietary simulation tool of the German Aerospace Center (DLR).

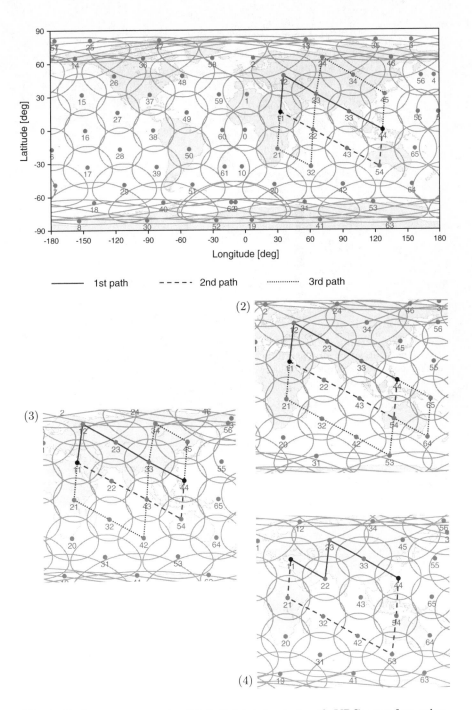

Fig. 7.17. Instantaneous sets of link-disjoint shortest-path VPCs over four subsequent time intervals ($\Delta t = 2$ min) for satellite pair S_{11}–S_{44}

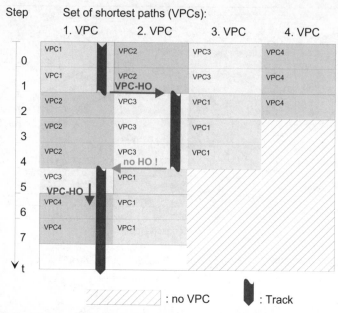

Fig. 7.18. System period optimization approach for one terminating satellite pair. VPC handover situations are indicated. The same VPC numbers and grayscale levels identify one unique physical path

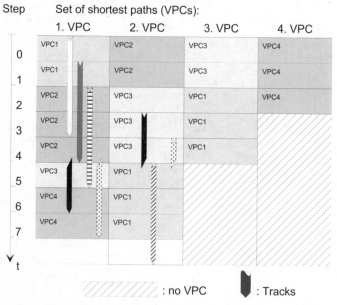

Fig. 7.19. Sliding-window optimization approach. The same VPC numbers and grayscale levels identify one unique physical path. Sliding window length: four steps

handovers, that any connection starting in the respective time interval will use in the operational phase of the system. In other words, active connections set up in different intervals will use different VPCs (and consequently different VPs on all single links, even if physically the same!). This indeed requires an increased number of VPIs; more specifically, the *average* number of VPIs required per ISL grows proportionally with the sliding window size.

Figure 7.20 illustrates a performance comparison of the two proposed optimization approaches in terms of HO delay jitter distribution experienced by a fixed length 2 minute call in the Iridium ISL subnetwork. The simulated calls are equally distributed over time and all possible OD pairs. The results emphasize the superior performance of the sliding-window approach over the period-based optimization scheme.

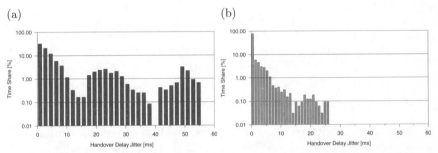

Fig. 7.20. Distribution of VPC-HO related delay jitter for a 2 minute call from the first satellite S_0 in Iridium, averaged over all possible last satellites and one system period. *Optimization strategy:* (a) minimize number of VPC-HOs during one system period; (b) minimize HO delay jitter within a 2 minute sliding window.

Figure 7.21 provides the distribution of VPC-HO delay jitter for a typical telephone service, i.e. showing a negative exponentially distributed call holding time with mean $1/\mu = 3$ min. The length of the sliding window for HO delay jitter minimization is 5 minutes. Note that the shown distribution contains only those calls which are affected by a handover during their lifetime; with the given simulation parameters, these are in fact less than 20% of all calls. Altogether this results in a very low percentage of calls that have to cope with an HO delay jitter larger than 20 to 30 ms. However, if one wants to guarantee optimal performance in the light of ATM quality of service, the remaining delay jitter values – which are minimized within the topology-specific limitations – represent critical requirements for any hitless path switching functionality. Comparing Fig. 7.20(b) with Fig. 7.21, one can see significantly higher jitter values in the realistic traffic scenario. This is due to the fact that in the former theoretical case the optimization window equals the fixed (!) call duration, whereas in the latter case 20% of the calls extend beyond the optimized period and are then exposed to path handovers.

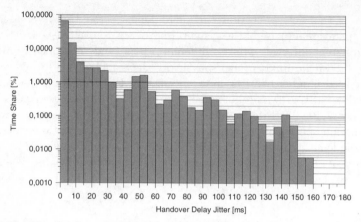

Fig. 7.21. Distribution of VPC-HO related delay jitter for a typical telephone service (negative exponentially distributed call holding time with mean $1/\mu = 3$ min), averaged over all first/last satellite pairs and one system period. *Optimization strategy:* minimize HO delay jitter within a 5 minute sliding window (i.e. 80% of calls end within optimized track)

7.9.3 On-line Adaptive ISL Routing

With connection-oriented communications, the task of on-line routing is to select a route at call setup, which is then only changed in case of a forced path handover. If the traffic load (on links or nodes) is taken into account as one of the decision criteria for route selection, then the on-line routing becomes traffic adaptive. Working within the presented off-line routing framework, the task of on-line routing is reduced to selecting one of the predefined alternative paths. Therefore it becomes less complex but cannot exploit the full potential of adaptivity. Prior to a proper integration of the two concepts it is therefore reasonable to determine the maximum theoretical gain from operating an on-line adaptive routing without any restrictions. This is the scope of this section, extracting the major results from an in-depth study [WM97]. The integrated approach is discussed in the context of ISL network dimensioning in Sect. 12.2.

For the purpose of numerical studies, first an on-line but non-adaptive routing scheme has been implemented which is based on a version of the Dijkstra shortest-path algorithm – the *Moore–Dijkstra algorithm (MDA)* – using ISL lengths as link costs. Therefore the algorithm determines the shortest routes with minimum delay. This scheme serves as a reference for the performance comparison with traffic adaptive routing.

Traffic adaptive routing in the ISL scenario is mainly motivated by the aim of using the on-board ISL capacity as efficiently as possible by smoothing peaks in the ISL loads. Strong time variations in the ISL load can become particularly critical considering the expected high data rates in broadband systems, since all satellites and ISLs have to be designed for the worst case.

The study of adaptive routing schemes has shown that efficient implementations work in a *distributed* or *decentralized* manner. According to [Sch87], a distributed traffic adaptive routing procedure generally consists of the following three components: (i) a measurement process that collects the required information on the actual traffic situation in the network, (ii) an update protocol that distributes (e.g. by flooding) this information to all nodes, and (iii) the shortest path computation to be performed in every single node, using the updated traffic data built into the respective link cost function.

The interval Δt_u between successive updates of the routing tables, in relation to the typical duration of connections, is essential for a reasonable trade-off between performance gain and complexity. Typically, there is some lower bound for Δt_u below which only little performance improvement can be achieved.

In the considered implementation of on-line traffic adaptive routing, the *distributed Bellman–Ford (DBF) algorithm* (see Chap. 5.2 in [BG87]) is used for the path search. The link cost is a function of link traffic alone, thereby investigating the maximum potential with respect to worst-case link (WCL) load reduction through traffic adaptivity.

The investigations have been performed for Iridium in a global telephony scenario including a call model, addressable market data characterizing the subscriber distribution and source traffic demand, a daily user activity model, and a global traffic flow model describing the destinations of generated calls. Details of the traffic modeling and further references can be found in [WM97]. All simulations have been performed over the complete Iridium system period $T_s = 5$ days in order to guarantee that all possible pairs of source traffic patterns and orbital patterns are considered. For Δt_u a value of 20 seconds has been used, which is in a reasonable relationship to the holding times of simulated voice calls and clearly below the discretization step width of the topology.

Figure 7.22(a) displays the pronounced asymmetry of ISL loads with the non-adaptive MDA routing scheme, always using shortest-delay paths. This is mainly due to the combination of strong geographic source traffic variations and the bounded ISL topology. Note that with 106 maintained ISLs a uniform distribution across the network would imply roughly that 1% of the total network traffic be carried on each link. This is far from being the case: the 10 most loaded links carry roughly half of the network traffic. In contrast to this, Fig. 7.22(b) proves that the traffic adaptive DBF scheme clearly achieves a more even traffic distribution in the network: the 10 most loaded links carry roughly 20% of the total network traffic.

For the following ISL load values it should be noted that all figures are normalized to the total source traffic. By doing this, the influence of daily activity variations is excluded and the effects of the routing schemes can be exposed more clearly.

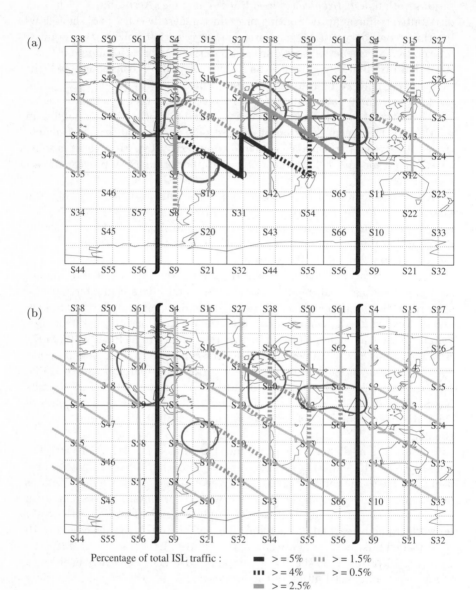

Fig. 7.22. ISL traffic distribution with (a) the MDA and (b) the DBF scheme at $t = 300$ min. The seam appears over the Eastern USA and India/China. High source traffic areas are highlighted

Figure 7.23(a) illustrates that the DBF scheme can reduce the maximum ISL traffic by roughly 30% in the average. On the other hand, one observes a constant increase of 30 to 40% in the average ISL traffic load. This is due to the fact that reducing peak loads on single ISLs is often achieved by shifting parts of the traffic from the shortest paths to alternative paths with more hops. According to its link cost metric, at a given time step the DBF scheme prefers previously less loaded links, whereas the load is decreased on the critical ones. Thus the group of ISLs carrying medium traffic loads grows compared to MDA, Fig. 7.23(b).

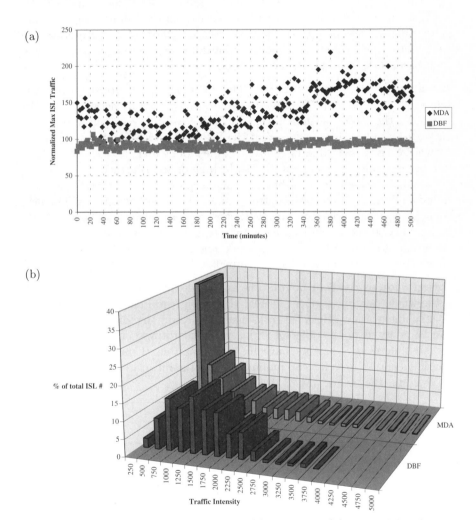

Fig. 7.23. MDA/DBF scheme comparison: (a) WCL load, (b) ISL load distribution

7.10 Integration of Terrestrial and Satellite Mobile Networks

If S-PCNs are to complement terrestrial PCNs, the networks have to be integrated to a certain degree.

Integration at Terminal Level (Dual-Mode Terminals). Terminal integration is the simplest way of providing combined usage of the networks. Basically, a dual-mode terminal contains a terrestrial and a satellite terminal. The mobile user has a separate number in both networks. Incoming calls reach the mobile user via the network chosen from the calling user. For outgoing calls the mobile user chooses the network which is currently available with better quality and takes into account different service charges. There is no network handover during a connection.

Integration at Network Level. The combination of terrestrial and satellite networks is more efficient if they are integrated at network level. In Fig. 7.24, the mobile switching centers (MSC, S-MSC) can mutually access the registers of both networks, and the S-PCN appears as an international network with its own country code. Now, the mobile user can have one unique number regardless of the network to which he or she is attached, and a network handover becomes possible. For this purpose, an active dual-mode terminal must also monitor the control channels of the other network. Further, some kind of interworking unit (IWU) between terrestrial MSC and S-MSC is necessary. If the networks use common registers, the S-PCN appears as an extension of a terrestrial mobile network, and a subscriber will get a country code depending on the home base [Hub96].

Integration at System Level. The highest degree of integration is reached if the whole satellite segment is treated as an integral part of total coverage (system integration). Both networks may use a common radio interface, which can be adapted to the different propagation conditions. Also, the protocols for both network segments have to be compatible. This approach, which considers a satellite cell as an umbrella cell of the integrated network, will be followed for the future UMTS [Del95].

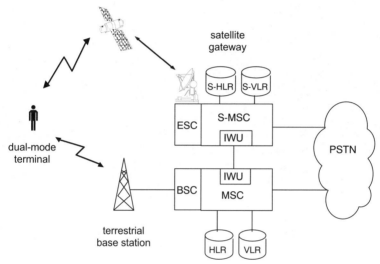

Fig. 7.24. Network integration with dual-mode terminals and mutual register access

8. Satellite Technology

8.1 Satellite Subsystems

A communications satellite essentially consists of two main functional units, the communications payload and the bus (also called the platform). The bus provides all the necessary electrical and mechanical support to the payload and consists of several subsystems [Ric99]:

- The attitude and orbit control system (AOCS) stabilizes the spacecraft and controls its orbit.
- The propulsion system provides the necessary velocity increments and torques to the AOCS.
- The thermal control system maintains the temperature of various subsystems within tolerable limits.
- The structure provides the necessary mechanical support during all phases of the mission.
- The telemetry, tracking, and command (TT&C) subsystem transmits the status of various subsystems of the satellite to the satellite control center and accepts satellite-related commands from the control center. The TT&C also provides tracking support to ground stations.
- The electrical power supply system provides the necessary DC power.

In addition, GEO satellites usually are equipped with perigee- and apogee-kick motors injecting the satellite from the LEO parking orbit via the geostationary transfer orbit (GTO) into the GEO orbit.

Table 8.1 contains an example of a mass budget for a LEO satellite. The main parameters are

- the payload mass,
- the dry mass (total mass of the satellite in orbit without propellant),
- the begin-of-life (BOL) mass (dry mass plus propellants), and
- the launch mass (BOL mass plus propellants for orbit transfer and insertion).

For GEO satellites with kick motors, the BOL mass is the dry mass including the kick motors and the propellants for the satellite. The launch mass is the BOL mass plus the fuel for the kick motors [MB93].

Table 8.1. Mass budget of a LEO satellite (Teledesic) [GKOH94]

Satellite launch mass budgets	Mass (kg)
Structure	87
Mechanisms	52
Cabling	22
TT&C incl. command & data handling	9
Temperature control	37
Attitude/orbit determination and control	12
Propulsion (without propellant)	20
Power	239
Communication payload	144
Contingency (20%)	125
Satellite dry mass	747
Propellant:	
Orbit transfer and insertion	11
Orbit maintenance and gap filling	7
De-orbit retro maneuvers	3
Contingency and margin	27
Propellant mass	48
Satellite launch mass	795

Table 8.2 presents an example of the power budget of a LEO satellite. The main characteristics are the power of the solar cells (DC power), the power required for the payload, and the radiated RF power, which for the considered satellite approximately amounts to 600 W. More detailed information on the mass and power budgets of a satellite can be found in [Ric99, GM93, Pri84].

In this book we will concentrate on the communications payload. For more information on the satellite bus we refer to the literature such as [MB93, Ric99].

Communications Payload. The communications payload comprises antennas for receiving and transmitting the signals from and to the feeder and service links, and repeaters processing the received forward and return link signals.

Two types of repeaters are used, transparent repeaters and regenerative repeaters. A transparent repeater (also called a bent-pipe repeater) only amplifies the uplink signal and translates it to the downlink frequency. A regenerative repeater, in addition to amplification and frequency translation, performs demodulation, baseband processing, and re-modulation.

The main tasks of the payload will be explained for the forward link, Fig. 8.1:

− Signal reception: The feeder uplink signal from the gateway is received via the feeder link antenna.

Table 8.2. Power budget of a Teledesic LEO satellite [GKOH94]

Category	Solar array power (W)	Eclipse battery power (W)
Communications payload	3000	3000
Engineering housekeeping	51	76
Power subsystem		
Shunt regulators	30	30
Power distribution loss	30	30
Power/charge controller	60	60
Battery charging (maximum)	2168	0
Contingency (20%)	1068	634
Total power requirement	6407	3830
Power capabilities:		
Solar array		
Begin of life	11 595	N/A
End of life	6626	N/A
Battery depth of discharge (maximum)		
Average satellite	N/A	17%
Worst case	N/A	31%

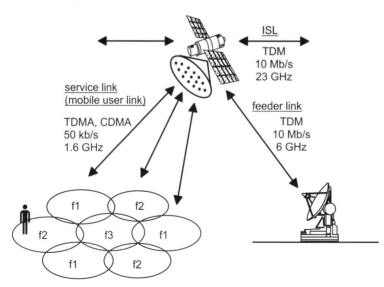

Fig. 8.1. Main tasks of the communications payload and typical link parameters

- Channelization: the received signal is divided up and fed into several spot beam transponders (channels). For an FDM feeder uplink, this can be achieved by a suitable filter bank. For a TDM feeder uplink, a demultiplexer working in the time domain is required.
- Demultiplexing: In a regenerative repeater, single channels or groups of channels are extracted from the received feeder link signal.
- Routing/switching: The extracted signals are routed into the wanted spot beams or are switched to an appropriate ISL.
- Frequency translation: The feeder link frequencies are translated into the corresponding mobile link frequencies.
- Power amplification: The signals are amplified to the power level required for transmission.
- Beam-forming: The transmit signals for the service link are distributed onto the radiating antenna elements or feeds, in order to transmit them within the corresponding spot beams.
- Transmission: The signals are transmitted via the service link antenna.

8.2 Antenna Technology

The multiple spot beams for the mobile link are formed either by a reflector antenna with multiple feeds or with a phased-array antenna requiring a complex feed network to generate multiple beams. GEO satellites tend to use multifeed reflector antennas, while LEO/MEO satellites use phased-array antennas. New GEO systems with a large number of spot beams have either to use a large antenna aperture (ACeS) or to resort to higher frequencies (EuroSkyWay). Earth-fixed cell systems with regional frequency assignment require agile digital beam-forming.

Inter-orbit ISLs require steerable antennas, which have to track the neighboring satellites, at least during a part of the orbit, cf. Fig. 2.22.

For the feeder link, usually either reflector antennas or horn antennas are used.

8.2.1 GEO Antennas for Mobile Links with Spot Beams

Geostationary satellites have a relatively small view angle approximately amounting to $\pm 9°$. For this scenario, multifeed reflector antennas can be used [VTSC+95]. The feed elements can be horn radiators or dipole radiators.

As an example for this approach, Fig. 8.2 shows the Garuda GEO satellite of the ACeS system, with multifeed reflector antennas for the mobile up- and downlink [NBA97]. Figure 8.3 shows the transmit feed array, consisting of 88 L band cup-dipole radiators.

For beam-forming, the spot beam signals are distributed onto the feed elements of the mobile link antenna. Each beam is formed by adequate weighting

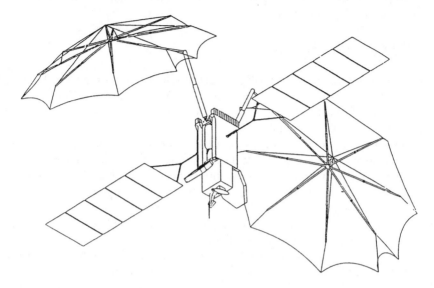

Fig. 8.2. The Garuda satellite [NBA97]

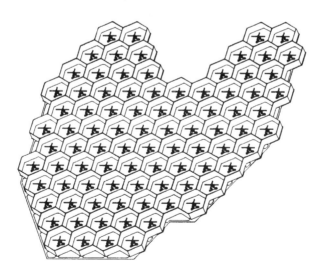

Fig. 8.3. Transmit antenna feed model of the Garuda satellite [FRM97]

of the amplitude and phase of a group of feed elements. In the Inmarsat-3 satellites, central beams are formed by 3 to 7 feed elements, fringe beams require up to 16 feed elements.

8.2.2 LEO/MEO Antennas

As opposed to GEO satellites, LEO/MEO satellites must cover a much larger range of nadir angles. The view angle ranges from $\pm 20°$ for MEO satellites to $\pm 60°$ for LEO satellites. For this scenario, planar phased-array antennas represent a suitable technology.

Phased-array antennas consist of a large number of radiators, which are regularly positioned on a plain structure forming the antenna aperture. The distance between these antenna elements typically is around 0.6λ. The radiator elements are realized e.g. as patch antennas or as dipole radiators.

For a receive antenna, the signals received by the antenna elements are individually phase shifted, such that an RF signal arriving from a certain angle produces co-phased signals. These signals are summed up to constitute the receive signal for a spot beam. By weighting the amplitudes of the signal components (tapering), the shape of the spot beam characteristic can be determined. Each spot beam uses a separate phase-shifting and tapering pattern.

For a transmit antenna, the reverse process is used, as shown in Fig. 8.4 for the Globalstar transmit antenna.

Similar to reflector antennas, the two-sided half-power beamwidth for a phased-array antenna with tapered aperture distribution is

$$2\theta_{3dB} \approx 70° \cdot \lambda/D \tag{8.1}$$

with D denoting the array diameter. As an example, Fig. 8.5 shows an ICO satellite under construction, illustrating two phased-array antennas for the mobile user link, one for transmission and one for reception. Figure 8.6 gives a closer look at the array of dipole radiators. Another example of a phased-array antenna is the Iridium satellite antenna shown in Fig. C.10. Iridium uses the same frequency band for the mobile up- and downlinks. Thus, the same antenna can be used for transmission and reception. In order to serve the large range of nadir angle, the Iridium satellite contains three separate phased-array antennas, each of which generates 16 satellite-fixed spot beams.

In active phased-array antennas, each element of a receive antenna is equipped with a low-noise amplifier (LNA) and each element of a transmit antenna is equipped with a high-power amplifier (HPA). An advantage of this approach is that solid-state power amplifiers (SSPAs) can be used, because each radiator needs only a small transmit power. SSPAs exhibit good linearity and – as well as LNAs and phase shifters – can be realized as microwave integrated circuits and integrated into the planar antenna structure. Due to the fact that each spot beam is generated by a large number of antenna

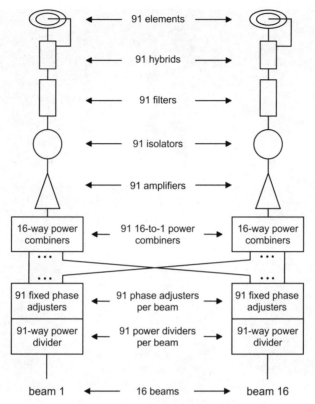

Fig. 8.4. Beam-forming for the Globalstar phased-array transmit antenna at 2.5 GHz [Die97]. No amplitude tapering is used

elements, the failure of an element or the corresponding LNA or HPA causes negligible degradation, thus improving antenna reliability.

For adaptive phased-array antennas required for earth-fixed spot beams (and for electronically steered terminal antennas), the attenuators for the tapering and the phase shifters must be electronically controllable. As an example for such a phase shifter, Fig. 8.7 shows the principle of a 3-bit switched line length phase shifter [Pat98]. The effective delay of this phase shifter is controlled by inserting or bypassing different line lengths with diode switches.

Isoflux Antenna. The variation of the signal attenuation with the slant range can be compensated with the Isoflux design which tries to achieve a constant power flux density at the ground throughout the satellite footprint. This can be approximated by designing the fringe beams with a higher antenna gain.

Fig. 8.5. ICO satellite under construction (ICO Global Communications)

8.3 Payload Architecture

As mentioned before, satellites with transparent transponders just frequency-shift and amplify the received uplink signal before it is retransmitted into the downlink (e.g. Globalstar).

In satellites with regenerative transponders, the received multiplex signal is demodulated, decoded, and demultiplexed. Various analog, digital, or hybrid techniques for demultiplexing (channelization) are possible, with digital solutions requiring complex and powerful signal processing on board the satellite. Then, channels have to be routed to spot beams, and if ISLs are used, a switching function must be implemented in the satellite. Before transmission, the signals are multiplexed, coded, and modulated. With regenerative repeaters, the feeder link and the (mobile) user link can be optimized separately with regard to multiplexing, modulation, and error control, enhancing the signal quality as well as the bandwidth and power efficiency of the system.

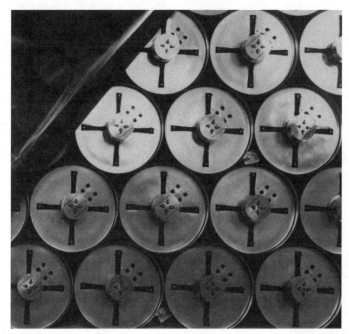

Fig. 8.6. ICO dipole radiators in a phased-array satellite antenna (ICO Global Communications)

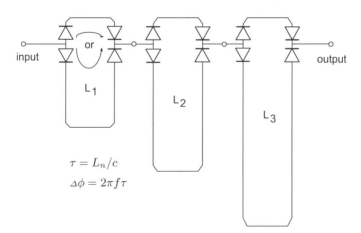

$$\tau = L_n/c$$
$$\Delta\phi = 2\pi f\tau$$

Fig. 8.7. 3-bit switched line length phase shifter

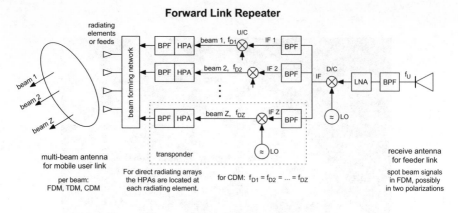

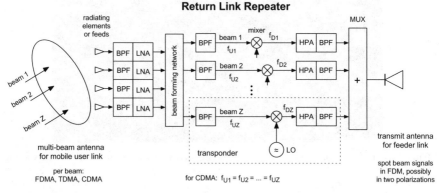

Fig. 8.8. Payload architecture of a satellite with spot beams and transparent transponders (principle).
BPF = bandpass filter;
LNA = low-noise amplifier;
HPA = high-power amplifier;
LO = local oscillator;
D/C = downconverter;
IF = intermediate frequency;
U/C = upconverter

Figures 8.8 and 8.9 illustrate basic transparent and regenerative payload architectures.

Let us first consider the transparent repeater: We assume that the feeder link signals are multiplexed in FDM, whereas the spot beam signals in the service link may be multiplexed in FDM, TDM, or CDM.

In the forward link repeater, the signal from the feeder link receive antenna is fed into a low-noise amplifier (LNA) through a bandpass filter (BPF). The BPF attenuates all out-of-band signals, e.g. from ground stations of other satellite systems. The low-noise amplifier stage typically provides 20 dB amplification of the weak received signals, followed by further amplification.

Then, the signal is downconverted to the intermediate frequency (IF), and a bandpass filter bank is used to channelize the FDM signal multiplex into the signals corresponding to the spot beams of the service link.

For each spot beam a transponder is implemented, which translates the input signal to the wanted spot beam frequency and provides for power amplification. A bandpass filter is used at the output of the mixer to remove the unwanted out-of-band frequencies generated in the mixing process (not shown). Additionally, each transponder contains electronically switchable attenuators to control the transponder gain. The attenuators are controlled through tele-commands [Ric99]. The IF signals are upconverted to the spot beams' radio frequencies and fed into the high-power amplifiers. To minimize intermodulation noise, the HPAs are operated in the linear part of their characteristic. Bandpass filters reject unwanted signal components generated through the nonlinear power amplification. Finally, the transponder output signals are fed into the beam-forming network, which appropriately distributes the signal onto the feeds or radiating elements of the transmit antenna, such that each transponder signal is radiated into the corresponding spot beam.

Channelization reduces the number of carriers passing simultaneously through the high-power amplifier, thereby minimizing intermodulation noise. Also, channelization reduces the required output power of the high-power amplifiers which is technologically limited.

For high output powers, travelling wave tube (TWT) amplifiers are commonly used, in particular at higher frequency bands. For direct radiating antenna arrays, one HPA is located at each radiating element. In this case, the required HPA output power is substantially reduced, and solid-state amplifiers can be used which exhibit better linearity and reliability.

In the return link repeater, the signals from the multiple-spot beam receive antenna elements or feeds are bandpass filtered and fed into low-noise amplifiers (LNA). Then, the signals corresponding to the single spot beams are reconstructed and channelized via the beam-forming network.

In each spot beam transponder, the signal is converted to the feeder downlink frequency where the spot beam signals are transmitted in FDM. The frequency conversion may be performed in two steps (dual-conversion) using an appropriate IF. Then, the transponder signals are high-power amplified, bandpass filtered, and multiplexed onto the feeder link.

In a regenerative return link repeater the spot beam signals are demultiplexed into single channels and demodulated, either by using a filter bank demultiplexer followed by demodulators, or by a multicarrier demodulator. For each signal, a decoder can be used to correct transmission errors which have occurred in the mobile user link. For the transmission in the feeder link, the decoded signals may be time multiplexed, re-coded, re-modulated, and

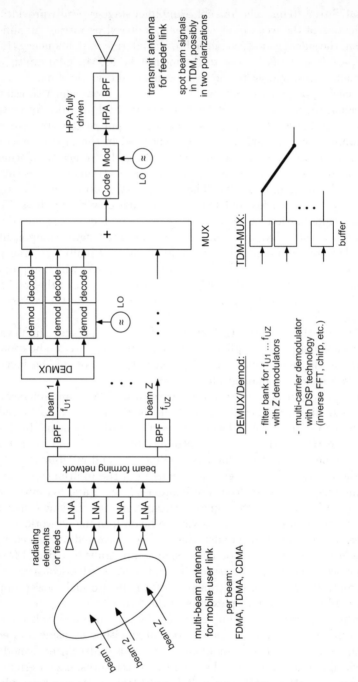

Fig. 8.9. Return link repeater architecture of a satellite with spot beams and regenerative transponders (principle). Satellite without on-board switching

amplified. Because the feeder link is a fixed link, less coding overhead may be applied, and a higher-level modulation scheme as well as a more bandwidth-efficient TDM multiplexing may be used. In this way, the service link can be optimized for the more challenging mobile communications, whereas the feeder link can be optimized with regard to bandwidth requirement.

9. Regulatory, Organizational, and Financial Aspects

Apart from the technical aspects, there are a number of other issues related to S-PCN systems which have to be solved before a system can successfully be operated:

Frequency bands for the mobile and feeder links have to be allocated. For each system, a license must be granted for implementation and operation of the space segment. Further, a license for establishing a gateway station is required by the gateway operator from the respective country. A license for providing the service must be granted to the service provider by the respective country. Connection agreements with terrestrial fixed networks, and roaming agreements with terrestrial mobile networks, must be set up.

A number of political issues between system proponents and various countries must be cleared up concerning the cost of licenses and airtime, ownership of gateways, type approval and free circulation of terminals, and service accessibility.

Finally, a huge amount of costs must be financed during system construction. The service charging policy is an important instrument to penetrate the markets in developed and less developed countries.

9.1 Allocation of Frequency Bands

The ITU (International Telecommunication Union) is an organization associated to the United Nations. The radio communication sector of the ITU (ITU-R) performs the following tasks:

- Global assignment of frequency bands. Frequencies are allocated to types of services, not to systems.
- Establishment of technical rules for the efficient and economic use of the spectrum. Such radio regulations have the status of a treaty and represent international law.
- Coordination of orbital slots in the GEO arc.

The radio regulations are reviewed and updated approximately every other year in the frame of World Radiocommunication Conferences (WRCs), e.g. WRC-97 and WRC-2000.

Situation of the Frequency Allocation after WRC-97. A rough overview of the frequency bands for satellite mobile radio systems has been given in Chap. 1. The ITU frequency allocations distinguish three regions of the earth:

- Region 1: Europe, Africa, the Middle East, Northern Asia
- Region 2: the Americas
- Region 3: South-East Asia, Australia.

Without claiming completeness, Tables 9.1 through 9.5 list frequency allocations for satellite systems for low-rate data, voice, and multimedia communications. Figure 9.1 illustrates the current ITU allocations for the satellite component of IMT-2000/UMTS. Tables 9.6 through 9.8 show frequency allocations for aeronautical platforms, feeder links, and intersatellite links. The frequency allocations are listed in Article S5 of the ITU Radio Regulations.

The following topics are planned to be negotiated at the WRC-2000:

- review of allocations for IMT-2000/UMTS
- review of power-flux density limits for the Ka band
- evolution of domestic rules (USA, Europe, etc.) in order to achieve a homogeneous worldwide regulatory status of the Ka band.

Table 9.1. Global allocations for MSS data LEOs ("little-LEOs") in the VHF/UHF band

Frequency band	Direction	Service	Status	System
137–138 MHz	down[a]	MSS[c]	prim./sec.	Orbcomm
148–149.9 MHz	up[b]	MSS	primary	Orbcomm
149.9–150.05 MHz	up[b]	LMSS	primary	Orbcomm
399.9–400.05 MHz	up[b]	LMSS	primary	
400.15–401 MHz	down[a]	MSS	primary	LEO One

[a] Downlink bands: threshold power-flux density for coordination: -125 dB(W/m^2/4kHz).
[b] All uplink bands are subject to coordination under S9.11A [Gou96].
[c] MSS includes aeronautical, land mobile, and maritime mobile satellite services.

Table 9.2. Allocations for mobile user links for MSS voice LEOs ("big-LEOs") in the L and S bands

Frequency band	Direction	Status	Region	System
1610–1626.5 MHz[a]	up	primary	global	Globalstar
1613.8–1626.5 MHz	down	secondary	global	Iridium
1980–1990 MHz	up	from 2000	region 1+3	ICO
		from 2005	global	IMT-2000(S)
1990–2010 MHz	up	from 2000	global	ICO
				IMT-2000(S)
2010–2025 MHz	up	from 2000	USA+Canada	ICO
		from 2002	region 2	IMT-2000(S)
2160–2170 MHz[b]	down	from 2000	USA+Canada	IMT-2000(S)
		from 2002	region 2	
2170–2200 MHz[b]	down	from 2000	global	ICO
				IMT-2000(S)
2483.5–2500 MHz[c]	down	primary	global	Globalstar

[a] 1610–1626.5 MHz uplink: max. EIRP-power density -15 or -3 dB(W/4kHz), depending on the subband.
[b] 2160–2200 MHz downlink: threshold power-flux density for coordination: down to -141 dB(W/m^2/4kHz), $\varepsilon = 5°$.
[c] 2483.5–2500 MHz downlink: threshold power-flux density for coordination: down to -144 dB(W/m^2/4kHz), $\varepsilon = 5°$.

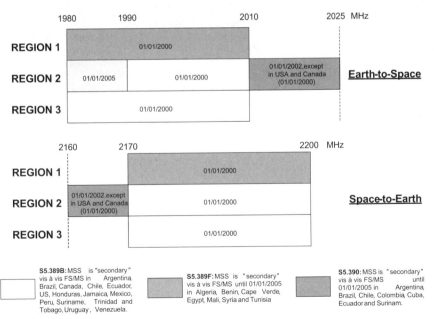

Fig. 9.1. Current ITU allocations for the satellite component of IMT-2000/UMTS

Table 9.3. Global allocations for mobile user links for L band GEO-MSS systems

Frequency band	Direction	Status	System
1525–1559 MHz[a]	down	primary/secondary	Inmarsat, ACeS
1626.5–1660.5 MHz	up	primary/secondary	Inmarsat, ACeS

[a] 1525–1530 MHz downlink: Threshold power-flux density for coordination: down to -146 dB(W/m^2/4kHz), $\varepsilon = 5°$

Table 9.4. Global allocations for user links for Ku and K/Ka band GSO and NGSO FSS systems (global allocations after WRC-97) [Dum98]

Frequency band	Direction	Resolutions, remarks	System Examples
10.7–11.7 GHz	down		SkyBridge
12.75–14.8 GHz	up		SkyBridge
17.7–18.6 GHz	down	Res. 130, 131, 538	Spaceway
			EuroSkyWay
18.6–18.8 GHz	down	Res. 131, S22.2	Spaceway
		only for GSO systems	EuroSkyWay
18.8–19.3 GHz	down	Res. 46/S9.11A, 132	Teledesic
		mainly for NGSO systems	
19.7–20.2 GHz	down	Res. 130	Spaceway, Astrolink
27.5–28.6 GHz	up	Res. 130	Spaceway
28.6–29.1 GHz	up	Res. 46/S9.11A, 132	Teledesic
		mainly for NGSO systems	
29.1–29.5 GHz	up	Res. 46/S9.11A, S22.2	Spaceway
		for GSO systems	
29.5–30.0 GHz	up	Res. 130	Spaceway, Astrolink

– Res. 46/S9.11A (pre WRC-97): sets equal rights to NGSO-FSS and GSO-FSS systems (first come / first served). Requires coordination between GSO- and NGSO-FSS systems.
– Article S22.2 (pre WRC-97): protection of GSO systems against NGSO systems.
– Resolution 130: allows sharing between NGSO-FSS and GSO-FSS systems without coordination. GSO-FSS systems are protected through hard limits for NGSO-FSS systems with regard to power flux density (PDF). item Resolution 131: sharing between NGSO-FSS and terrestrial FS (fixed services). FS is protected through PDF limits for NGSO-FSS systems.
– Resolution 132: extends Res. 46/S9.11A to the Ka band.
– Resolution 538: sharing between NGSO-FSS and BSS (broadcast satellite service) feeder links. BSS feeder links are protected through PDF limits for NGSO-FSS systems.

Table 9.5. Global allocations for user links for V band (EHF band) GSO and NGSO systems

Frequency band	Direction	Systems	Service	System
37.5–39.5 GHz	down	GSO	FSS	Orblink
39.5–40.5 GHz	down	GSO and NGSO	MSS and FSS	
47.2–50.2 GHz	up	GSO	FSS	Orblink

Table 9.6. Global allocations for user links for aeronautical platforms

Frequency band	Status	System
47.2–47.5 GHz	primary	SkyStation
47.9–48.2 GHz	primary	SkyStation

Table 9.7. Global FSS allocations intended for MSS and FSS feeder links

Band	Frequency band	Direction	System type	System examples
C band	3400–4200 MHz	down	GSO	Inmarsat, ACeS
C band	5091–5250 MHz	up	NGSO	Globalstar, ICO
C band	5850–6700 MHz	up	GSO	Inmarsat, ACeS
C band	6700–7075 MHz	down	NGSO	Globalstar, ICO
Ku band	15.43–15.63 GHz	up/down	NGSO	Ellipso
Ka band	17.7–19.3 GHz	down		Teledesic
Ka band	19.3–19.7 GHz	up/down	NGSO	Iridium (down)
Ka band	27.5–29.1 GHz	up	NGSO	Teledesic
Ka band	29.1–29.5 GHz	up	NGSO	Iridium

Table 9.8. Global frequency bands for intersatellite links between LEO satellites

Frequency band	Status	System
22.55–23.55 MHz	primary	Iridium
59.3–64 GHz	primary	Teledesic

9.2 Licensing/Regulation

Various licenses must be obtained before a satellite system can be operated:

- System license: For each system, a license must be granted for the implementation and operation of the space segment (satellite constellation) and for the frequency allocation for the system. This license is filed for by the system proponent and is granted by a regulatory authority of one country in coordination with the ITU.
- A license for establishing and operating a gateway station is required by the gateway operator from the respective country.
- A license for providing the service must be granted to the service provider by the respective country. Also, the respective frequency band must be allocated for the system in each country.
- The terminal supplier or the service provider must achieve a type approval of the terminals for each country.
- Mutual acknowledgement of national licenses is required.
- Interconnection agreements with terrestrial fixed networks must be established.
- Roaming agreements with terrestrial mobile networks must be set up, allowing cellular users to roam into the satellite network and vice versa.

The regulation of satellite services is handled very differently around the world, representing a difficulty for global satellite systems. Establishing a global regulatory framework is a very important aspect for such systems.

9.2.1 Granting a System License

In order to receive a system license, a detailed procedure must be followed:

- Advanced publication and detailed publication of the planned system. These procedures are required for the filing with the ITU (through a national regulatory authority) and are used to inform other authorities and system operators.
- Publication in the Weekly Circular of the ITU with the possibility to comment.
- Clearing of objections and coordination with other system operators. In this respect, systems which were applied for earlier have a higher priority.
- Entry into the Master International Frequency Register and frequency allocation.

In the last few years, a common problem was the filing of "paper satellites" which were used only to gain orbital positions. As a counter-measure, the Administrative Due Diligence has been introduced, which is a procedure controlling the actual usage of an allocation. This procedure requests the periodical publication of information concerning the system, the operator, and a time schedule.

9.2.2 Licensing in the USA

In the USA, the Federal Communications Commission (FCC) is responsible for licensing. The FCC is concerned with system licenses as well as with service licenses and has established the following requirements for obtaining a big-LEO system license:

- global continuous coverage with voice service (except the poles) for 75% of the time
- continuous coverage of the USA with voice service for 100% of the time
- guaranteed financing
- cooperation with radio astronomy
- Ka band feeder links.

In January 1995, three "big-LEO" systems were licensed by the FCC:

- Globalstar
- Iridium
- Odyssey (subsequently abandoned).

In July 1997, two more "big-LEO" systems were licensed by the FCC:

- Ellipso
- Constellation.

Also, a number of "little-LEO" systems (data systems) were licensed:

- Orbcomm
- VITA
- E-Sat
- Final Analysis
- Leo One USA.

The license for a little-LEO system requires that the system uses the frequency bands 137–138 MHz, 148–149 MHz, and 400.15–401 MHz in time-sharing mode. This means that the satellites are periodically switched off when they overlap with existing satellites or ground stations.

In order to prevent the operation of terminals in countries where the MSS service is not licensed, the USA intends to license only (handheld or mobile) terminals with position determination capability.

In May 1997 a number of Ka band GEO systems received a license from the FCC. Among them were Spaceway, Orion Network, EchoStar, and Ka-Star. In 1997 the Ka band LEO system Teledesic also received a license from the FCC.

9.2.3 Licensing in Europe

Europe-wide licensing is a task of the corresponding departments of the
CEPT (Conference Européene des Administrations des Postes et Telecommunications), which is a committee of European post and telecommunication
administrations. CEPT is composed of representatives from communications
regulatory agencies of more than 40 countries all over Europe. The major
committees of CEPT are:

- the European Radiocommunications Committee (ERC)
- the European Committee of Telecommunications Regulatory Authorities
  (ECTRA)
- the European Radio Office (ERO)
- the European Numbering Office (ENO).

In July 1997 CEPT decided on a harmonized authorization concept and a
concept of the harmonized use of the spectrum for S-PCN systems. The next
step is the implementation of these decisions in the CEPT member countries.

The EC (European Commission) envisages the development of a regulatory structure for the MSS, which will consider the following aspects:

- selection of satellite operators (space segment)
- frequency allocation and licensing of service providers
- licensing of terminals
- licensing of gateway operators
- internetworking and interoperability all over Europe
- numbering
- fair connection conditions to the existing terrestrial networks
- cooperation with developing countries.

The EC does not intend to grant licenses by itself. The fulfillment of the EC
requirements will, however, be necessary for applying for national licenses
in the corresponding countries. In Germany licenses will be issued by the
Regulierungsbehörde für Telekommunikation und Post (RegTP).

Standardization in Europe is overseen by the European Telecommunications Standards Institute (ETSI), which is composed of representatives from
private companies and regulation authorities.

9.2.4 Common Use of Frequency Bands by Several Systems

The allocated frequency bands will be used by more than one system. While
CDMA systems can share a band, TDMA systems must use separate band
segments. Common use of frequency bands must be considered for user links,
as well as for feeder links.

Common Use of Frequency Bands for the Mobile User Links of S-PCN Systems. For CDMA systems (Globalstar, Ellipso) the common use of a frequency band is possible (band-sharing). For TDMA systems (Iridium) the splitting of frequency bands is necessary (band-splitting). CDMA and TDMA systems cannot co-exist in the same frequency band (mutual exclusivity).

USA. For the 1610–1626.5 MHz band, the FCC proposed to use the lower 11.35 MHz for the uplink of CDMA systems and to reserve the upper 5.15 MHz for the TDD uplink and downlink of the Iridium TDMA system, Fig. 9.2. This was defined in a Notice of Proposed Rulemaking, NPRM:

- 1610–1621.35 MHz (= 11.35 MHz) for CDMA systems: Globalstar, etc.
- 1621.35–1626.5 MHz (= 5.15 MHz) for TDMA: Iridium.

The 2483.5-2500 MHz band was proposed for the downlink of CDMA systems.

Europe. The CEPT envisages taking over the allocation of the USA, but to reserve:

- 1618.25–1621.35 MHz (= 3.1 MHz) for a regional European CDMA-LEO system
- 1621.35–1622.375 MHz (= 1.025 MHz) for a regional European TDMA-LEO system [MSN96].

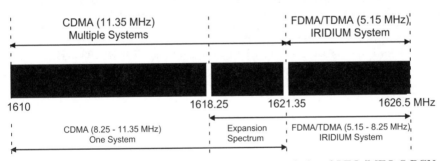

Fig. 9.2. L band frequency allocation for mobile user links of LEO/MEO S-PCN systems

9.2.5 Global Licensing and Political Aspects

A number of global licensing aspects and political aspects are relevant for S-PCN systems:

- Allocation of international numbering plans. In 1996, the following country codes were allocated:
  - ICO: +8810

- Iridium: +8816 and +8817
- Globalstar: +8818.
- S-PCN systems need landing rights in all countries they intend to serve. These must be negotiated country by country.
- The USA has a considerable political advantage because it was the first to begin the process of licensing and frequency allocation (through the FCC). Hence, the USA has a strong influence on global regulation.
- In order to allow global use of (dual-mode) terminals, the cross-border operation of terminals has to be agreed upon, and international roaming agreements must be set up. This may be hindered by national interests.
- The licensing of patents (intellectual property rights) is another important issue.
- According to international law each country can determine its own regulation strategy. National sovereignty with respect to the licensing of a system must be maintained.
- System providers must prevent the usage of their systems in countries where the system is not licensed.
- Some countries suspect discrimination by the USA with regard to system access (Iraq, Cuba, etc.).
- African and Asian countries fear that the S-PCN systems will draw off traffic from their national telecommunication networks. To prevent unauthorized traffic and bypassing of national telecommunications networks through satellite call-back services, these countries are interested in transparent call records and reliable data on unauthorized traffic.

Some political problems have been discussed at the World Telecommunications Policy Forum (WTPF) initiated by the ITU in 1996. Here, some conflicts between system proponents and developing countries arose [CI996]: System proponents are interested in low cost of licenses, foreign ownership of gateways, type approval and free circulation of terminals, and transparent regulation. On the other hand, developing countries are interested in low cost of airtime, opportunity for domestic ownership of gateways, and unrestricted service accessibility.

The GMPCS MoU. The GMPCS MoU (Global Mobile Personal Communications by Satellite Memorandum of Understanding) is a non-binding agreement of administrations, GMPCS system operators, GMPCS service providers, and GMPCS terminal manufacturers to cooperate in the development of type approval and marking of terminals, licensing and free circulation of terminals, and access to traffic data.

9.3 Financing and Marketing of S-PCN Systems

Huge costs must be borne during system construction. Typically, about half of the cost is invested by strategic partners, and the rest is financed by credits. The service charging policy is an important instrument for penetrating the markets in developed and less developed countries. On the one hand, the service charge (charge for airtime) must be attractive for the user; on the other hand, the system cost must be recovered within the amortization period. To obtain a rough estimate of the service charge to be envisaged, the following parameters are taken into account [CGH+92, GKOH94]:

- I [US-$] is the net investment for the system, including the following costs:
  - development
  - production, including redundancy (spare satellites, etc.)
  - installation of the space segment (satellite launches including spare satellites) and the ground segment
  - initial system operation.
  Typical values for the investment range between $I \approx$ US-$ 1 billion for regional GEO systems (e.g. ACeS) and US-$ 9 billion for Teledesic.
- k is the internal annual interest rate. Typical values are $k = 15 \ldots 30\%$ [GKOH94].
- T [years] is the period of time until amortization. Typical values are $T = 3 \ldots 8$ years.
- C_s [Erl] is the global system capacity (number of duplex channels). Typical values are $C_s = 20\,000 \ldots 100\,000$ Erl.
- U_s is the global system utilization. Typical values are $U_s = 5 \ldots 15\%$ (for LEO/MEO systems).

With these parameters, a lower limit for the envisaged charge per minute of airtime can be established [CGH+92]:

$$\text{service charge} = \frac{I(1+k)^T}{(365 \cdot 24 \cdot 60)TC_sU_s} \quad \text{in US-\$/min.} \quad (9.1)$$

This equation does not take into account that investments are required before start of service and that subscriber numbers build up only gradually.

Table 9.9 gives some examples of required service charges, which have been derived from Eq. (9.1). It can be seen that the results are within a realistic order of magnitude. Especially, it appears that broadband satellite systems can provide services for a very low charge.

The system utilization can be calculated from the following parameters:

N_u = number of users (subscribers) in the system.
D_a = average user activity in minutes per month. Typical values are $D_a = 100 \ldots 150$ min/month [CGH+92].
A_u = average user activity in Erl. Typical values are $A_u = 2 \ldots 4$ mErl.

With these parameters, the system utilization is given by

$$U_s = \frac{N_uA_u}{C_s} = \frac{N_uD_a}{\left(\frac{365}{12} \cdot 24 \cdot 60\right)C_s} \,. \quad (9.2)$$

Table 9.9. Required service charges for some satellite systems

System:	Globalstar	Iridium	Teledesic
I	US-$ 2.5 billion	US-$ 5 billion	US-$ 9 billion
k	22%	22%	22%
T	3 years	3 years	3 years
C_s	65 000 channels	86 000 channels	$7.2 \cdot 10^6$ at 16 kb/s
A_s	10%	10%	5%
Service charge:	US-$ 0.44/minute + terrestrial charges	US-$ 0.67/minute	US-$ 0.03/minute

The service revenues should allow recovery of the system cost, and after some time to break even and to achieve a positive cost balance. Figure 9.3 illustrates the development of the (simplified) cost balance for a LEO satellite system (Teledesic).

9.4 Operation of S-PCN Systems

A number of partners must cooperate in the successful operation of a (global mobile) satellite system. The system operator

− finances the system
− monitors and maintains the system
− sells airtime to gateway operators.

The gateway operators

− obtain the national license for setup and operation of the GWs
− procure and operate the GWs
− sell airtime to service providers (service wholesalers)
− negotiate connection agreements with operators of local terrestrial networks and international transit networks
− pay charges for terrestrial lines within S-PCN connections
− receive charges from the terrestrial operator for S-PCN connections initiated from the terrestrial network.

The service providers

− obtain national licenses for the operation of mobile and fixed terminals
− buy services from the gateway operators
− sell services to the users (service retailers)
− take care of marketing, billing, and customer service.

National/regional service providers (wholesalers, retailers) are subject to the regulatory conditions of the corresponding countries.

Finally, the clearinghouse

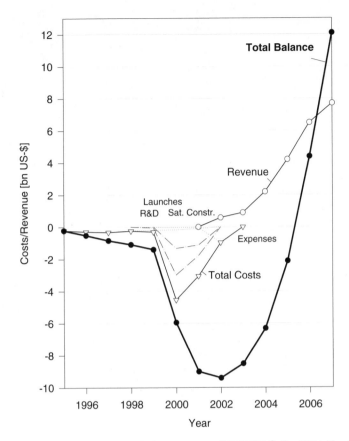

Fig. 9.3. Cost balance for the Teledesic system [GKOH94]. In 1994 the begin of operation was scheduled for 2001.

- produces bills and distributes the revenues among system operators, GW operators, and service providers
- establishes the numbering plan.

Part III

Satellite Systems for Broadband Multimedia
Communications

10. Multimedia Communications in Satellite Systems

10.1 Types of Broadband Communication Networks

10.1.1 Traditional Circuit-Switched Networks and the Packet-Switched Internet

The PSTN, ISDN, and also narrowband cellular systems can be considered traditional networks. Such networks are circuit-switched and in the past have been used for the bulk of voice and data communications. For some years now, however, the Internet has played an important and explosively growing role in data communications and a variety of applications related to the World Wide Web. Traditional networks and the Internet differ in some basic characteristics with regard to the kind of data transfer and the communication procedures, as follows.

Traditional networks are connection-oriented. In advance of the communication, a path through the network (a connection) is established, according to the address of the communication partner. The routing in the network nodes (switches) is then determined, and the availability of network resources along the connection path is checked. In this way, the quality of the connection can be guaranteed throughout the communication session. All data packets follow the same path through the network. Traditional networks have been developed mainly for the continuous transmission of data, typically for voice and video communications. These data are transmitted without any essential storage in the network.

The Internet is a connectionless network and is optimized for bursty transmission of data. Transmitted data packets contain the full address of the recipient. They are transported separately through the network, and therefore no connection setup is necessary. In each network node (router or server), a data packet is temporarily stored, the full address of the packet is analyzed, and the packet is then forwarded through an appropriate outgoing link. Depending on the current network load, packets belonging to a certain recipient can be transported via different routes and therefore can arrive in a different order as that sent. Usually, no quality of service (QoS) can be guaranteed for a communication session.

10.1.2 New Multimedia Satellite Systems Using New Satellite Orbits

Within the next few years, a large number of new satellite systems for broad-band multimedia communications will be developed, mainly for fixed terminals, but to a lesser degree also for portable and mobile terminals. Tables C.7 and C.8 give an overview of such systems, and some of these, such as Astrolink, EuroSkyWay, SkyBridge, Spaceway, and Teledesic, are discussed in more detail in App. C.

These systems will allow video telephony, Internet access, and other (interactive) multimedia applications with asymmetric and time-varying data rates up to a few tens of Mb/s. The user terminals can be fixed terminals (USATs), laptop terminals, or mobile terminals. Compared to terrestrial systems, broadband satellite systems are attractive because they can set up worldwide broadband networks with direct end user connectivity, have a relatively short time to market, and are especially suited for broadcast services and asymmetric services.

Some of the new systems will use geostationary satellites, which have been used in the past for fixed and mobile communications, because of their stationary position and their large coverage area. In particular, systems for regional coverage will use geostationary satellites, since their coverage can be precisely adapted to the envisaged service area of the system. In addition, GEO satellites are efficient in providing broadcast services.

Other new systems will use LEO or MEO satellite orbits, which are suitable for systems with global coverage. In particular, LEO orbits exhibit small signal attenuation and because of their low propagation delay are advantageous for systems providing interactive multimedia communications. Teledesic, the most advanced of these systems, will be based on a LEO satellite constellation with intersatellite links and will be seamlessly integrated into the Global Information Infrastructure (GII).

It has become increasingly evident that such a seamless interoperability of satellite networks and terrestrial networks is necessary for achieving the full potential of the emerging GII. In the light of this, the Internet TCP/IP protocol suite and the high-speed ATM protocol architecture which dominate terrestrial networks are also of great importance for satellite networks. Therefore, in this chapter we will review the fundamentals of these protocols as well as the implications of applying them to satellite networks.

10.2 Multimedia Services and Traffic Characterization

In this section we focus on two traffic types that are likely to be the dominating ones in evolving multiservice satellite networks: (i) video traffic imposing the highest requirements in terms of delay and delay jitter for real-time variable bit rate services, and (ii) self-similar traffic revealing high burstiness and thus introducing new challenges predominantly to resource management and dimensioning, and to traffic shaping and protocol adaptations as soon as some level of QoS is required.

10.2.1 Video Traffic and MPEG Coding

Whereas text and data files must be transmitted without losses, for voice, image, and video lossy compression techniques can be used, substantially reducing the amount of data to be transmitted. Moreover, such compression techniques can live with the occasional loss of transmitted packets causing only graceful degradation of quality.

JPEG is a standard for coding still images, developed by the Joint Photographic Experts Group. This coding scheme is based on the two-dimensional discrete cosine transform (DCT) essentially performing a spatial frequency analysis of all 8×8 arrays of image pixels. Quantization, proper sequencing, and entropy coding of the frequency coefficients can then substantially reduce the amount of information to be transmitted, e.g. corresponding to a 24 to 1 compression ratio.

In video, each still picture of the sequence can be compressed similar to JPEG. Moreover, since the differences between adjacent still pictures are generally small compared to the amount of information in a single still picture, additional gain can be achieved by encoding the differences between adjacent still pictures (interframe compression). This is used in the MPEG standard for video and associated audio compression, developed by the Moving Picture Experts Group. With MPEG, video signals can be compressed to a bit rate of about 1.5 Mb/s with good quality, or to a bit rate of 384 kb/s with video conferencing quality.

MPEG interframe compression is based on motion compensation, assuming that a portion of an image in one frame will be very similar to an equal-sized portion in any nearby frame. From the different locations of a block in a previous and in the present frame, the location in the next frame can be predicted. Using two reference frames, one in the past and one in the future, even more efficient bidirectional interpolation can be achieved. A more detailed description of image and video coding can be found in [Sta98].

The availability of variable bit rate (VBR) services in ATM networks offers freedom for the traffic sources to work with time-varying bit rates, allowing the use of variable rate video coding and layered video coding. On the other hand, the problems of cell loss and cell delay jitter must be taken into account in the terminals.

Actually, as soon as video compression is used, the resulting bit rate varies in time. Usually, the frame generation rate is constant (e.g. 30 frames/s in the NTSC format). To maintain constant-quality pictures, the encoder must generate more bits during high-action scenes. Especially, for high-quality video, variable-rate coding is used, resulting in widely varying frame sizes.

For transmission over fixed bit rate lines, the fluctuating information rate must be converted into a fixed bit rate, e.g. by inserting an output buffer between the encoder and the network. To ensure that the buffer is not overflowing or underflowing, a feedback signal is provided to the encoder, telling it to reduce or increase the information rate. For networks providing VBR services (e.g. ATM), the output of the encoder can be fed directly into the network, resulting in a variable rate video encoder [Pry95].

Statistical multiplexing of VBR video streams can provide substantial gain in required bandwidth: if several sources are multiplexed onto a single trunk line of an ATM network, the required trunk bit rate will be much lower than the sum of the single peak bit rates. Temporally smoothing the VBR video stream to a certain extent is another means of reducing the required peak bandwidth.

The following image or video compression standards are in common use:

- H.261 for ISDN video teleconferencing, 64 kb/s to 2 Mb/s,
- H.262 for ATM/broadband video conferencing,
- H.263 low bit rate video codec for use in POTS teleconferencing at rates of 14.4–56 kb/s,
- JPEG for still images,
- MPEG-1 for recorder-quality video at 1.2 Mb/s and for storing movies on CD-ROM,
- MPEG-2 for compression and transmission of full-motion video (HDTV, VoD) at rates of 4–60 Mb/s, and
- MPEG-4 for wireless multimedia communications and for medium-resolution video conferencing at 4 kb/s to 4 Mb/s; MPEG-4 is object-oriented.

MPEG video streams can advantageously be put on top of ATM by using the AAL 5 adaptation layer, which is very efficient for VBR traffic [Tan96].

10.2.2 Self-Similar Traffic

The utilization of network resources under the constraint of QoS can only be optimized by using a realistic characterization of the traffic. Traditionally, traffic is described by Markov processes (e.g. Poisson arrival of calls/packets). Multimedia is better characterized as self-similar[1] traffic, which is traffic that is bursty on many or all time scales [ST99]. Examples of self-similar traffic are

- Ethernet traffic (having similar statistical properties at a range of time scales from milliseconds to days or even weeks),
- compressed video traffic, and
- Web traffic between browsers and servers.

---

[1] Self-similarity is the property associated with fractals. Self-similar time series exhibit bursts at a wide range of time scales.

Statistical multiplexing of self-similar traffic streams does not result in smoothing of traffic. The correlation structures of aggregated processes (processes averaged over non-overlapping subblocks of size m) do not degenerate as $m \to \infty$. The autocorrelation function decays only slowly (is heavy-tailed). The most important feature of a self-similar process is that the variance of the arithmetic mean decreases more slowly than the reciprocal of the sample size.

Self-similarity, meaning scale-invariant burstiness, introduces new complexities into network dimensioning and into the optimization of network performance. The provision of QoS at high resource utilization is difficult for this kind of traffic.

10.3 ATM-Based Communication in Satellite Systems

10.3.1 Principles of ATM

It is expected that the volume of multimedia communications will show an exponential growth in the next few years. A suitable transport scheme for multimedia information is the asynchronous transfer mode (ATM), which is foreseen as the transfer (transmission and switching) scheme for the future broadband ISDN (B-ISDN). ATM combines the strengths of circuit switching and packet switching techniques, and is therefore capable of transporting all kinds of information and supporting each type of service. ATM can be mainly characterized by the following features:

- *Fixed length packets*: The information is transported in fixed length packets called cells. Each ATM cell consists of 53 bytes: the first 5 bytes constitute the cell header and the remaining 48 bytes contain the user data.
- *Virtual connections*: ATM allows multiple logical connections to be multiplexed over a single physical interface. All cells belonging to one connection follow the same route through the network (connection-oriented transfer). The cells arrive at the destination in the same order as generated by the source. Moreover, they exhibit low transfer latency and small delay jitter. No fixed transmission capacity is permanently allocated to a connection, rather sources use the capacity as demanded (virtual connection). All cells belonging to one connection can be recognized by virtual connection identifiers assigned at call setup time. The use of multiple identifiers for one "bundled" connection allows the support of multimedia services.
- *Statistical multiplexing*: The traffic sources can generate cells asynchronously and at different rates. In a multiplexing device (e.g. a switch), after buffering, the cells from different sources and services are multiplexed in a statistical manner, according to their arrival times and service classes. Hence, there is no fixed time relation between the input and output cells of the multiplexer. Statistical multiplexing allows the network resources to be

flexibly and efficiently shared among all connections, favoring time-varying bit rates and multimedia services.

- *Service integrated and independent network:* With ATM, all kinds of speech, video, and data services can be provided by a single network. Also non-native ATM streams, originating for instance from TCP/IP- and MPEG-based applications, can be transported in an efficient manner. Moreover, future services with still unknown characteristics will be supported without any modifications of the network.
- *Quality of service negotiation:* During call setup a traffic contract is negotiated between the source and network, containing the relevant traffic parameters of the requested service. Once the network accepts the call, it must meet certain QoS requirements (cell loss ratio, cell delay variation, etc.) throughout the lifetime of the connection. The negotiation between the source and network allows, on the one hand, estimation of the network resources needed by the source, and ensures, on the other hand, that the resources are not underutilized.
- *Reduced control functionality:* ATM is designed for high-rate physical links with very low bit error rates, therefore, no error control is applied in the ATM layer.

ATM is described in detail in, for example, the textbooks by De Prycker [Pry95], Tanenbaum [Tan96], and Stallings [Sta98]. In the following we summarize the fundamentals of ATM, utilizing some material from these references.

ATM Protocol Architecture. Figure 10.1 shows the ATM protocol stack for an interface between user and network.

The physical layer (PHY) is mainly responsible for the transmission of the ATM cells. Typical data rates are 155.52 Mb/s and 622.08 Mb/s. The ATM network layer mainly performs switching/routing and multiplexing. This layer is common to all services. The ATM adaptation layer (AAL) is mainly responsible for adapting service information to the ATM stream; thus it is service

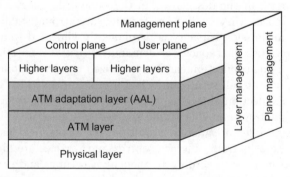

Fig. 10.1. ATM protocol architecture

Convergence	Convergence sublayer (CS)	AAL
Segmentation and reassembly	Segmentation and reassembly sublayer (SAR)	
Generic flow control Cell VPI/VCI translation Cell multiplex/demultiplex Cell rate decoupling (with UNASSIGNED cells: ATM-Forum)		ATM
Cell rate decoupling (with IDLE cells: ITU-T) Header error check (HEC) generation/verification Cell scrambling/descrambling Cell delineation (using HEC) Path signal identification Frequency justification Frame scrambling/descrambling Frame generation/recovery	Transmission convergence (TC)	PHY
Bit timing Line coding Physical medium dependent scrambling/descrambling	Physical medium (PM)	

Fig. 10.2. Layers and sublayers of the ATM reference model and their functions

dependent. Figure 10.2 lists the functions of the ATM reference model layers and sublayers.

Virtual ATM Connections. ATM is connection-oriented. This means that at connection setup an end-to-end route through the network (i.e. a series of links and nodes) is determined, which is used by all data cells belonging to that connection. With this procedure, a unidirectional *virtual channel (VC)* is assigned to the virtual connection in each link between two network nodes lying on the route; the series of VCs constituting the end-to-end route is called a *virtual channel connection (VCC)*. Each VC of the VCC and each cell using the VC are uniquely labeled by a *virtual channel identifier (VCI)* which usually changes from link to link; thus, a VCC is determined by a number of VCIs belonging to the links on the route and a routing table in each node on the route, translating the VCIs between the incoming and outgoing links. Accordingly, each link contains a number of VCs, mapped to different connections through their VCIs.

When a group of VCCs use the same route along a part of their end-to-end routes, the respective VCs in a commonly used link can be combined into a single *virtual path (VP)* identified by a *virtual path identifier (VPI)*. The common part of the route is characterized by a series of VPs and VPIs constituting a *virtual path connection (VPC)*. Thus, each link of a connection is characterized by a VCI and optionally by a VPI. This concept of aggregating parallel VCs into a single VP substantially reduces the complexity of switching in the network nodes and eases the setup of a new VCC. While each virtual end-to-end connection is identified by a temporary VCC, VPCs may be defined between network nodes in a more static manner.

Translation and switching of VCs and VPs is done in VC switches and VP switches (also called cross-connects), see Fig. 10.3. Note that VP switches do not change VCIs.

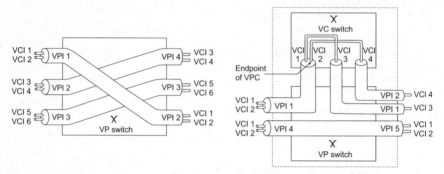

Fig. 10.3. VP switching and VC switching in ATM. Source: [HH91]

The ATM Cell Format. Figure 10.4 shows the structure of an ATM cell, including VPI and VCI in the header. It should be noted that the VPI and VCI are in principle different for every link on the route.

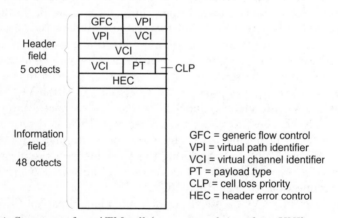

Fig. 10.4. Structure of an ATM cell (user–network interface, UNI)

The GFC field can be used to control the cell flow at the user–network interface (UNI), e.g. for traffic of different QoS. The VPI constitutes a routing field for the network, whereas the VCI is used for routing to and from the end user. The PT field indicates the type of information carried in the cell, e.g. distinguishing user data and management cells. The CLP bit indicates

the priority of the cell (CLP = 0 meaning higher priority). In the case of
network congestion cells with CLP = 1 may be discarded. Also, cells with
CLP = 1 can be used to transport data beyond the pre-negotiated capacity.
The HEC field is calculated from the remaining 32 bits of the cell header. It
can be used to detect errors in the header and in some cases to correct them.

ATM Service Categories. An ATM network can simultaneously transport
different types of traffic, such as real-time voice and video traffic or bursty
TCP traffic. In order to cope with such different traffic requirements, the
following service categories have been defined by the ATM Forum:

- real-time services
  - constant bit rate (CBR)
  - real-time variable bit rate (rt-VBR)
- non-real-time services
  - non-real-time variable bit rate (nrt-VBR)
  - available bit rate (ABR)
  - unspecified bit rate (UBR).

Real-time services include video and audio streams that do not allow a
lack of continuity. Also, interactive services put tight constraints on delay.
Typically, any delay above a few hundred milliseconds becomes noticeable
and annoying.

The *CBR service* is used by applications requiring a fixed data rate to
be continuously available during the connection lifetime. In addition, a relatively tight upper bound on transfer delay must be guaranteed. Examples of
CBR applications include uncompressed audio and video, and telephony. The
network guarantees that the required capacity is available and also polices
the incoming CBR traffic to ensure that the subscriber does not exceed its
allocation. However, no error or flow control is applied.

In contrast to CBR services, *rt-VBR services* are transmitted with a time-
varying data rate. Rt-VBR is intended for applications that require tightly
constrained cell delay and delay variation, but accept occasional cell loss.
A typical example is interactive compressed video (e.g. video conferencing)
represented by a sequence of image frames of varying size. A VBR connection
is defined in terms of a sustained rate for normal use and a faster burst rate
for occasional use at peak periods. Statistical multiplexing of rt-VBR services
provides some gain in bandwidth efficiency.

Non-real-time services relate to bursty traffic not having tight constraints
on delay and delay variation.

The *nrt-VBR service* is similar to rt-VBR, except that there is no delay
variation bound specified. Also, a certain low cell loss ratio is allowed. With
this service, the end system specifies a peak cell rate, a sustainable or average
cell rate, and a measure of burstiness. The nrt-VBR service can for instance
be used for applications with critical response-time requirements like airline
reservations or banking transactions.

Unused network capacity can be made available to the *UBR service*, which can tolerate variable delay and some cell losses. No amount of capacity is guaranteed for this service. TCP-based traffic (e.g. email) is a typical example for this "best-effort" service. UBR cells use the remaining network capacity. If congestion occurs, UBR cells will be discarded with no feedback to the sender.

ABR is a service for bursty sources that provides higher quality than UBR. For ABR services the bandwidth range should be known roughly, to allow the definition of a peak cell rate and a guaranteed minimum cell rate. Any unused network capacity above the minimum cell rate is shared in a fair and controlled way among all ABR sources, up to their peak cell rates. ABR is the only service for which the network applies congestion feedback for source rate control.

ATM Traffic Descriptors. As indicated above, a number of traffic descriptors are used to characterize the different ATM service categories and to allocate network resources at connection setup, Fig. 10.5:

Attribute	ATM layer service category				
	CBR	rt-VBR	nrt-VBR	UBR	ABR
PCR and CDVT	specified				
SCR, MBS, CDVT	n/a	specified		n/a	
MCR	n/a				specified
peak-to-peak CDV	specified		unspecified		
maxCTD	specified		unspecified		
CLR	specified			un-specified	network-specific
feedback	unspecified				specified

Fig. 10.5. ATM service category attributes

- The *peak cell rate (PCR)* defines an upper bound on the traffic that can be submitted by a source on an ATM connection.
- The *sustainable cell rate (SCR)* defines an upper bound on the average rate of a VBR ATM connection.
- The *maximum burst size (MBS)* is the maximum number of cells that can be sent continuously at the PCR. There must be sufficient gaps between bursts so that the overall rate does not exceed the SCR.

- The *minimum cell rate (MCR)* of an ABR connection defines the minimum rate requested from the network.
- The *cell delay variation tolerance (CDVT)* represents a bound on the delay variability due to the slotted nature of ATM, the physical layer overhead, and cell multiplexing. Only if the delay variation introduced by the source traffic conforms to the CDVT value can a certain cell delay variance be guaranteed by the network.
- The *maximum cell transfer delay (maxCTD)* defines the maximum acceptable delay for a connection. Cells exceeding this threshold must be either discarded or delivered late.
- Cells having a delay below maxCTD may exhibit a peak-to-peak cell delay variation (*peak-to-peak CDV*). CDV is usually negotiated during connection setup.
- The *cell loss ratio (CLR)* is the ratio of lost cells to the total transmitted cells on a connection.

Whereas the first five parameters specify the source traffic (traffic parameters), the last three parameters define the quality of the requested service (QoS parameters).

ATM Adaptation Layer. The ATM adaptation layer (AAL) is required to support information transfer protocols which are not based on ATM. Typical services provided by the AAL are the segmentation and reassembly of higher-layer data blocks, flow control, and error control. ITU-T has defined four AAL service classes covering a broad range of requirements. The classification is based on requirements for timing and constant bit rate and distinguishes connection-oriented and connectionless services, see Fig. 10.6.

	class A	class B	class C	class D
timing relation between source and destination	required		not required	
bit rate	constant	variable		
connection mode	connection-oriented			connectionless
AAL protocol	type 1	type 2	type 3/4	
			type 5	

Fig. 10.6. Service classification for AAL [Sta98]

To support these classes of service, different AAL protocols have been defined. *AAL type 1* deals with real-time CBR sources. The main responsibility of the AAL is to pack the source bits into cells for transmission and to unpack them at reception. No error detection is applied; however, missing cells are reported to the application.

AAL type 2 is intended for applications such as compressed video and audio requiring timing but not requiring CBR.

The services provided by *AAL type 3/4* may be connectionless or connection-oriented and may be message mode or streaming mode. In the message mode, each call from the application injects one message into the network.

AAL type 5 was introduced to provide a streamlined transport facility for connection-oriented higher-layer protocols. Due to its reduced overhead, AAL type 5 is highly efficient. Moreover, it is very flexible since it can provide unicast and multicast, reliable or unreliable services, and supports both, message mode and streaming mode. AAL type 5 is becoming increasingly popular and in the future may represent the most frequently used AAL type.

ATM Traffic and Congestion Control. Several traffic parameters can be negotiated during connection setup, including peak rate and burstiness. The network may use different strategies to deal with congestion and to manage existing and requested VCCs [Sta98]. In order to prevent congestion, the network may simply deny new requests. Additionally, cells may be discarded if negotiated parameters are violated or if congestion becomes severe.

After preparatory work by the ITU-T, the ATM Forum has published a set of traffic and congestion control capabilities in its *Traffic Management and Specification Version 4.0* [ATM96].

ATM networks have special requirements for congestion control, because they carry a mix of real-time and non-real-time traffic streams with widely differing bit rates and burstiness. In addition, cell transmission time is extremely short compared to propagation delays across the network, rendering reactive congestion control quite inefficient.

Cell delay variation is an important parameter for CBR services. A certain amount of delay variation is compensated at the receiver; however, cells with exceeding delay will be discarded. Since ATM is based on fixed size cells and minimum processing overhead at the switches, the only reason for cell delay variation in the network could be congestion. This is minimized through admission control and negotiation at connection setup.

Random cell delay variation can occur at the traffic sources, however, due to the processing in the ATM layers and due to the interleaving of cells from different ATM connections.

ITU-T and the ATM Forum have defined several traffic control functions to avoid congestion or to minimize congestion effects:

- resource management using virtual paths as a long-term action,
- connection admission control (CAC) being activated at each connection setup,
- usage parameter control (UPC),
- selective cell discard, and
- traffic shaping.

Whereas the first two functions act over somewhat longer terms, the last three functions react immediately to cells.

Connection admission control (CAC) is the first means of protecting the network from excessive load. At connection setup, the user must specify the requested service in terms of service category (CBR, rt-VBR, nrt-VBR, ABR, UBR) and traffic descriptors (PCR, SCR, MBS, MCR, CDVT, CDV, max-CTD, CLR). The network accepts the new connection only if it can provide the requested resources while at the same time maintaining the agreed QoS of existing connections.

Once a connection has been accepted, the usage parameter control (UPC) function of the network monitors the connection to determine whether the traffic conforms to the negotiated characteristics. UPC is performed for CBR, rt-VBR, and nrt-VBR traffic, without using any feedback to the source. In more detail, UPC comprises two separate functions:

– control of peak cell rate and the associated CDVT
– control of sustainable cell rate and the associated burst tolerance.

Cells which do not conform to these criteria can be tagged by UPC. The compliancy of a cell as well as the setting of the CLP bit in the cell can be taken into account by UPC to decide if the cell should be discarded.

The UPC traffic policing function can be supplemented by a traffic shaping policy, which smooths out a traffic flow and reduces burstiness.

Typical non-real-time applications such as file transfer or Web access do not have well-defined traffic characteristics. With UBR, such applications can share unused network capacity in a relatively uncontrolled way. As congestion arises, cells will be lost and must be retransmitted.

With ABR, feedback is provided to the sources to adjust the load dynamically and thus avoid cell loss and share the capacity fairly. This feedback is provided periodically by resource management cells which are inserted into the source traffic and returned by the destination end system. In case of congestion, these cells are modified by ATM switches or the destination end system and enable the source to adjust its rate of cell transmission. In this way, switches can also take account of a fair capacity allocation to different connections. The time delays inherent in providing this feedback dictate the use of buffers along a connection's path, absorbing excess traffic generated prior to the arrival of the feedback at the source. For ABR sources that adapt their transmission rate to the provided feedback, a low CLR is guaranteed [CLS96].

10.3.2 Implications for ATM-Based Satellite Networks

ATM is already being successfully introduced into WANs and MANs, and will be used in the future broadband ISDN. Further, ATM is going to be introduced also in terrestrial radio networks (wireless ATM, mobile ATM).

Finally, ATM will have as well a great impact on future broadband satellite systems, which will be seamlessly integrated with terrestrial broadband networks in order to achieve the full potential of the emerging GII. In the light of this, it is advantageous for the satellite networks to adopt the ATM transmission scheme and implement ATM switches on board the satellites.

However, compared to multimedia communications in fixed ATM networks, multimedia communications via satellites are characterized by special constraints, as illustrated in Fig. 10.7:

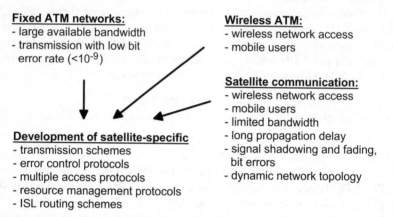

Fixed ATM networks:
- large available bandwidth
- transmission with low bit
  error rate ($<10^{-9}$)

Wireless ATM:
- wireless network access
- mobile users

Satellite communication:
- wireless network access
- mobile users
- limited bandwidth
- long propagation delay
- signal shadowing and fading,
  bit errors
- dynamic network topology

Development of satellite-specific
- transmission schemes
- error control protocols
- multiple access protocols
- resource management protocols
- ISL routing schemes

Fig. 10.7. Specific constraints and requirements for ATM-based satellite networks

- Because of the special network topology and the dynamic nature of non-geostationary satellites, multimedia communications over satellites require specific network and protocol architectures.
- In addition to fixed user terminals, portable and mobile terminals may be used for satellite communications. However, the satellite link budget requires the usage of directive terminal antennas and a line-of-sight connection to the satellite. Moreover, the bit rate will be limited for portable and mobile terminals. These constraints will restrict the application scenarios and services for mobile and portable satellite multimedia communications.
- Compared to fixed links, the bit error rate of the satellite channel is higher and can be time-varying, especially for mobile applications. Moreover, the satellite link exhibits substantial propagation delay. This requires the development of robust transmission schemes and error control protocols.
- The satellite uplink is simultaneously used by a number of terminals (shared medium). This requires the introduction of a multiple access scheme, which in spite of the substantial propagation delay must be suitable for real-time services. Especially the time-varying usage of transmission capacity for services with variable bit rates requires a highly sophisticated multiple access scheme.

- A LEO satellite network with intersatellite links (ISLs) exhibits a time-varying topology requiring a dedicated routing scheme. Significant criteria for the development of appropriate schemes are the CLR and the delay jitter caused by ISL route changes and satellite handovers.
- The available transmission bandwidth and the on-board capacity of the satellites are limited. This requires the development of specific methods of resource management. For ISL-based LEO systems the ATM connection admission control (CAC) function of the access satellite has to be combined with the routing scheme in the dynamic ISL subnetwork. Moreover, since the total link capacity is divided into several radio carriers, a carrier assignment strategy (CAS) has to select the radio carrier to be used by the terminal.
- The majority of distributed applications is based on the TCP/IP protocol suite. In order to support such applications as well as new developments in areas like electronic commerce, etc., the suitability of TCP/UDP/IP and related protocols (e.g. for multicast or QoS provisioning) must be investigated.
- For multimedia satellites a new technology for the ATM satellite switch payload has to be developed. In addition to cell switching, the payload must include various protocol and management functions for medium access control, CAC, resource management, and routing.
- New antenna technologies for satellites and terminals must be developed. Earth-fixed spot beams require active satellite antennas with digital beam-forming. User terminal antennas must track non-geostationary satellites and provide for a seamless handover between succeeding satellites. For portable and mobile terminals, the antennas must acquire a satellite and compensate for the motion of the user.

Some of these issues will be discussed in more detail in Chap. 11.

10.4 Internet Services via Satellite Systems

10.4.1 Principles of TCP/IP

The Internet is based on the TCP/IP protocol suite, where the communication task is organized in five layers [Sta98]:

- application layer
- host-to-host, or transport layer
- internet layer
- network access layer
- physical layer.

Figure 10.8 shows the protocol stack for a network operating on TCP/IP. The physical layer is concerned with the transmission medium.

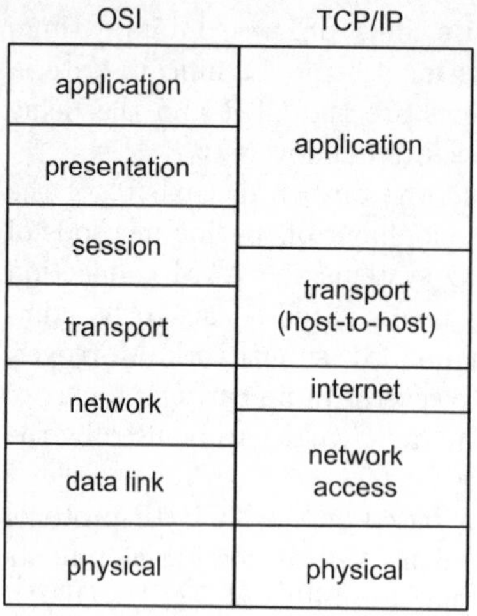

OSI	TCP/IP
application	application
presentation	
session	
transport	transport (host-to-host)
network	internet
data link	network access
physical	physical

Fig. 10.8. A comparison of the TCP/IP and OSI protocol architectures [Sta98]

The network access layer is concerned with the exchange of data between an end system and the network to which it is attached. Different standards exist for circuit switching, packet switching (e.g. X.25), LANs (e.g. Ethernet), and others. The protocol layers above the network access layer need not be concerned with the specifics of the network being used for transmission. The network access layer routes data across a single network.

By using IP, the internet layer allows data to traverse multiple interconnected networks. IP provides routing functions across multiple networks and is implemented in end systems as well as in routers connecting two networks.

In the transport layer, TCP is used to achieve reliable host-to-host data transport. This comprises the delivery of data without losses and in the same order as sent. TCP is implemented only in the end systems.

The application layer supports various user applications (e.g. file transfer). A separate module is needed for each type of application.

10.4.2 Internet Protocol (IP)

The Internet is formed of a large number of different subnetworks interconnected by routers. IP is an unreliable datagram-oriented network layer protocol, whose function is to permit data traffic to flow seamlessly between different types of subnetworks (Ethernet, ATM, frame relay, etc.). It can be used as an end-to-end transmission protocol between systems, and accounts for the routing of datagrams.

IP resides in the terminal devices and in routers which connect different subnetworks and function as switches in the network, routing datagrams (packets) towards their destination, based on an address field contained in the datagram. Routers must further translate between different addressing schemes and handle different packet sizes of the subnetworks. To overcome these differences, datagrams may have to be broken into smaller packets (fragmentation) and reassembled when they reach their destination.

IP does not guarantee delivery, or that packets will arrive in the proper sequence. (Packets can get out of order since they may follow different paths through the network, thereby encountering different amounts of delay.) Packets can fail to be delivered for several reasons. If the network becomes congested one or more routers may become overloaded and their buffers may begin to overflow. Rather than simply discarding all newly arriving packets, the routers are programmed to discard packets in a random fashion to prevent buffer overflow. Also, packets may arrive in duplicate. In sum, IP is engineered to make the best effort to deliver a message but does not guarantee to do so.

IPv4. The existing IP is known as IPv4 (RFC 791). Among other information, the IP header contains information regarding

- the type of service, indicating requirements for throughput, reliability, and delay
- the total length of the IP block
- the time span for which a datagram may "live" in the network
- an 8 bit check sum over the IP header
- source address
- destination address.

The maximum length of an IP datagram is 65 535 bytes.

The type-of-service (TOS) field of the IP header consists of two subfields: a 3 bit precedence subfield and a 4 bit TOS subfield. The TOS subfield provides guidance to the IP entity (in the source or router) on selecting the next hop for this datagram. The precedence subfield provides guidance on the relative allocation of router resources for this datagram.

The TOS field is set by the source to indicate the type or QoS that should be provided. A router can respond in three different ways to the TOS value:

- Select a proper route. For example, a satellite link is avoided if the datagram requests minimized delay.
- The router can request a type of service from the next subnetwork that closely matches the requested TOS (e.g. ATM networks support different types of service).
- A router may give preferential treatment in queues to datagrams requesting minimized delay, or a router might attempt to avoid discarding datagrams that have requested maximized reliability.

The precedence subfield is set to indicate the degree of urgency or priority to be associated with a datagram. The precedence indicator ranges from the highest level of *network control* to the lowest level of *routine*. A router should implement a precedence-ordered queue service and in case of congestion can discard datagrams of low precedence.

IPv6. IPv6 (RFC 1883) uses a new addressing scheme which overcomes the limitations of IPv4 (128 bit addresses instead of 32 bit addresses). Also, IPv6 supports higher transmission speeds, real-time services, and a mix of data streams. Finally, IPv6 supports packet sizes up to more than 4 billion bytes.

The IPv6 header contains a 4 bit priority field enabling the source to identify the desired transmit and delivery priority of each packet relative to other packets from the same source. First, packets are classified as being part of traffic either for which the source is providing congestion control, or for which the source is not providing congestion control (e.g. real-time traffic). Second, packets are assigned one of eight levels of relative priority within each classification. For congestion-controlled traffic, the priorities range from uncharacterized traffic to Internet control traffic having the highest priority. For traffic without congestion control, the priority ranges from "most willing to discard" to "least willing to discard". This can be used to distinguish packets of different relevance in video and audio services. In case of congestion, less important packets can be discarded.

In addition to unicast addresses, IPv6 also allows multicast addresses. A packet sent to a multicast address is delivered to all node interfaces identified by that address.

10.4.3 Transport Control Protocol (TCP)

TCP is a full-duplex connection-oriented end-to-end transport protocol for packet-switched networks, guaranteeing reliable and lossless delivery of data (RFC 793). Therefore, TCP can be used on top of a broad range of communication networks.

TCP may divide the data to be transmitted into several data segments. The size of a segment is negotiated between sender and receiver at the start of transmission, and may be limited by features of the network, but typically might be 1000 bytes.

TCP Segment. A TCP segment consists of a header and an (optional) data field. The TCP segment header contains

- the source port (source TCP user),
- the destination port (destination TCP user),
- a sequence number related to the first byte of the segment,
- an ACK field (piggybacked acknowledgement) containing the sequence number of the next byte that the TCP entity expects to receive,
- the length of the TCP header,
- the number of bytes which can be accepted by the sender, and
- a 16 bit check sum for error detection.

The source and destination ports identify the applications at the source and destination systems that are using the respective connection. The sequence number and the acknowledgement number support flow control and error control.

TCP Error Control. It is the function of TCP residing in the end devices (computers) to ensure the proper delivery of a complete message. To this end, TCP partitions the data stream to be transmitted into segments and assigns a unique sequence number to each byte of information. The receiver keeps track of these sequence numbers to ensure the complete reception of the data up to a particular byte number. At the receiver, the sequence numbers are used to arrange the received data into the correct order, in this way compensating for errors caused by different delay times.

In order to detect transmission errors, a check sum is added to each data segment. At the receiver, corrupted data are detected by comparing the received check sum with a local check sum derived from the received data.

For computing the check sum, a pseudo-header is added to the TCP data segment, containing the source and destination addresses, the length of the data segment, and information concerning TCP. This allows the detection of misdelivered but otherwise error-free segments.

The receiver sends back acknowledgements (ACKs) to indicate that the data have been received completely and free of error, up to the sequence number returned in the ACK. If no ACK comes back to the sender, the respective data are retransmitted after time-out.

Retransmission Strategy. Two events necessitate the retransmission of a segment. First, the segment may arrive in error and be discarded after error check. Second, the segment may fail to arrive at all. In both cases, no ACK will be issued (TCP foresees no negative ACK), and a retransmission is required. A retransmission takes place if the retransmission timer expires before the segment is acknowledged. The timer can be set to a fixed value, somewhat larger to a typical round-trip delay, or the timer can be adapted according to the average observed time taken to acknowledge data segments.

There are several possibilities with regard to the policies of data acceptance, retransmission, and acknowledgement.

Data segments can be accepted

- if they arrive in order. This requires the retransmission of everything that was sent since the last acknowledged datagram (go-back N ARQ).
- if they are within the receive window. Only the missing bytes are retransmitted (selective-repeat ARQ).

There are different ways of timer usage:

- One retransmission timer can be set for the entire queue. If it expires, the segment at the front of the queue is retransmitted.
- One retransmission timer can be set for the entire queue. If it expires, all segments in the queue are retransmitted.
- One timer can be maintained for each segment.

Data can be acknowledged immediately, or piggybacked onto the next outbound data segment.

TCP Flow Control. TCP uses a sliding-window mechanism, decoupling the acknowledgement of received data units from the granting of permission to send additional data units. A receive window and a send window at each of both end systems take care of these procedures. Concerning flow control, the ACKs returned to the sender indicate the number of bytes that it is allowed to transmit before waiting for the next acknowledgement.

The amount of data the receiver grants to the sender to transmit depends on the available receive buffer space. A conservative approach is to allow only new segments up to the available space. With this, the potential throughput S can be expressed as

$$S = \begin{cases} R & W \geq RD \\ W/D & W < RD, \end{cases} \tag{10.1}$$

where W is the window size in bits, R the transmission rate in b/s, and D the round-trip delay, including delays in different subnetworks and in routers.

The maximum window size is $65\,535$ bytes = $524\,280$ bits. For high-speed long-delay connections, a window scaling factor can be used to improve throughput. For multiplexed connections, the throughput is divided up among them. Due to the error control mechanism, the throughput is decreased for error-prone transmission.

TCP Congestion Control. The purpose of congestion control is to limit the total amount of data entering the Internet to the amount that it can carry.

TCP can only use the sliding-window flow and error control mechanism to detect, avoid, and recover from network congestion. The rate at which a TCP entity can send data is determined by the rate of incoming ACKs to previous segments. If ACKs return slowly, TCP does not know if this is due to transmission errors, a slow receiver, or network congestion. Therefore, the TCP sliding-window mechanism must be used in a way that takes into account the need for congestion control. That means, regardless of the cause, the sender is obliged to assume that the problem is congestion and institute a congestion control algorithm.

Techniques for TCP congestion control through retransmission timer control are to

- adapt the retransmission timer (RTO) according to an RTT (round-trip time) variance estimation,
- increase (e.g. double) the RTO each time the same segment is retransmitted (exponential RTO backoff), and
- not use the measured RTT for a retransmitted segment (which is ambiguous, because it is not known if the RTT results from the first or second transmission) to update the RTT variance estimation (Karn's algorithm).

Techniques for TCP congestion control through window management are the following:

– *Slow start*: TCP makes use of a congestion window, measured in segments. The total allowed window is

$$awnd = \min[credit, cwnd],\qquad(10.2)$$

where *awnd* is the allowed window in segments, *cwnd* the congestion window in segments, and *credit* the unused credit in segments, granted in the most recent ACK.

When a new connection is opened, the TCP entity initializes $cwnd = 1$, in order not to flood a possibly congested network with too much data. That is, TCP is allowed to send only one segment and then must wait for an acknowledgement before transmitting a second segment. Each time an acknowledgement is received, the value of *cwnd* is increased by one, up to some maximum value. In this way, *cwnd* actually grows exponentially, and TCP opens up its window until the flow is controlled by the incoming ACKs rather than by *cwnd*.

– *Dynamic window sizing on congestion*: In order to cope with congestion during a running connection, a slow start threshold is set to half the current congestion window, *cwnd* is reset to $cwnd = 1$, and the slow-start process is performed until *cwnd* arrives at the threshold. After that, *cwnd* is increased only by one for each RTT. This introduces a linear increase as opposed to the initial exponential increase. This strategy is illustrated in Fig. 10.9.

– *Fast retransmit*: The RTO that is used by a sending TCP entity will generally be noticeably longer than the actual RTT. A consequence is that if a segment is lost, TCP may be slow to retransmit. Fast retransmit takes advantage of the following rule: if a TCP entity receives a segment out

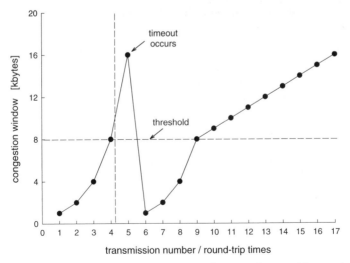

Fig. 10.9. Example of the TCP slow-start and congestion avoidance algorithms [Sta98]

of order, it must immediately issue an ACK for the last in-order segment that was received. TCP will continue to repeat this ACK with each incoming segment until the missing segment arrives to "plug the hole" in its buffer. The arrival of a duplicated ACK can function as an early warning system to tell the source TCP that a segment has been lost and must be retransmitted. To be sure, it is recommended to wait until three duplicate ACKs to the same segment have returned. Then it is highly likely that the following segment has been lost and should be retransmitted immediately, rather than waiting for a time-out.

- *Fast recovery*: In case of a fast retransmit, the TCP should take some congestion avoidance measures. It could use the slow-start congestion avoidance procedure used when a time-out occurs. However, this approach may be unnecessarily conservative, so the fast recovery technique was proposed: retransmit the lost segment, cut *cwnd* in half, and then proceed with the linear increase of *cwnd*. This technique avoids the initial exponential slow-start process.

UDP. UDP is another transport-level protocol used in the Internet. UDP provides a connectionless service for application-level procedures. It does not guarantee delivery, preservation of sequence, or protection against duplication. Because UDP is connectionless, it needs only a minimum protocol mechanism. It is suitable for transaction-oriented applications.

10.4.4 TCP/IP in the Satellite Environment

TCP/IP was developed without taking into consideration its performance over very high-speed (fiber optic) links, long-delay links (GEO satellite), or error-prone links (mobile satellite in particular). These problems have been addressed, e.g. in [PR98]. In the following, an excerpt of that text is used.

Influence of TCP Flow Control. A problem for links via geostationary satellites that involve a response time of approx. 0.5 seconds is that TCP will not allow for more data to be sent beyond a certain window size before receiving an acknowledgement. This window size is currently set to 64 kbytes and is limited by the fact that only 16 bits are available in the IP header to describe the packet size. Note that no two datagrams with the same IP identifier can be sent within a window. This limits the throughput to 2^{16} bytes divided by the response time (round-trip delay) of the end-to-end circuit, cf. Eq. (10.1).

For a GEO path with a round-trip delay of 600 ms the maximum throughput is approximately 840 kb/s [PS97]. Figure 10.10 shows the effect of the RTT on throughput as a function of window size. Because TCP resides in the users' computers the only way to "spoof" it is to place at the forwarding earth station a terminal device that acknowledges receipt of data segments as if it were the distant receiver.

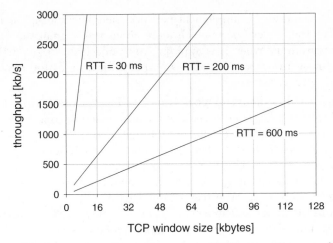

Fig. 10.10. Maximum throughput for a single TCP connection as a function of window size and round-trip time (RTT)

The Internet Engineering Task Force (IETF) has been at work recommending changes to TCP/IP to overcome this and other limitations inherent in the current design. An increase in the size of the window to 2^{30} bytes (window scaling) is proposed (RFC 1323) which would raise the throughput over a GEO satellite link to about 15 Gb/s. Since even at Q/V band, satellite frequency assignments are likely to have no more than 3 GHz bandwidth, this will probably not impose any limitation for the foreseeable future.

If many connections share the satellite link, a correspondingly higher total throughput can be achieved.

Influence of TCP Retransmission Strategy. TCP ensures the complete delivery of data over a link by retransmitting anything for which it does not receive an acknowledgement. For accepting only data received in order, TCP retransmits everything that was sent since the last acknowledged datagram. This ARQ scheme is clearly inefficient in a situation where many bytes in a packet were received correctly and only one or two arrived corrupted. In these situations it is preferable to retransmit only the corrupted information, i.e. use a selective acknowledgement.

A relatively straightforward modification to TCP that goes some way toward remedying its current shortcomings has been approved by the IETF. This permits the acknowledgement of datagrams received correctly, but out of order. This new feature has been termed selective acknowledgement (SACK).

With the introduction of this TCP option, the TCP sender is able to better manage which segments are retransmitted, as it has more information. With SACK, instead of the receiver returning the highest in-order segment received, it informs the sender about all the segments it has received. Up to three consecutive blocks of missing segments can be detected (RFC 2018).

This allows the sender to implement network-friendly retransmission. In addition, since the sender knows much more about the state of the network it can safely determine when it is appropriate to inject new segments into the network during recovery. This allows better utilization of the network and therefore better performance.

Influence of Slow-Start Algorithm for TCP Congestion Control. The TCP slow-start mechanism causes a transmission to commence with the sending of a single segment of information (datagram). Once this is acknowledged, two segments are sent, then four, eight, etc., up to a limit given by the maximum window size. This slow-start algorithm will cause the throughput on long delay links to rarely reach its maximum. It is particularly troublesome when transmitting Web pages formatted by HTTP, since TCP treats each item in the image as requiring a separate transmission sequence.

By beginning slow start by sending more than one segment, transfer time can be reduced by several RTTs. This has been shown to be especially effective in the satellite environment. An IETF proposal is to commence by sending four segments.

In addition to starting with a larger number of segments, the NASA Lewis Research Center is investigating alternative methods for generating and utilizing acknowledgements that will provide a more rapid speed-up during a slow start. This will be especially useful in the long-delay satellite environment, but should benefit all networks including terrestrial networks.

Influence of TCP Dynamic Window Sizing on Congestion. The TCP congestion control algorithm requires that the sending rate is immediately reduced, and is increased only linearly (by one segment at a time) above a certain threshold. On long-delay circuits the consequences of this congestion algorithm are particularly severe since it now takes an inordinately long time to reach maximum throughput. This is also potentially very severe for satellite circuits with their higher error rates, since any loss is interpreted as being caused by congestion. The best means of avoiding this error loss problem appears to be to operate the link with sufficient (concatenated) coding to ensure very low BER.

For example, the COMSAT Link Accelerator for IP, CLA-2000/IP, invokes a link-error-dependent amount of Reed–Solomon outer coding, resulting in a very low TCP packet error ratio. This dynamic adaptive coding method, coupled with data compression, considerably improves the throughput of applications such as file transfer protocol (FTP) running over TCP. Similarly, the combination of TCP link-level retransmissions with forward error correction at the data link level is discussed in [PAP+95].

Further methods to improve TCP/IP performance in satellite networks are [Mat99]:

– To use data segments with maximum size (maximum transmission units, MTUs). This size can be detected at connection setup by using the path MTU discovery technique.

- To use explicit congestion notification, either backward to the data source, or forward to the recipient which in turn notifies the source.
- To use a fair rate sharing between connections exhibiting different RTT. Without this, long-delay connections get an unduly low capacity share.
- To open more than one connection in parallel.

Another way of limiting the effects of satellite links on TCP (e.g. retransmissions) is to split the TCP connection into a fixed and a satellite part [BB95]. Also, the application of the fast retransmit and fast recovery procedures [CI95] can improve TCP on satellite links. The influence of the long propagation delay of a satellite link on TCP is alleviated by the SSCOP (service-specific connection-oriented protocol) transport layer protocol, which is designed for a large bandwidth–delay product. In order to make SSCOP work in a connectionless network environment such as IP, some modifications have been proposed resulting in the satellite transport protocol STP [HK97].

QoS in IP over Satellite. The basic problem is how to guarantee bandwidth on request and how to control delay and packet losses.

The resource reservation protocol (RSVP) is an existing approach, reserving end-to-end capacity during connection setup by means of hop-by-hop resource reservation protocols.

The differentiated services approach aims to offer voice, video, and multimedia using IP. The basic idea is to use the IPv4 TOS header field or the IPv6 differentiated traffic byte to determine the per-hop treatment for different packets.

Existing Systems. Several satellite systems are already providing Internet services, such as Hughes' DirecPC (with a telephone-line return path) or Astra, Gilat, and Web-Sat with two-way satellite links.

10.4.5 IP over ATM in the Satellite Environment

Because a large number of applications exist for TCP/IP, the TCP/IP layers can be advantageously used on top of the ATM layer to transfer the corresponding services via an ATM network.

This approach is attracting much attention, although it is inefficient since certain functionality is present in the ATM network layer as well as in the IP layer. The source host first establishes an ATM connection to the destination host and then sends independent IP packets over it, Fig. 10.11. Although the ATM layer delivers these packets in order, the TCP layer contains the full mechanism for reordering out-of-order packets.

IP over ATM typically uses the UBR or ABR service categories of ATM. While UBR is expected to yield low throughput and high unfairness over a satellite link, ABR will provide at least the negotiated MCR. Due to limited buffer space in switches and the large RTT of satellite links, ABR traffic may encounter problems related to congestion control [Mat99].

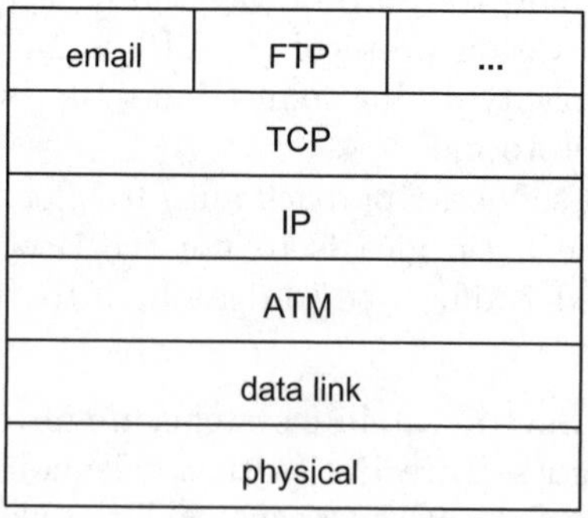

email	FTP	...
TCP		
IP		
ATM		
data link		
physical		

Fig. 10.11. Running TCP/IP over an ATM subnet [Tan96]

11. ATM-Based Satellite Networks

11.1 System Architecture

Figure 11.1 shows a simplified concept of a multimedia satellite system. It is based on a satellite constellation which uses an ISL subnetwork for the transmission of long-distance traffic. The end users can access the system over the air interface. The satellite network is connected via fixed earth stations (gateways) to the terrestrial broadband ISDN and other terrestrial fixed networks.

Because of the crowded situation in the low-frequency bands around 2 GHz and the large bandwidth required for broadband applications, future multimedia satellite systems will operate at higher frequencies. Possible frequency bands are the Ku band (11/14 GHz downlink/uplink), the Ka band (20/30 GHz), and the V band (40/50 GHz).

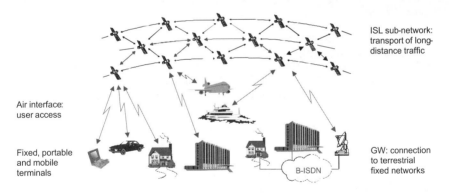

Fig. 11.1. System concept of an ISL-based LEO satellite system for multimedia communications and its connection to terrestrial fixed networks

Due to their non-geostationarity, LEO satellite constellations are particularly suited for systems with global coverage. However, because of the smaller coverage area of a LEO satellite, a larger number of satellites are required. At the start of operation, the full satellite constellation must be available. Further on, satellite handovers are necessary, the signal delay varies with time, and Doppler shifts occur. In principle, for global LEO satellite systems a large number of gateway stations are necessary, because every satellite must stay in contact with the ground segment at all times. However, if ISLs are used, the gateway stations can be contacted indirectly over the ISLs. Thus, the number of gateways can be reduced, and their position can be chosen freely. Long-distance connections can be routed via ISLs, saving expensive terrestrial lines. In systems without ISLs, usually transparent satellites are used, performing frequency translation and amplification. In systems with ISLs, switching functions have to be implemented on board the satellites, in order to route the signals through the ISL network.

For the satellite payload a number of aspects must be defined when establishing the system architecture. One important area is the kind of on-board switching (adoption of ATM switching, input/output buffering, number of ports, switching capacity). Another area is the design of the satellite antenna (number and pattern of spot beams, adoption of hopping beams, adoption of earth-fixed beams, etc.).

Further, it must be determined how the satellites share the available frequency band and how the spot beams of a satellite use the bandwidth available for the satellite (frequency planning, frequency reuse).

User Terminals. Several types of user terminals may be considered:

− fixed terminals
− portable terminals
− mobile terminals.

These can be

− terminals for individual use
− terminals for user groups.

The terminal classes will also be different with regard to the supported services. Especially, mobile and portable terminals may support fewer services (and with lower bit rates) than fixed terminals. Also, business users may require different service profiles compared to residential users.

Because of the high data rates, terminals must use directional antennas. Fixed terminals for GEO satellites may use fixed antennas; for LEO satellites they require tracking antennas with two beams to prepare for satellite handovers. Mobile terminals must additionally compensate for user motion. Antenna steering can be mechanical, electronic (phased arrays), or can use hybrid techniques.

11.2 Services

Figure 11.2 shows a schematic application scenario of a multimedia satellite system. The user applications may be based on TCP/IP, ISDN, native ATM, or MPEG protocols and formats, and via protocol conversion (ATM adaptation layer, AAL) generate ATM traffic corresponding to the categories of constant bit rate (CBR), variable bit rate (VBR), available bit rate (ABR), or unspecified bit rate (UBR). In the ATM layer, these traffic streams are multiplexed into a single stream of ATM cells. The transmission of the multiplexed ATM traffic via the satellite air interface requires a special modem at the terminal and on board the satellite, implementing a satellite-specific physical layer (S-PHY), a medium access control layer (S-MAC), and a data link control layer (S-DLC). On board the satellites, ATM switching is used to route the ATM cells into the appropriate ISL or downlink. The interworking with fixed terrestrial networks is provided by gateway stations.

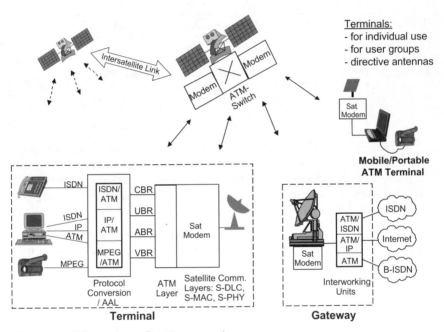

Fig. 11.2. Schematic application scenario

Popular user services are video telephony, video conferencing, and video on demand. Internet applications and services such as WWW browsing, email, and file transfer show completely different characteristics compared to real-time services and are of increasing importance. Table 11.1 shows a mapping of envisaged user services onto ATM service categories.

Table 11.2 maps the ATM service categories onto satellite service classes that were proposed for the ITU-R S.ATM recommendation [ATM98]. In particular, in satellite systems the CBR and rt-VBR ATM services will be jointly treated as a stringent class service, probably with a constant (peak) rate allocation in order to guarantee limited cell delay variation.

Table 11.1. ATM service categories and corresponding user services

ATM service categories	Corresponding user services
Constant bit rate (CBR)	Circuit emulation services
Real-time variable bit rate (rt-VBR)	Compressed audio and video (MPEG-2, MPEG-4)
Non-real-time VBR (nrt-VBR)	Transaction processing, frame relay transport
Available bit rate (ABR)	LAN emulation, file transfer, video on demand
Unspecified bit rate (UBR)	File transfer, email, fax, telnet

Table 11.2. Service classes proposed for ITU-R S.ATM

ITU-R S.ATM recommendation	Corresponding ATM service categories
Class-1 (stringent class)	CBR, rt-VBR
Class-2 (tolerant class)	nrt-VBR, ABR
Class-3 (bilevel class)	VBR and ABR high-speed data
Class-4 (unspecified class)	UBR

Since the various ATM services imply very distinct source and connection traffic characteristics, it is essential to take these features into account for the design of the communication protocols. The various services can be described, depending on their category, by a subgroup of the following parameters and attributes [ATM96], cf. Sect. 10.3.1:

- Traffic parameters: PCR (peak cell rate), CDVT (cell delay variation tolerance), SCR (sustainable cell rate), MBS (maximum burst size), MCR (minimum cell rate).
- QoS parameters: peak-to-peak CDV (peak-to-peak cell delay variation), maxCTD (maximum cell transfer delay), CLR (cell loss ratio).
- Other attributes: support of feedback and priorities, BT (burst tolerance), etc.

A typical range of user bit rates and radio carrier capacities are as follows:

- Bit rate per user: 16 kb/s to 2 Mb/s
- Uplink carrier capacity: 2, 16, 32, 64 Mb/s
- Downlink carrier capacity: 32, 64 Mb/s.

The uplink capacity within a spot beam is divided into a number of carriers, with a high carrier capacity favoring the efficiency of statistical multiplexing.

Internet services based on TCP/IP are one of the most important services in future multimedia networks. When providing such services via ATM-based satellite networks, it must be taken into account that the requirements for

flow and congestion control for the satellite part of the TCP connection differ
from these for the TCP connection within the fixed network. In Sect. 10.4.4
such issues are discussed and some new developments introduced.

11.3 Protocol Architecture

A multimedia satellite system might be considered an ATM-based meshed
sky network [TIA98] with dynamic network topology, including end user access, ISLs, satellite–gateway connections, and network interconnection via
satellites.

In this configuration, several interconnected satellites form an in-orbit
ATM network. This ATM satellite network performs ATM switching, traffic
and congestion control, and QoS management equivalent to terrestrial ATM
networks. The in-orbit ATM switches use NNI signaling for the ISL communication, and UNI or NNI signaling for communication with the ground
stations.

The protocol reference model for the fully meshed ATM-Sat network is
shown in Fig. 11.3, indicating the satellite-specific protocol layers S-PHY, S-
MAC, and S-DLC that have been inserted below the ATM layer. The resource
and mobility management functions can be implemented at the gateway earth
station or on board the satellites and can be invoked by ATM UNI or NNI
signaling via If.a or If.c. Alternatively, internal signaling channels between
terminals, gateways, satellites, and an NCC (network control center) can be
used (see the thin lines in Fig. 11.3).

In the following sections, we discuss some important functionalities of multimedia satellite systems, as indicated in the protocol architecture, Fig. 11.3.

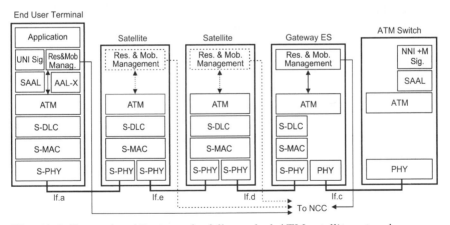

Fig. 11.3. Protocol architecture of a fully meshed ATM satellite network

11.4 ATM Resource Management

The task of resource management is to provide a fair allocation of system resources to the users. In ATM multimedia satellite networks, the available up- and downlink bandwidth and the switching capacity in the satellites are limited. This requires the application of appropriate resource management functions which can be classified into two groups: ATM resource management functions and radio resource management functions (see Sect. 11.6).

ATM resource management functions take care of call management (including call setup, call termination, call monitoring, routing, etc.), traffic control, and congestion control. In particular, as with wireline networks, also in satellite networks a connection admission control (CAC) function is required, which decides whether a new connection can be accepted without unduly decreasing the quality of ongoing connections.

Time-critical functions such as the switching of cells, queueing, flow control, and scheduling of time slots should be installed on-board the satellites. Connection admission control and radio resource management can (e.g. partly) be placed in the network control center.

In particular, if the network shall provide services with different constant and time-varying bit rates, the issue of resource management is not trivial. For the assessment of the corresponding functions, the availability of a suitable source traffic model is very important: On the one hand, the traffic behavior of different services must be modeled at connection and cell level, on the other hand, the distribution of the users onto the satellite spot beams has to be considered [WHS97].

11.4.1 Connection Admission Control and Usage Parameter Control

A main function of traffic control is *connection admission control (CAC)*, which decides if a sufficient amount of free capacity is available in the network to accommodate a new connection without decreasing the QoS of existing connections below their negotiated level. CAC takes into account the bandwidth demand of the new connection, the free capacity available in the network, and the measurement results of the actual cell rates of the existing connections (provided by usage parameter control, UPC). Also, predefined priorities must be supported when allocating the resources, which is implicitly done by respecting a set of traffic descriptors for each connection.

For ISL-based LEO systems, the customary CAC function known from terrestrial fixed and wireless ATM networks has to be extended to the ISL subnetwork, since in addition to the resources in the satellite uplink, the availability of resources in the ISL subnetwork must also be checked. Accordingly, the CAC function can be performed in two steps:

1. *Uplink-CAC*: Since it can be expected that the severest bandwidth constraint appears in the satellite uplink, in a first step, the uplink-CAC

checks for available resources in this radio channel. This check can be based on bandwidth requirements resulting from specified traffic parameters of the new connection, and, if available, on UPC measurements of the already existing connections.

2. *ISL-CAC*: If the uplink-CAC was successful, the routing strategy proposes a set of paths through the ISL subnetwork that could be used by the new connection. A selected path must be checked by the ISL-CAC, to verify whether each satellite in the path has the required capacity (the ISL capacity is considered sufficiently large). If this is not the case, another path is tested. If no path is found which fulfills the requirements, the call must be rejected.

Another function of traffic control is *usage parameter control (UPC)*, which must ensure that no traffic source sends more data than negotiated during the call setup.

There are several alternatives for the placement of the CAC and UPC functions in an ISL-based LEO system. CAC can be implemented on board the satellite as well as in the gateway station. The CAC in the satellite implies higher demands on the on-board processing, but allows independence from the earth segment. This can be an important aspect, since the satellites are moving and the gateway station serving a given satellite changes accordingly.

Taking into account the substantial round-trip delay in satellite systems, the parameter policing function of UPC can be implemented in the terminals, to quickly detect and react to the possible misbehavior of a source with regard to the pre-negotiated traffic parameters. This supports the interception of unauthorized traffic already at its origin.

Additionally, a monitoring function of UPC can be envisaged on board the satellite, to perform measurements of the ATM traffic being transported over the uplink radio channel. These measurements can be used to support the feedback mechanisms that recognize congestion, and to react, for example, by triggering traffic shaping in the terminals. Moreover, real-time traffic statistics gained from the UPC measurements can be made available to the CAC as well as to the MAC time slot scheduler (Section 11.5.1), in order to support their actions.

11.4.2 Congestion Control, Traffic Shaping, and Flow Control

Further aspects of traffic management are compiled in [ZA99]:

Congestion Control. Congestion denotes a situation of high network load, leading to a possible buffer overflow in the network. Several methods exist to counteract network congestion: whereas network congestion can be avoided by appropriate network design and configuration (open-loop congestion control), closed-loop congestion control uses feedback from the network to decrease network traffic, thus helping the network to recover from congestion. When designing closed-loop congestion control for a satellite network, it must be taken into account that the substantial round-trip delay may slow down the reaction to network congestion.

Closed-loop congestion control can work at end systems, reducing the source rate according to the level of network congestion. To this end, feedback from the network or the buffer occupancy at the traffic source can be used. The rate of ABR/UBR traffic sources can be directly reduced. Rate scaling of VBR source traffic can be based on multiple-rate coding or layered video coding, as well as on frame or block dropping. Such methods lead to graceful degradation of audio or video quality.

Congestion control can also take place in the network, using random cell discard, cell priority packeting with selective cell discard, or VC priorities.

Traffic shaping is another task related to traffic control. Traffic shaping at an end system with a bursty source is achieved by buffering the generated cells in order to smooth the traffic flow. This is necessary if the source peak rate cannot be borne by the satellite uplink or the network. An example is the leaky bucket traffic shaper. Traffic shaping may also be applied in network nodes in order to achieve a better utilization of network resources.

Flow control regulates the traffic rate between sender and receiver, acting in a point-to-point manner. For a short round-trip delay, it can work with end-to-end feedback; for a long round-trip delay, hop-by-hop flow control can be applied.

11.5 Multiple Access for ATM Satellite Systems

The ATM cells generated by independent traffic sources within a satellite spot beam are transmitted over a common satellite uplink. Therefore, a MAC protocol is needed, providing a set of rules for controlling the usage of this shared medium. There are three main approaches for the uplink MAC: frequency-division multiple access (FDMA), time-division multiple access (TDMA), and code-division multiple access (CDMA).

The MAC protocol should provide support for CBR, VBR, ABR, UBR, and GFR service categories, and in principle should expand the statistical multiplexing of wired ATM into the radio environment. Real-time services put the highest demand on the MAC protocol. In particular, the time-varying usage of transmission capacity by rt-VBR services constitutes a challenge to the MAC. A fair and dynamic allocation of transmission capacity to the ATM virtual connections regarding their individual QoS parameters is crucial for the overall system performance.

For the development of multiple access schemes for ATM satellite networks the long signal propagation delay (especially to geostationary satellites) must be considered, together with high and variable bit rates as well as different QoS requirements for the service classes.

Figure 11.4 shows the different realization aspects of statistical multiplexing in (satellite) radio networks as opposed to fixed networks.

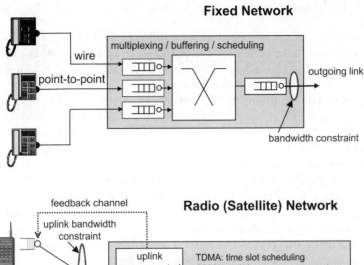

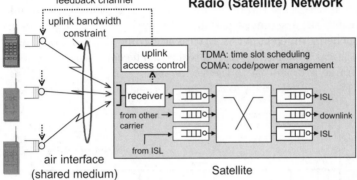

Fig. 11.4. Statistical multiplexing in fixed networks and in radio (satellite) networks

In fixed networks, each traffic source is connected to the switch by wirelines with sufficient capacity. It is assumed that the access links from the sources and the input buffers are dimensioned in such a way that they do not impose any constraints on the traffic, so that the sources can transmit anytime up to their peak cell rate. Moreover, no mutual influence between the access signals is present. The scheduling function, which schedules the traffic according to pre-negotiated QoS parameters, is realized within the ATM switch. The bandwidth constraint in fixed networks appears at the outgoing link from a multiplexing point, after buffering was applied.

In radio networks, especially in satellite networks, a severe bandwidth constraint is present in the shared uplink to the satellite or base station, respectively. Moreover, due to the shared medium, the radio signals transmitted by the terminals may interfere with each other. Therefore, at the receiving satellite, there must be an uplink access control function, which regulates the medium access and, via a feedback channel, broadcasts the corresponding information to the terminals.

11.5.1 TDMA-Based Multiple Access

For multimedia satellite networks with different and time-varying bit rates TDMA schemes with multiple carriers (MF-TDMA) are suitable where each TDMA time slot typically contains an ATM cell plus a satellite-specific header for error control and inband signaling.

To meet the requirements of different ATM services, the TDMA frame can be divided into signaling slots, reserved data slots, and random access slots, where the assignment can be adapted to the current traffic situation [BÖW+98], Fig. 11.5. Thus, the access scheme can be realized as a combination of random access with inherent cell losses and reservation-based access with longer response times. Usually, slotted Aloha is used for random access; the reservation-based access protocol can operate with a reservation subframe and random access by the terminals. The organization of the uplink frame structure and the assignment of data slots is performed by a scheduling algorithm, queueing different ATM services according to their priorities. In this way, statistical multiplexing can be realized. The current configuration of the TDMA frame is broadcast to the terminals.

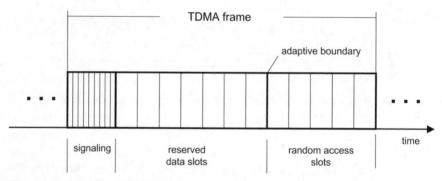

Fig. 11.5. Adaptive TDMA scheme for multimedia satellite networks

The reservation slots are used for services with constant or variable bit rates. Data slots with random access will primarily be used by services without QoS (UBR). However, they can also be used in a PRMA-like scheme (advanced PRMA, [BÖW+98]) for real-time services with variable bit rate (rt-VBR). Signaling slots are used for service requests and for requesting the dynamic allocation of data slots for VBR services without real-time requirements (nrt-VBR).

The realization of the VBR (especially rt-VBR) services is the most challenging problem for TDMA techniques. When the bit rate increases, new resources may be obtained by contention-based access such as reservation Aloha or packet reservation multiple access (PRMA). But such techniques alone cannot guarantee the QoS parameters, particularly the cell loss ratio [Bos99, WBO+98].

Contention-free reservation techniques are one alternative, where the resources are requested and reserved before data transmission. The request can be transmitted either via piggybacking (implicit reservation) or via a dedicated request channel, constituted for example by signaling slots (explicit reservation). A combination of both techniques could be used to achieve satisfying results also for rt-VBR services.

The uplink scheduler, which is part of the MAC protocol, is the entity that decides on the reservation of slots for a connection and their location in the frame, according to the pre-negotiated QoS parameters. The scheduler also includes the UPC function which gathers information about already existing connections, such as the service class, the QoS parameters, and actual traffic statistics. The main criterion for the scheduling decision is the priority of the service, with the highest priority assigned to the CBR services, followed by the rt-VBR, nrt-VBR, ABR, and UBR services. The terminals are informed about the slot reservation via the downlink broadcast channel.

For the radio interface, the scheduling is a particular challenge, since it has to be performed prior to data transmission, and the traffic patterns cannot always be predicted exactly. The problem increases with propagation delay, especially for rt-VBR services, because here also a limit for cell delay variation has to be guaranteed.

11.5.2 CDMA-Based Multiple Access

Besides TDMA schemes, also CDMA schemes (MF-CDMA) are investigated for multimedia satellite systems. In contrast to TDMA having a hard-limited capacity, CDMA exhibits graceful degradation with increasing number of simultaneously transmitted user signals. Therefore, CDMA schemes can carry VBR services without a priori resource allocation, and the statistical multiplexing of ATM can be directly implemented in the air interface. This feature is especially advantageous when considering the long signaling delay in satellite links.

Moreover, no feedback of the frame structure is required, thus avoiding additional delays which are undesired for real-time services. However, the disadvantages of CDMA are its high complexity, especially for high bit rates, the requirement for a fast and exact power control, and a substantial backoff of the satellite transponder because of the time-varying CDMA signal power. Also, in CDMA, the spreading codes must be allocated, and the transmit powers must be tightly controlled.

For providing services with different and time-varying bit rates, several CDMA methods are possible:

– In multi-chip rate schemes, the same spreading factor is used for services with different bit rates, resulting in different chip rates.

- In multisequence schemes, more than one spreading sequence can be used for a service in order to allow different bit rates.
- Multi-code rate schemes vary the channel code rate, as realized in the Globalstar system [Glo95], or only vary the spreading factor, as proposed in the wideband-CDMA approach for UMTS, cf. Sect. 5.11.1.
- T/CDMA is a combination of CDMA and TDMA and has also been proposed for the UMTS air interface, cf. Sect. 5.11.2.

The capacity of CDMA is limited by the interference between user signals. Unfortunately, the interference between satellite spot beam cells decreases more slowly than between terrestrial radio cells, thus producing more multiple access interference [Lut97]. On the other hand, a multibeam satellite antenna detects all users in all spot beam cells. This fact favors the usage of multiuser detection schemes [Ver86], which aim to use the available information from all users simultaneously, instead of decoding each user separately and treating the other user signals as noise, cf. Sect. 5.9.3.

The main drawback of multiuser detection schemes is their high complexity, which grows exponentially with the number of users. Sub-optimum interference cancellation schemes, as proposed for example in [AGR98], may reduce the receiver complexity. These schemes are based on the idea that each user is decoded individually, but the information of all previously decoded users is used to improve the current decoding step. This process is repeated until near optimal decoding is achieved.

In the approach by Rimoldi and Urbanke [RU96], the data stream of the individual users is split into different components. To each component a distinct channel code rate is assigned. The component using the lowest code rate can be decoded individually, treating all other components as noise. After successful decoding, the signal of the component is subtracted from the overall signal and then the next component can be decoded.

In summary, TDMA and CDMA both have their advantages and drawbacks. Problems with TDMA are:

- a complicated scheduling of time slots, which must take into account the different QoS requirements of various traffic sources
- the delay due to signaling for dynamic bit rate adaptation
- the hard limitation of overall bandwidth.

Problems with CDMA are:

- the high complexity of the CDMA receiver on board the satellite, especially for high bit rates
- the transponder backoff, which is necessary because of the non-constant signal amplitude
- the required power control, which must be fast and precise.

11.6 Radio Resource Management

In satellite systems, radio resource management is especially important, because the bandwidth of the uplink is limited.

In satellite systems utilizing multi-frequency access schemes (MF-TDMA, MF-CDMA), the total link capacity is divided up among several radio carriers. This requires a carrier assignment strategy (CAS) to manage the mapping of the connections onto the uplink radio carriers.

Within an uplink carrier, radio resource management is closely related to the MAC functions such as the scheduling of time slots in TDMA, or the allocation of spreading codes in CDMA. Some relations between the resource management groups relevant for satellite networks are outlined in Fig. 11.6.

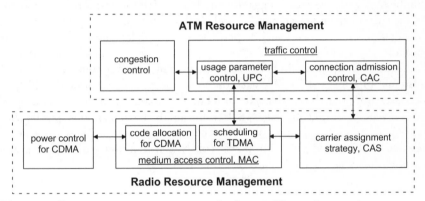

Fig. 11.6. Resource management groups in the satellite environment

Carrier Assignment Strategy (CAS). During call setup, the CAS supports CAC, proposing a radio carrier for the new connection if it can be accepted in the network. In MF-TDMA systems supporting frequency hopping, a connection can exploit free capacity that is available in different carriers. The cooperation of the CAS with a dynamic reservation-based MAC protocol favors the provision of VBR connections and the exploitation of statistical multiplexing.

A second issue is the distribution of the ATM service categories among the radio carriers. Research in bandwidth allocation in terrestrial networks has proposed concepts like hierarchical virtual partitioning [MZ96] and service separation [BDM97]. These ideas postulate that the statistical multiplexing gain of ATM can be better exploited if each ATM traffic stream is multiplexed with traffic streams of the same class. These concepts can be extended to satellites, taking into account the constraint of the subdivision of the uplink bandwidth into carriers. Two types of CAS can be envisaged for the ATM satellite environment:

1. *Complete sharing CAS*: Each carrier can carry all traffic classes. The management is very simple, and statistical multiplexing gain can be achieved among all traffic classes.
2. *Complete partitioning CAS*: The pool of carriers is divided into regions, each one carrying only a certain traffic class. The management is more complex, and statistical multiplexing is performed only within each traffic class. The multiplexing gain can be improved if the boundary between the carrier regions is set dynamically, depending on the load per traffic class in the network.

Bandwidth Allocation and Power Control. Bandwidth allocation is required to achieve a fair bandwidth sharing between a number of connections, by means of deterministic or statistical multiplexing. Bandwidth allocation can be dynamic, e.g. when renegotiating bandwidth for ABR traffic according to the current demand. Dynamic bandwidth allocation can be feedback-based (depending on network load) or prediction-based (depending on connection traffic).

Another topic of radio resource management is power control for CDMA systems. In standard CDMA with only one bit rate, every user should have the same power at the receiver. In an ATM environment, however, an adjusted strategy is advantageous, e.g. ABR traffic might have a smaller power than real-time traffic. In contrast to this, an interference cancellation scheme would require a different approach to power control, since a user with higher power actually has a smaller influence on other users.

11.7 Error Control

Originally, ATM was designed for highly reliable transmission media like cable or optical fiber, where error control plays only a minor role. In contrast to this, the bit error rate of the satellite channel is high and can be time-varying, especially for mobile applications. Therefore, additional error control is inevitable when transmitting ATM cells via a satellite channel. Moreover, the error control methods should adapt to the time-variant conditions of the satellite channel.

For the conception of error control it is necessary to consider the format of the ATM cells, the different QoS requirements of the ATM service categories, and the specific user application requirements. The ATM service requirements relevant for error control are defined in terms of:

– cell loss ratio
– probability of undetected errors
– cell delay and delay variation.

ATM cells include an 8 bit header error control (HEC) which is capable of detecting errors in the cell header and also of correcting one single bit error, if the receiving side is in the correction mode. After an error has been detected and possibly corrected, the receiver switches from correction mode to detection mode, where only error detection and not error correction is performed. This prevents a false error correction in case of bad channel conditions. If no error has been detected, the receiver switches back to correction mode. All cells with detected and uncorrected errors in the cell header are discarded (cell loss). Therefore, higher-layer automatic repeat request (ARQ) is necessary for services that are not sensitive to delay but sensitive to data loss.

For ATM over satellite, HEC is not sufficient to meet the QoS requirements for the different ATM service categories. Forward error correction (FEC), ARQ, or combinations of both (hybrid ARQ) must be considered. Moreover, even with additional error control, some of the ATM QoS requirements, which have been defined for fixed networks, have to be relaxed for ATM over satellite.

The error control functions should be distributed between the S-DLC, S-MAC, and S-PHY layers, and the AAL. Another issue is the integration of adaptive error control with higher protocol layers. New concepts in this context are hierarchical coding, integration of source and channel coding, and the use of link-level ARQ protocols for real-time services.

Various error control schemes can be implemented in different protocol layers, see Fig. 11.7:

- In the physical layer, variable-rate convolutional coding might be appropriate.
- In the data link control layer, block coding and ARQ may be suitable.
- In the ATM adaptation layer, early packet discard may prevent the transmission of useless ATM cells.

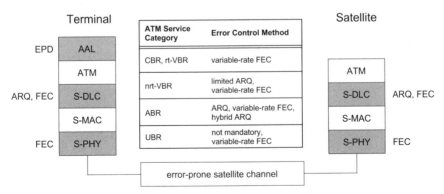

Fig. 11.7. Error control in ATM-based multimedia satellite systems

Different ATM service categories require the application or combination of different error control methods:

- CBR and rt-VBR traffic have a strict delay constraint. Therefore, only FEC in the S-DLC and S-PHY layers may be applied, but no ARQ. Error detection should filter out erroneous packets at S-DLC level to reduce unnecessary traffic in the ATM switch. The absence of ARQ means that cell loss is possible. Cell loss ratio will be the evaluation criterion.
- nrt-VBR traffic also has some delay constraint. Therefore, error detection and FEC in the S-DLC and S-PHY layers and additional limited (time-constrained) ARQ can be used, if some buffering delay can be tolerated. The time limitation of ARQ means that cell loss is possible. Cell loss ratio and cell buffering time will be the evaluation criteria.
- ABR traffic has no delay constraint, but no cell loss is allowed. Therefore, FEC must be combined with ARQ. FEC may be applied in the S-DLC and S-PHY layers, and selective repeat ARQ for each virtual connection could be used in the S-DLC layer. This means that no cell loss occurs, but the delay varies.
- UBR traffic has no constraints at all; cell loss and delay variations are allowed. FEC in the S-DLC and S-PHY layers can optionally be combined with ARQ in the S-DLC layer.

12. Network Dimensioning

Network dimensioning is generally closely coupled with network routing in the network design or synthesis process. Whereas routing has to determine the path of connections, given the topology and the capacity of transmission and switching equipment, the task of network dimensioning is to determine these capacities assuming a particular routing method [Gir90]. Given the end-to-end traffic matrices, network dimensioning is usually treated as an (iterative) optimization problem with specific network costs as the target function and some constraints on the quality of service (delay, blocking, etc.) to be met by the network.

This chapter focusses on two dimensioning problems which are relevant for future broadband satellite systems. The first part is devoted to the capacity dimensioning of a regional GEO satellite system with multiple spot beams, where the characterization and modeling of multiservice aspects are specifically addressed. The second part focusses on the ISL segment of broadband LEO satellite systems.

12.1 Spot Beam Capacity Dimensioning for GEO Systems

12.1.1 Motivation and Approach

Currently developed or planned satellite communication systems reveal a clear trend toward providing an extended service spectrum to the end user. Some well-known candidates of such "multimedia" systems are the LEO systems Teledesic and SkyBridge (Alcatel), and the GEO system EuroSkyWay (Alenia Aerospazio), see App. C.

In this section, we will denote such systems as multiservice systems, providing a set of different services for various applications. This terminology does not presuppose "true" multimedia end user terminals, but also includes systems providing voice, fax, and data in an alternate manner.

It is obvious that one of the crucial tasks in the multiservice scenario is to develop a reliable and tractable method to quantitatively estimate the bandwidth demand for a specified system. This is important for both sys-

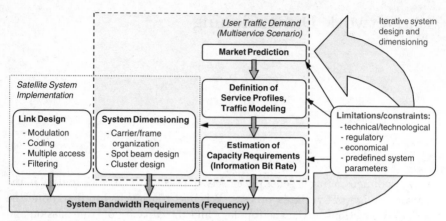

Fig. 12.1. Estimation of bandwidth requirements in an iterative system design process

tem designers dimensioning their systems, and regulation/licensing bodies allocating the spectrum to different systems and competitors.

This section outlines a methodology to estimate the capacity requirements and the corresponding bandwidth demand for a multiservice satellite system. Considering the usually iterative system design process as illustrated in Fig. 12.1, we restrict ourselves to one iteration step and assume the link design parameters to be fixed. We integrate the results from several research projects and from the earlier publications [WL98a] and [WL98b].

12.1.2 Market Prediction

Being a complete research area in itself, market prediction will not be treated in general terms here; rather we summarize a straightforward approach to estimate the (European) satellite UMTS (S-UMTS) market using a closed-form market prediction model as presented from different perspectives in [HS99], [HSD+98], and [WHS97].

Figure 12.2 provides an overview of the adopted approach. Based on the fundamental assumption that S-UMTS will be complementary to the terrestrial UMTS (T-UMTS), three key market segments can be identified: (i) land services to private or business users in rural or remote areas without T-UMTS coverage, to international business travelers, and to both commercial vehicles and private cars; (ii) maritime services to passenger and cargo ships, cruise and research vessels; (iii) aeronautical services providing business applications and in-flight entertainment supported by a kind of on-board LAN. According to the market segment, the service profile, the bit rate requirements, and the pricing policy, user terminals can be classified into mobile and portable ones, the latter ranging from handheld to the laptop type, and into those used by individuals and those used by user groups (e.g. a group of

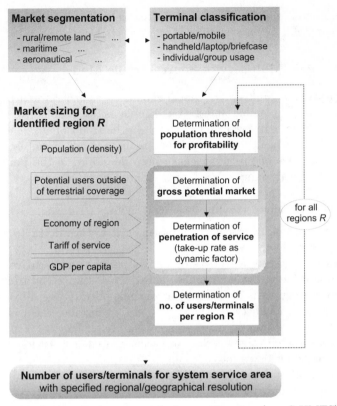

Fig. 12.2. Market prediction approach for multiservice (e.g. S-UMTS) satellite systems. GDP = gross domestic product

passengers on board an aircraft using a common terminal via user interfaces built into the seats).

The heart of the approach toward determining the size of the S-UMTS market consists of three distinct steps aiming at the determination of the number of users/terminals per identified region R of the service area. First, a population density threshold is assessed, below which the implementation of satellite service provision becomes profitable compared to a terrestrial cellular solution. For profitable regions and market segments the gross potential market (GPM) is then determined as the group of potential users (permanently or part-time) outside of terrestrial coverage. From this, the resulting service penetration (i.e. the percentage of GPM people subscribing to the service) is determined taking into account the affordability (ratio between GDP per capita and tariff) and the predicted take-up rate of the market in form of a logistic model. By performing these core steps subsequently for all identified regions of the projected service area, subscriber numbers with a specified regional/geographic resolution can be derived. Together with appropriate user activity and traffic modeling, this is the key input to the capacity dimensioning of a target system.

12.1.3 Generic Multiservice Source Traffic Model

Assumptions. A multiservice model should be as general and parametric (thus scalable) as possible and also as simple as acceptable. We propose a "linear" model on call level that is based on the traditional Poisson model, which has been extended to account for specific features of multiservice traffic. Essential parts of this model have been developed in the framework of the network dimensioning subtask of the European ACTS project SECOMS [LLV97].

The basic modeling assumptions are as follows:

1. The model and all calculations based on it are restricted to the worst case or busy hour, respectively.
2. Call-level modeling with Poisson arrival process and negative exponentially distributed holding time is applied to all service and traffic types.
3. A call in progress is modeled by a constant bit rate. For all variable bit rate services, an effective bit rate is used, for instance the average value as a simple solution.
4. The constant/effective bit rates are accumulated over services and users. Blocking is not taken into account.
5. The correlation between the different services of one user are captured in a linear scaling factor.

The model is restricted to on-demand services initiated by multiservice users and cannot cover broadcast services. For a combined on-demand/broadcast service system, the respective traffic requirements can simply be summed up.

Service Profiles and Traffic Parameters. The system provides up to S services s, with specific subsets being available via U types of user terminals u. These service profiles can be described by an $S \times U$ indicator matrix

$$\Delta = \begin{pmatrix} \delta_{11} \cdots \delta_{1u} \cdots \delta_{1U} \\ \vdots \ddots \vdots \ddots \vdots \\ \delta_{s1} \cdots \delta_{su} \cdots \delta_{sU} \\ \vdots \ddots \vdots \ddots \vdots \\ \delta_{S1} \cdots \delta_{Su} \cdots \delta_{SU} \end{pmatrix} \qquad \delta_{su} = \begin{cases} 1 & s \text{ available via } u \\ 0 & \text{otherwise.} \end{cases} \tag{12.1}$$

Terminals may be classified as *individual* or *group* terminals, the latter ones being used by a group of people.

According to the ITU definition a service consists of elementary *service components* (e.g. the components *voice* and *video* for the service *video telephony*). Each unidirectional component of a service s is specified by the traffic parameters listed in Table 12.1. Distinct components of the same service are separately taken into account with their specific parameter values, although this is not explicitly indicated by indices. In fact only maximum bit rate and burstiness are component specific, whereas application frequency and call holding time have unique values for all components of the same service s.

Table 12.1. Traffic parameters specifying a unidirectional service component of a service s

Parameter	Notation	Unit	Explanation/comment
Application frequency	λ_s^*	1/s	Long-term average call arrival rate
Mean call holding time	$1/\mu_s$	s	
Maximum bit rate	R_s	b/s	
Burstiness	b_s	–	Average divided by max. bit rate ($= 1$ for CBR, < 1 for non-CBR)

The traffic parameters from Table 12.1 allow the calculation of the *long-term average* bit rate requirements for a single unidirectional service component. Further parameters are required to capture the *busy hour*, the *group terminal*, and the *multiservice* effects. Therefore, in Table 12.2 four linear scaling factors are introduced which will be referred to as *multiservice parameters*.

Table 12.2. Specific parameters (scaling factors) for the multiservice scenario

Parameter	Notation	Unit	Range
Busy hour factor	$m_{\mathrm{bh},s,u}$	–	≥ 1
Multiservice correlation factor	$m_{\mathrm{ms},s,u}$	–	$0 \ldots 1$
General scaling and adaptation factor	$m_{\mathrm{sa},s,u}$	–	$0 \ldots 1$
Group terminal factor	$m_{\mathrm{gt},u}$	–	≥ 0

Busy Hour Factor. The busy hour factor is the ratio between the maximum (busy hour) and the average traffic demand. Including the busy-hour effect is standard for dimensioning single-service networks. However, in the multiservice scenario the busy hour factor will generally depend on both the service and user terminal type.

Multiservice Correlation Factor. This parameter captures the multiservice behavior and can be illustrated by the following example. Voice telephony, video telephony, and video conferencing make up a service class (symmetric conversational services). In the multiservice scenario the classical "voice-only" user now has the choice between the above-mentioned three alternative services, and it can be assumed that the latter two replace a certain conversation demand which was formerly served by voice telephony only. The total activity can be expected to remain constant.

General Scaling and Adaptation Factor. This factor describes other influences on the expected source traffic volume, such as the forced shaping of traffic in cases where the nominal information bit rate of a user terminal does not allow the maximum bit rate assumed for a service. This situation is more likely for services revealing a high burstiness, like those related to computer interconnection and Internet applications. Another aspect is that certain services in the profile of a terminal type may not be used by all users.

Group Terminal Factor. The group terminal factor takes into account the shared use of services within the group. For instance, the factor can be used to adjust a predefined total terminal traffic while preserving the correct service shares.

Mathematical Notation. The first three factors are service and terminal dependent, which again requires three $S \times U$ matrices of parameter values for a complete description, one for each factor type f, $f \in \{\mathrm{bh, ms, sa}\}$:

$$\mathbf{M}_f = \begin{pmatrix} m_{f,1,1} \cdots m_{f,1,u} \cdots m_{f,1,U} \\ \vdots \ \ \ddots \ \ \vdots \ \ \ddots \ \ \vdots \\ m_{f,s,1} \cdots m_{f,s,u} \cdots m_{f,s,U} \\ \vdots \ \ \ddots \ \ \vdots \ \ \ddots \ \ \vdots \\ m_{f,S,1} \cdots m_{f,S,u} \cdots m_{f,S,U} \end{pmatrix} \qquad m_{f,s,u} \in \begin{cases} [1,\infty) & f = \mathrm{bh} \\ [0,1] & f = \mathrm{ms} \\ [0,1] & f = \mathrm{sa}. \end{cases}$$

(12.2)

The group terminal characteristics are captured in a vector with U values:

$$\mathbf{m}_{\mathrm{gt}} = \begin{pmatrix} m_{\mathrm{gt},1} \cdots m_{\mathrm{gt},u} \cdots m_{\mathrm{gt},U} \end{pmatrix} \qquad m_{\mathrm{gt},u} \in [0,\infty). \qquad (12.3)$$

12.1.4 Calculation of the Spot Beam Capacity Requirements

Source traffic from on-demand services obviously drives the capacity requirements not only on the *source uplink*, but on all links in the system. First, each connection contributes to the *destination downlink* in the spot beam of the respective partner. For capacity calculations there will be a distinction if the partner is a mobile user or a fixed earth station, since in general they use separate frequency bands. The connection also contributes in the return direction to the *destination uplink* and *source downlink*, with a potentially different capacity demand, depending on the asymmetry of the service. In the remainder of this subsection, all considerations are restricted to the source uplink, which is also referred to as the *user* or *service uplink*.

Contribution of On-Demand Services. For the calculation of capacity demand from on-demand services, individual and group terminals have to be treated differently.

The traffic contribution (in terms of bit rate) of a unidirectional service component s from an *individual terminal u* is

$$A_{s,u} = m_{\text{bh},s,u}\, m_{\text{ms},s,u}\, m_{\text{sa},s,u}\, \frac{\lambda_s^*}{\mu_s} b_s R_s, \tag{12.4}$$

and the resulting cumulative traffic sent from a multiservice terminal is

$$A_u = \sum_s A_{s,u} = \sum_s \delta_{su} m_{\text{bh},s,u}\, m_{\text{ms},s,u}\, m_{\text{sa},s,u}\, \frac{\lambda_s^*}{\mu_s} b_s R_s. \tag{12.5}$$

The cumulative traffic sent from a *group terminal u* is assumed to be constant at a certain percentage p of the specified maximum information bit rate of the respective terminal, $R_{\text{max},u}$:

$$
\begin{aligned}
A_u &= m_{\text{gt},u} \sum_s \delta_{su}\, m_{\text{bh},s,u}\, m_{\text{ms},s,u}\, m_{\text{sa},s,u}\, \frac{\lambda_s^*}{\mu_s} b_s R_s \\
&\overset{!}{=} p\, R_{\text{max},u} \qquad p \le 1.
\end{aligned}
\tag{12.6}
$$

According to this formula, $m_{\text{gt},u}$ is used to adjust the terminal traffic to this predefined value, while the service split w.r.t. the bit rates is preserved.

Given the number of (individual or group) terminals u in the spot beam i, $N_u(i)$, the cumulative busy-hour source traffic $A(i)$ in spot beam i can be calculated as

$$A(i) = \sum_u N_u(i) A_u. \tag{12.7}$$

Contribution of Broadcast Services. For broadcast services, the related traffic demand does not depend on user behavior and can therefore not be treated using the developed multiservice source traffic model. Rather, the cumulative traffic from all the programs of a sending station is added to the on-demand load, once for the feeder uplink in the sending spot beam and once for the service downlink in each receiving spot beam.

12.1.5 System Bandwidth Demand Calculation

In Chap. 6, the required bandwidth of (MF-)TDMA and (MF-)CDMA systems is calculated for a single constant bit rate service, such as classical telephony. Here we present an extension (i) to the multiservice case and (ii) to the case where the system dimensioning parameters (carrier/frame organization, spot beam layout, cluster design) have to be adapted to a given market demand.

The following discussion considers on-demand services in an MF-TDMA system; the application to the cases of broadcast services and/or MF-CDMA systems is straightforward.

With the notation of Chaps. 5 and 6, the single-service system bandwidth B_{ss}, equaling the cluster bandwidth B_{cl}, can be derived from Eqs. (6.16), (6.17), and (6.51) as

$$B_{ss} = B_{cl} = K_s T N \left(1 + \beta + \frac{B_g}{R_B} \right) \left(1 + \frac{H + G}{n_s} \right) \frac{R_b}{r \log_2 M}, \qquad (12.8)$$

where K_s is the effective cluster size and the product $K_s T N$ is the number of channels per cluster.

In the multiservice case the term *channel* refers to a service composed of its specific service components. Moreover, the reference must be the *worst-case* cluster in terms of traffic load with respect to geographic and time variations. Thus, neglecting blocking, one can postulate that the service-specific number of channels in spot beam i should equal the average number of active users of the respective service s in the busy hour, $\tilde{N}_s(i)$. The worst-case cluster of spot beams is defined as $\mathcal{C} = \{c_1, \ldots, c_{K_s}\}$, c_n reflecting the IDs of the worst-case spot beams making use of the K_s non-overlapping frequencies $f_1, \ldots, f_{K_s}$. The number of required service-specific channels in this worst-case cluster, $\hat{N}_{cl,s}$, is

$$\hat{N}_{cl,s} = \sum_{i \in \mathcal{C}} \tilde{N}_s(i). \qquad (12.9)$$

The required multiservice system bandwidth B_{ms} equals the worst-case cluster bandwidth $\hat{B}_{cl}$, which can be calculated by modifying and extending (12.8) for the multiservice case to

$$B_{ms} = \hat{B}_{cl} = \left(1 + \beta + \frac{B_g}{R_B} \right) \left(1 + \frac{H + G}{n_s} \right) \frac{1}{r \log_2 M} \sum_{i \in \mathcal{C}} \sum_s \tilde{N}_s(i) R_{b,s}. \qquad (12.10)$$

The right-hand double sum term is the cumulative information bit rate in the service uplink of the K_s worst-case spot beams. With Eq. (12.7) the system bandwidth in the multiservice case can be expressed by

$$B_{ms} = \hat{B}_{cl} = \left(1 + \beta + \frac{B_g}{R_B} \right) \left(1 + \frac{H + G}{n_s} \right) \frac{1}{r \log_2 M} \sum_{i \in \mathcal{C}} A(i). \qquad (12.11)$$

12.1.6 Applied Spot Beam Capacity Dimensioning: A Case Study

The European Study System SECOMS. Serving as a first-phase study system for the commercial EuroSkyWay project [PFV+99], SECOMS is a European geostationary satellite system mainly operating in the Ka band and providing multimedia services via six different portable and mobile terminal types. For the network dimensioning, a set of 13 basic services has been defined. The user may communicate via a satellite terminal *SatT* to another SECOMS SatT or to a partner in the terrestrial networks via a gateway station *GTW*. Additionally, a SatT can receive broadcast or retrieval services from a service provider station *SPS*. Figure 12.3 shows the coverage of the 32 spot beams and a reference set of GTW/SPS locations.

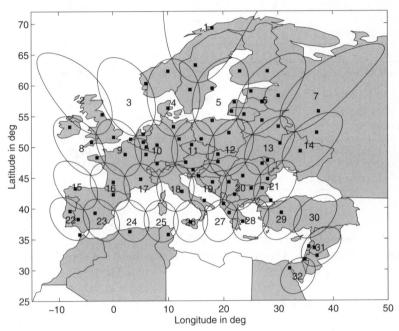

Fig. 12.3. SECOMS Ka band spot beam coverage and reference GTW distribution

Parameter Settings and Input Data. Basic inputs to the calculations are the subscriber numbers for the six different Ka band terminals in all served countries, which are highlighted in Fig. 12.3. These numbers have been derived using a detailed market prediction model, cf. Sect. 12.1.2, which delivers the terminal population data per country; these can then be mapped onto terminal figures per spot beam and directly used as numerical input $N_u(i)$ to Eq. (12.7).

Table 12.3 presents the six SECOMS Ka band terminals. Important considerations for the dimensioning process are the individual/group terminal classification and the values for maximum information rate and rate granularity, the latter corresponding to the selected MF-TDMA carrier and frame organization.

Table 12.3. Terminal types and their relevant characteristics/parameters

	SatT-A	SatT-B	SatT-C
Portable			
Case	Laptop	Briefcase	Briefcase
Use	Individual	Individual	Individual
Mobility during operation	No	No	No
Uplink information rate (granularity)	16–160 kb/s (16 kb/s)	16–512 kb/s (16 kb/s)	16–2048 kb/s (16 kb/s)
Downlink max. information rate	2.048 Mb/s	2.048 Mb/s	2.048 Mb/s
Mobile			
Mobile type	Car	Aircraft, ship, bus, train, truck	Aircraft, ship, bus, train, truck
Use	Individual	Group	Group
Mobility during operation	Yes	Yes	Yes
Uplink information rate (granularity)	16–160 kb/s (16 kb/s)	16–512 kb/s (16 kb/s)	16–2048 kb/s (16 kb/s)
Downlink max. information rate	2.048 Mb/s	2.048 Mb/s	2.048 Mb/s

The reference service profiles of all SECOMS Ka band terminals are displayed in Table 12.4. The services with ID numbers 5–8 form the group of broadcast services, whereas all others are on-demand services. The table values can be directly taken to form the indicator matrix Δ defined in Eq. (12.1).

Table 12.5 summarizes the numerical values for the basic traffic parameters introduced in Sect. 12.1.3. The values are based on original ITU data and earlier SECOMS project documentation [WHS97]. It should be noted that for some services only the dominating service components (in terms of bit rate contribution) have been taken into account. Moreover, for broadcast services the dimensioning has been restricted to *non-interactive* broadcast, although in the table figures are displayed for the respective return links.

For the busy hour factor one unique value has been specified for each of the three terminal types A, B, and C: $m_{bh,A} = 2.2$, $m_{bh,B} = 2.1$, and $m_{bh,C} = 2.9$.

Numerical values for the multiservice correlation factors have to reflect that the extended service profile of a terminal may lead to increased traffic volume generated by the user. However, the traffic will grow less than "linearly" with the addition of new services due to the reasons described in Sect. 12.1.3.

To date, no complete systems or user communities exist from which reliable data about these effects can be derived. Therefore, the following prag-

Table 12.4. Considered services and reference service profiles for all SECOMS terminals

ID		Port. A	Mob. A	Port. B	Mob. B	Port. C	Mob. C
1	Telephony, telefax	1	1	1	1	1	1
2	Video telephony	0	0	0	0	1	1
3	Video conference	0	0	0	0	1	1
4	Video surveillance	0	0	1	1	1	1
5	TV broadcasting	0	0	0	0	0	1
6	Audio broadcasting	0	1	0	1	0	1
7	Document broadcasting	1	0	1	1	1	1
8	Vehicle inf. broadcast.	0	1	0	1	0	1
9	Videography	0	0	1	0	1	1
10	Database retrieval	1	0	1	1	1	1
11	Computer interconnection	1	0	1	1	1	1
12	Email and paging	1	0	1	1	1	1
13	File transfer	1	0	1	1	1	1

Table 12.5. Basic traffic parameters

	λ_s^*	$1/\mu_s$	R_{return}	R_{forward}	b
Telephony, telefax	1/h	3 min	64 kb/s	64 kb/s	0.35
Video telephony	2/day	5 min	1150 kb/s	1150 kb/s	1.0
Video conference	1/day	60 min	1920 kb/s	1920 kb/s	0.33
Video surveillance	1/month	1 month	16 kb/s	1920 kb/s	1.0
TV broadcasting	2/day	2 h	16 kb/s	2 Mb/s	1.0
Audio broadcasting	2/day	2 h	16 kb/s	192 kb/s	1.0
Doc. broadcasting	2/day	1 h	16 kb/s	2 Mb/s	1.0
Veh. inf. broadcast.	2/day	10 min	16 kb/s	1.5 Mb/s	1.0
Videography	1/day	1 h	128 kb/s	1920 kb/s	0.33
Database retrieval	2/day	2 h	16 kb/s	2 Mb/s	0.05
Computer intercon.	2/day	30 min	2 Mb/s	2 Mb/s	0.001
Email and paging	5/day	1 s	16 kb/s	16 kb/s	1.0
File transfer	5/day	40 s	1 Mb/s	1 Mb/s	0.05

matic approach has been adopted, resulting in the reference values given in Table 12.6:

- The 13 SECOMS services are assigned to the following mainly "disjoint" groups:
  1. {telephony, video telephony, video conferencing},
  2. {video surveillance},
  3. {TV/audio/vehicle information broadcasting, videography},
  4. {document broadcasting, database retrieval}, and
  5. {computer interconnection, email, file transfer}.
- The normalized traffic intensity for a single user and a certain service group does not exceed a value of 1.
- For complete service groups (i.e. all services of this group are in the profile of a given terminal) the accumulated normalized traffic intensity is 1, whereas it is smaller for less complete groups.
- Potential impacts of the terminal type on the sharing between services of a group are considered to a limited extent where this seems reasonable.

Table 12.6. Reference set of multiservice correlation factors $m_{ms,s,u}$ for SECOMS

	Port. A	Mob. A	Port. B	Mob. B	Port. C	Mob. C
Telephony, telefax	1	1	1	1	0.6	0.6
Video telephony	0	0	0	0	0.3	0.3
Video conference	0	0	0	0	0.1	0.1
Video surveillance	0	0	1	1	1	1
TV broadcasting	0	0	0	0	0	0.4
Audio broadcasting	0	0.7	0	0.7	0	0.3
Vehicle inf. broadcast.	0	0.3	0	0.1	0	0.1
Videography	0	0	0.1	0	0.1	0.1
Document broadcasting	0.4	0	0.4	0.2	0.4	0.1
Database retrieval	0.6	0	0.6	0.6	0.6	0.6
Computer interconnection	0.2	0	0.3	0.2	0.3	0.2
Email and paging	0.5	0	0.4	0.6	0.4	0.6
File transfer	0.3	0	0.3	0.2	0.3	0.2

In order to include broadcast services, spot beam locations of eight sending stations, their cumulative bit rate, and the receiving spot beams related to each sender have been defined. Details can be found in [WL98b].

Numerical Results. Figure 12.4 shows the busy-hour traffic requirements in terms of accumulated information bit rate versus the spot beams. It can be seen in both histograms that the total traffic in the forward link (GTW up/SatT down) exceeds the respective one in the return link (SatT up/GTW

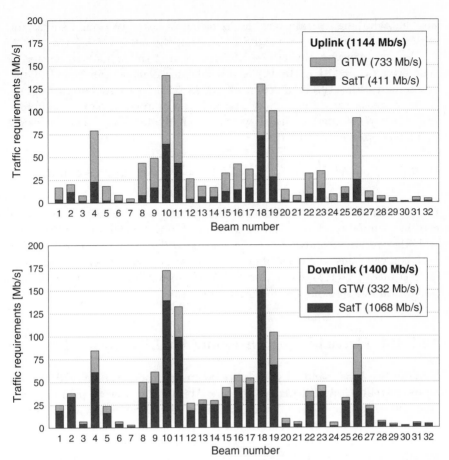

Fig. 12.4. Spot beam traffic distribution for SECOMS Ka band up- and downlink

down). This is due to the contribution of asymmetric components like retrieval and broadcast services; moreover, the nature of broadcasting (the dominating forward link traffic is sent in one but received in all or a number of spot beams) explains the fact that this asymmetry is more distinct on the downlink than on the uplink. Finally, all spot beam figures have been scaled proportionally so as to satisfy the technological limitation of 1.4 Gb/s gross capacity for the on-board switch. This limitation is directly applicable to the more critical downlink, resulting in a corresponding value of 1.14 Gb/s for the uplink. The four neighboring spot beams 10, 11, 18, and 19 reveal the highest traffic load. Due to the linearity of the approach all cumulative spot beam capacities can be easily given as either service or terminal splits.

Finally it is of interest to estimate the system bandwidth demand from the given traffic requirement figures based on the formulas derived in Sect. 12.1.5.

For the link design parameters the following values have been assumed for all services: $r = 0.75$, $M = 4$ (QPSK), $\beta + B_g/R_B = 0.5$, $(H+G)/n_s = 0.2$.

With an effective cluster size of $K_s = 4$, and the worst-case cluster of spot beams $\mathcal{C} = \{10, 11, 18, 19\}$, the bandwidth demand can be calculated according to Eq. (12.11). Since in SECOMS both feeder and service uplink traffic share the same resources, the formula can be applied to the total uplink traffic in the considered cluster, which is roughly 488 Mb/s. This yields an uplink system bandwidth of

$$B_{\mathrm{ms}} = \left(1 + \beta + \frac{B_g}{R_B}\right)\left(1 + \frac{H+G}{n_s}\right)\frac{1}{r \log_2 M} \cdot 488\,\mathrm{Mb/s} \approx 580\,\mathrm{MHz}.$$

(12.12)

Carrier granularity and the allocation of specific (e.g. exclusive) carriers to certain terminal types have not been included in the calculation; such effects will add overhead bandwidth to the calculated figure. A theoretical system with homogeneous traffic distribution – that is, a cumulative cluster information bit rate of $4 \cdot (1.14\,\mathrm{Gb/s})/32 \approx 143\,\mathrm{Mb/s}$ – would have required a system bandwidth of approximately 170 MHz only.

12.2 ISL Capacity Dimensioning for LEO Systems

Some future broadband LEO satellite communication systems will rely on an intersatellite link (ISL) trunk network. Within the network planning process, the routing and dimensioning tasks are in general closely coupled and essentially influence both the installation and operating costs of a system. The routing of global traffic flows over dynamic ISL network topologies has already been addressed in Sect. 7.9. For the capacity dimensioning of ISL networks one may use approaches known from terrestrial (ATM) networks to a certain extent, but has to take into account specific additional constraints like the time variance of the topology.

This section mainly summarizes the ideas and results of the earlier publications [WF99] and [WWFM99].

12.2.1 Topological Design of the ISL Network

The ISL network of polar satellite constellations exhibits two drawbacks with regard to connection-oriented routing: (i) the seam between counter-rotating orbits and (ii) the on/off switching of inter-orbit ISLs [WDV+97]. In [Wer95] it was shown that inclined Walker constellations provide the possibility to set up inter-orbit ISLs that can be maintained *permanently*. This is a highly desirable feature in the light of real-time connection-oriented services, and becomes even more striking when considering, for example, jitter-sensitive ATM cell streams.

M-Star [Mot96] was one of the first commercial system proposals to use inclined orbits and ISLs, Fig. 12.5. In the following we present a pragmatic approach to the ISL topology design in inclined Walker constellations in general, using M-Star as an example.

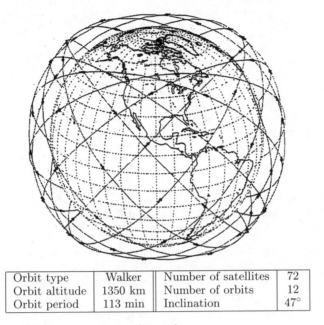

Orbit type	Walker	Number of satellites	72
Orbit altitude	1350 km	Number of orbits	12
Orbit period	113 min	Inclination	47°

Fig. 12.5. The M-Star constellation [Mot96]

A closer look at the planar projection of the constellation, in Fig. 12.6, facilitates the first step in the topology design, which is to identify potential ISLs. Due to the symmetry of the constellation it is sufficient to consider ISLs between satellite 0 and its eastward neighbors as an example. These are then applicable to all other satellite pairs correspondingly. Of course, the implementation of the intra-orbit ISLs 0–1 and 0–5 with both constant link distance and fixed pointing angles is obvious. Then one would intuitively envisage links toward the next neighbors on the adjacent plane, 0–6 and 0–11. Finally, the same procedure could be applied to partners on the next, second plane; here, the pairs 0–12 and 0–17 are selected because they are the ones with minimum phase difference.[1] This procedure is generally applicable to Walker constellations.

The feasibility of these envisaged links has to be proven taking into account the geometric and technological constraints for a specific constellation.

---

[1] From the snapshot in Fig. 12.6 the pair 0–16 also seems to be an attractive candidate; however, such "next, second partners" (in terms of phasing) have not been considered.

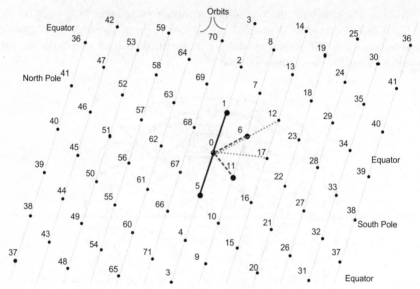

Fig. 12.6. Schematic view of the M-Star constellation at $t = 0$ and potential ISLs. The example for Sat 0 can be applied to all other satellites.

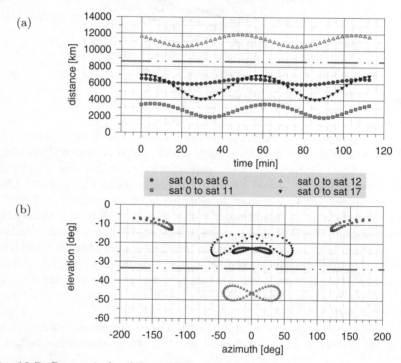

Fig. 12.7. Geometric feasibility of ISLs in M-Star: (a) time variation of ISL distance; (b) pointing diagram. The dash–dot line represents the bound with respect to earth shadowing

The diagrams in Fig. 12.7 display relevant geometrical data for this purpose. Both links toward the adjacent orbit show relatively little variation in distance and pointing angle (elevation and azimuth). This means that establishing these ISLs in permanent mode does not introduce severe problems. ISL 0–17 is less attractive but still possible, whereas earth shadowing prevents implementation of ISL 0–12.

For the following, a reference topology T1 is selected, with generic inter-orbit ISL 0–6 besides the intra-orbit ones, resulting in four bidirectional ISLs per satellite. In contrast to this, a topology T2 uses generic inter-orbit ISLs 0–6 and 0–11, thus yielding six ISLs per satellite. Figure 12.8 displays snapshots of T1 and T2, respectively.

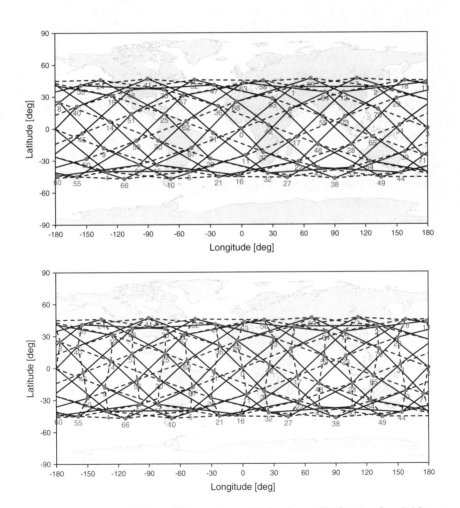

Fig. 12.8. Schematic M-Star ISL topologies T1 (top) and T2 (bottom): solid lines, intra-orbit ISLs; dashed lines, inter-orbit ISLs

12.2.2 ISL Routing Concept

As shown in Sect. 7.9, some gain can be expected from on-line traffic adaptive routing, or – in other words – from shifting a part of the connections to longer paths in order to balance the flows across the network. Network dimensioning is performed off-line prior to system operation, and it has specifically to incorporate the on-line routing decisions. This generally requires that the on-line routing works in a deterministic manner, so that it is predictable. In order to keep this tractable, the number of path options is usually restricted.

Off-Line Routing Framework. In Sect. 7.9, dynamic virtual topology routing (DVTR) has been presented as a general concept to provide connection-oriented communication in deterministically dynamic topology environments. Here, we recall this approach and restrict ourselves to ISL topologies which are based on permanent links only.

Off-line routing is embedded between topology design and capacity dimensioning, Fig. 12.9. The path search is performed by a K-shortest-path algorithm (KSPA) for every OD pair, where the single shortest path search task can be formulated as finding the least-cost path $p(s)$ for each time step s, cf. Eq. (7.9).

Performing this path search for all $s = \{0, \ldots, S-1\}$ establishes a discrete-time dynamic virtual ISL topology. It consists of an ordered set of K alternative paths for any OD pair at any step s. From these K alternatives, the $k \in \{1, \ldots, K\}$ best ones can be used for routing.

12.2.3 Network Dimensioning

Overall Approach and Assumptions.

1. *Target function.* The "classical" ATM network design process typically aims at minimizing some total network costs, where the key optimization areas are (i) the definition of the VPC topology, (ii) the VPC capacity assignment, and (iii) the routing rules for OD traffic.

   In satellite networks, each satellite and each ISL will face roughly the same worst-case traffic load at some time and hence have to be dimensioned with respect to this value; as a result, satellites in the constellation, including the ISL equipment, will be identical. Consequently the target function for the ISL network dimensioning problem is

   > Minimize the worst-case link (WCL) capacity, which is the maximum capacity required on any link at any time.

2. *Time-discrete approach.* Routing and dimensioning is performed in a time-discrete manner: we break down the ISL network dimensioning task into a number of independent dimensioning tasks, each minimizing the WCL capacity per step.

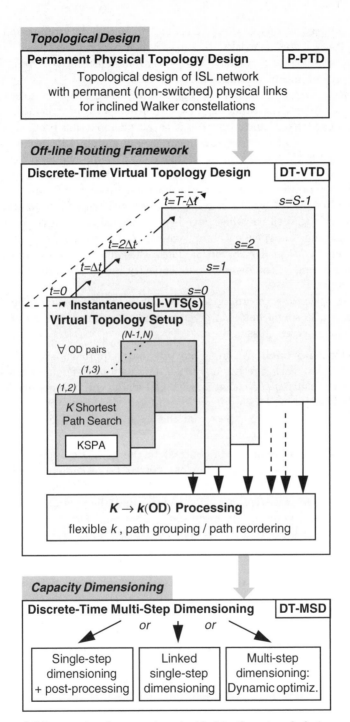

Fig. 12.9. Off-line routing framework embedded in the network design process

3. *Permanent virtual topology.* Overlaid on top of the time-variant physical ISL network is a VPC topology which is permanent over time. The VPC topology is predefined according to higher-priority criteria and not subject to optimization itself.

4. *Routing and VPC capacities.* The routing/distribution of given demand pair traffic on the available VPCs is either performed heuristically, according to fixed rules, or treated as an optimization problem which is formulated and solved using linear programming (LP) techniques. Assuming that a limited set of alternative VPCs may be used for splitting the traffic between a specific pair of end nodes, the main optimization parameters are then the splitting factors. Instead of Erlang traffic, we directly operate with given *OD demand pair capacities* or *bandwidths* assuming that these values have been previously calculated from the Erlang traffic – independently for each demand pair according to the *virtual trunking concept* [DH98, Sie95]. The VPC capacities result directly from the splitting of the demand pair capacity on all available VPCs of the OD pair.

5. *Link capacities.* According to the virtual trunking concept, the total capacity of a single link at a given step is determined by summing up all VPC capacities crossing it.

Heuristic Approach. A simple but pragmatic approach is based on observations made in earlier research [Bur97]. For a simple reference dimensioning we use *equal sharing (ES)*; that is, each OD traffic will be equally distributed on the k best paths, k being fixed for all OD pairs in the topology and over time. Intuitively, one can expect that the ES approach will lead to a decrease of WCL traffic load with increasing k.

Optimization Approach. In contrast to the heuristic approach, one may consider a dedicated optimization of a given target function, which is in our case the minimization of the overall WCL capacity. The WCL capacity is minimized for all steps s independently, and the overall WCL capacity is then the maximum of all minimized WCL capacities per step. Using the formulation of the network model and the dynamic routing concept presented in Sect. 7.9.2, we consider a given step without explicitly indexing with s.

We consider only permanent physical links in the ISL topology, which means that we have a fixed set of links l over all steps, $l \in \{1 \ldots L\}$; for topology T1, $L = 2N = 144$; for T2, $L = 3N = 216$. The offered capacity n_w per OD demand pair w is distributed among the k shortest paths p selected from the ordered set of existing paths, P_w, so that each path carries a certain share n_p of the total demand pair capacity,

$$n_w = \sum_{p \in P_w} n_p. \tag{12.13}$$

The required bandwidth n_l of a link l is obtained as the sum of the bandwidths n_p of all paths p containing this link,

$$n_l = \sum_{w \in W} \sum_{p \in P_w} \delta_l^p n_p, \tag{12.14}$$

where W denotes the set of all OD pairs and $\delta_l^p \in \{1, 0\}$ indicates if path p uses link l or not. Our objective to minimize the maximum required bandwidth on a single physical link can be expressed as a minimax optimization problem:

$$\max_l \{n_l\} = \max_l \left\{ \sum_{w \in W} \sum_{p \in P_w} \delta_l^p n_p \right\} \rightarrow \min, \tag{12.15}$$

subject to the (linear) constraint Eq. (12.13) for all demand pairs w.

In order to apply standard LP optimization techniques, the original minimax problem can be transformed into a smooth linear minimization problem by introducing a new scalar optimization variable $n_{\max}$, which is an upper bound on all n_l, and formulating the corresponding link capacity bounds as inequality constraints. Together with the set of equality constraints resulting from Eq. (12.13), the complete formulation of the LP optimization problem becomes

---

Minimize the WCL capacity,

$$\min_{\mathbf{n_p}} n_{\max}(\mathbf{n_p}), \tag{12.16}$$

subject to (i) the link capacity bounds

$$n_{\max} \geq \max_l n_l = \sum_{w \in W} \sum_{p \in P_w} \delta_l^p n_p \qquad \forall l \tag{12.17}$$

and (ii) the OD/path capacity requirements

$$n_w = \sum_{p \in P_w} n_p \qquad \forall w. \tag{12.18}$$

The optimization variables contained in the vector $\mathbf{n_p}$ are the shares of total OD capacity carried by each path belonging to the OD pair, or equivalently, the splitting factors that determine the OD capacity split into its correlated paths. Assuming for instance a fixed number of $k = 3$ alternative paths for all $N(N-1)/2$ OD pairs in our M-Star topology, we end up with a number of $kN(N-1)/2 = 7668$ optimization variables. In comparison, the number of equality constraints according to Eq. (12.18) equals the number of OD pairs, $N(N-1)/2 = 2556$, and Eq. (12.17) adds $L = 2N = 144$ upper bounds for the link capacities. Obviously, a larger k increases the optimization potential (and the computational complexity), whereas the number of constraints remains fixed for a given constellation (N) and ISL topology ($L = L(N)$).

So far, we have not introduced any specific constraint on the share of the total traffic that one path is allowed to carry. As a consequence, a single path may convey the complete offered OD traffic alone, whereas other paths may remain empty. In the following, we refer to this approach as *full optimization (FO)*. Although FO leads to the maximum possible WCL load reduction per step, there are some reasons – for example, consequences for operation in failure situations, potentially high load variations on single links from step to step, etc. – to introduce an additional linear constraint in the form of an upper bound α for the normalized share of the OD traffic one path is allowed to carry,

$$0 \leq n_p \leq \alpha n_w, \quad \alpha \in [1/k \ldots 1] \quad \forall p, \forall w. \tag{12.19}$$

This approach is referred to as *bounded optimization (BO)* with parameter α. The number of path capacity bounds is identical to the number of optimization variables.

All the above considerations on optimization are valid for a single step in the dynamic scenario, and are consequently applied to each single step independently.

12.2.4 Numerical Example

Scenario and Assumptions. All numerical studies presented in this section are based on a homogeneous traffic scenario, i.e. each OD demand pair generates traffic of one bandwidth unit over all steps. In the following, the terms traffic (load), capacity, and bandwidth are used synonymously. Restricting to such a scenario allows the isolation of topological impacts on the dimensioning process and on the dimensioning results.

Extensive dimensioning runs have been performed for both topologies T1 and T2, where $K = 18$ (for T1) and $K = 10$ (T2) best paths have formed the complete set resulting from KSPA. In all cases, $S = 50$ discrete steps have been considered, corresponding to a time step size $\Delta t = T/S = 113\text{min}/50 = 2.26$ min for the M-Star constellation.

Worst-Case Link (WCL) Traffic Load. First the performance of equal sharing (ES), full optimization (FO), and bounded optimization (BO) is compared in terms of the traffic load on the most loaded physical link (the WCL). The dimensioning has always been performed over all steps, and one observation is that the variation of the WCL load over time is very small. Already if all OD traffic is routed over the respective shortest path ($k = 1$), i.e. without any traffic optimization, the WCL load variation over the steps is less than 10%. And any optimization approach leads to a further reduction of this variation. Therefore we present only mean values in the following.

Figure 12.10 shows the WCL load results for the simple ES approach, comparing T1 and T2. As expected, the values for T2 are significantly lower due to the availability of many more links to distribute the traffic on. With growing k the WCL value decreases, but in both topologies "saturation"

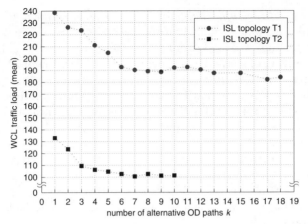

Fig. 12.10. Reduction of mean WCL traffic load through ES with growing k: performance comparison between topologies T1 and T2

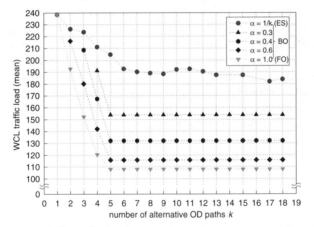

Fig. 12.11. Mean WCL traffic load versus k: performance comparison for ES, FO, and BO with different values of α; ISL topology T1

of the optimization is achieved for $k = 6$. The WCL load reduction of ES is then roughly 25%. Focussing on T1, Fig. 12.11 displays the performance results for the different optimization approaches. Comparing the two "extreme" approaches ES and FO, one observes that FO achieves an *additional* improvement of 45% compared to ES in the saturation zone. With $k = 5$ it is already possible to use the full optimization potential within T1. Comparing all curves as a whole, the distinct influence of the upper bound for path traffic, α, in the BO approach becomes obvious. As expected, the additional constraint reduces the optimization potential with respect to the WCL target value. However, with moderate values of α ($\alpha = 0.6$) the results come close to the FO ones.

Physical Link (PL) Traffic Load. In order to understand *how* the various traffic distribution and optimization approaches operate within the topology, it is helpful to study the load variation on fixed physical links (PLs) over the steps. Figure 12.12 displays the load over time for two selected PLs, one intra- and one inter-orbit link. In both cases we see pronounced load peaks when the respective link is over mid latitudes (i.e. the highest latitudes that the satellites reach with 47° inclination), whereas it is low near the equator. It is also obvious that the WCL is always an inter-orbit ISL. Concluding from both observations, WCL load reduction should typically work by shifting traffic away from those inter-orbit ISLs that are in the critical region at a certain step. Figure 12.13 confirms this expectation and illustrates the superior performance of the FO approach with respect to this shifting of traffic. For $k = 5$, FO leads to a constant PL load, which is the final limit for any optimization.

Another confirmation of these considerations is given by the PL load distributions over all PLs of a certain step, as displayed in Fig. 12.14 for step 0 (the shape of the curves being nearly identical for other steps). It is obvious how traffic from higher loaded links is "shifted" to lower loaded ones, and a complete balance is achieved for the inter-orbit ISLs when FO with $k \geq 5$ is used. In reality, of course, certain traffic is not just shifted from one link to another but OD traffic is shifted to other paths – sometimes to paths with more hops – resulting in an increase of both average path delay and average PL load in the network. This is the price to be paid for WCL load minimization. However, the increase in both values remains typically limited to less than 10% for FO.

Figure 12.15 summarizes pictorially some of the above-mentioned phenomena.

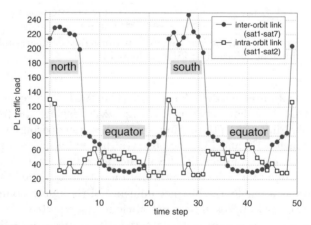

Fig. 12.12. Traffic load on selected PLs over one constellation period

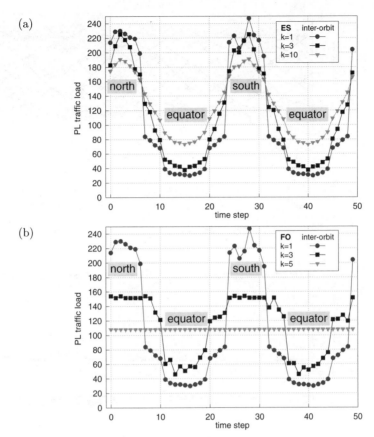

Fig. 12.13. Equalization of traffic load on a fixed inter-orbit PL through (a) the ES and (b) the FO approaches with varying k

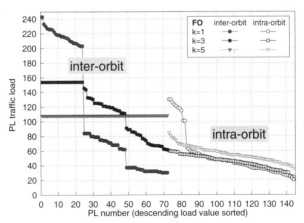

Fig. 12.14. PL traffic load distribution in the network for FO with selected k at step 0

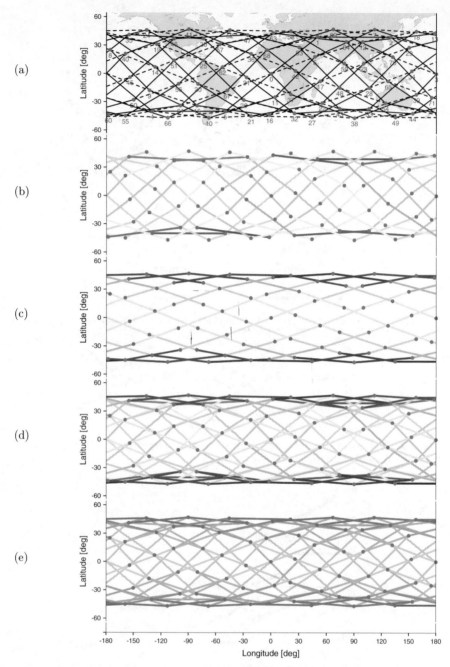

Fig. 12.15. Study of PL traffic distribution in the (a) M-Star topology T1 at step 0 (higher traffic load is illustrated in darker grayscale levels): (b)/(c) comparison of intra-/inter-orbit ISL traffic without optimization; (d)/(e) comparison of network traffic without/with optimization (ES with $k = 6$)

12.2.5 Extensions of the Dimensioning Approach

Path Grouping Concept. The number of alternative OD paths k has so far been identical for all OD pairs, being an arbitrarily fixed value. This may, however, lead to conflicts with the permanent VPC topology paradigm. Consider $k = 1$ and a satellite pair separated by one intra- and one inter-orbit hop as an illustrative example. Obviously, there are always two alternative shortest paths of similar length. Due to the variation of the inter-plane ISL distance, one or the other path will be the first one at a given step. If the alternative paths per step are strictly ordered according to their length, and always only the first one is chosen ($k = 1$), this effectively means switching VPCs, although it is not at all mandatory given the permanence of the physical ISL topology.

In [WWFM99] a *path grouping concept* has been presented which systematically forms groups of OD paths and then selects a specific k for each OD pair in such a manner that VPC switching is completely avoided, i.e. the same paths per OD pair are available for routing over all steps. Moreover, with this approach the WCL load is additionally decreased compared to the fixed-k case with only a negligible increase in average path delay.

History-Based Multi-Step Dimensioning. So far, the dimensioning procedure has considered isolated steps. This approach contains one systematic modeling deficiency: by assuming *independent* demand pair capacities in subsequent steps it does effectively neglect the "history" of single calls making up these instantaneous cumulative capacities. Specifically, in this approach it is implicitly assumed that calls are freely reroutable in each step according to the isolated optimization result. Consequently, the isolated step optimization results are usually "too good" as soon as any QoS requirements have to be fulfilled, as this in turn reduces the optimization freedom. Two specific call situations must be taken into account:

1. A call remains within the same OD satellite pair in the ISL subnetwork from one step to the next. In this case the call should stick to the OD path chosen in order to avoid uncontrolled path delay offset, increased call dropping probability, and unnecessary handover signaling.
2. A call has to be switched to a new OD pair from one step to the next because of a handover at the source and/or destination satellite. Being a "new" call for the new OD pair, the call could be freely routed (according to isolated optimization), but the new route selection could as well be deterministic, for instance according to the QoS requirement to minimize delay offsets during path switching.

In [WR99] a history-based multi-step dimensioning approach has been presented and compared to the isolated step approach, numerically assessing the reduction of the optimization potential through the introduction of call and path history. As a major conclusion, the "dimensioning error" in the isolated step approach turns out to be typically lower than 10%, which makes it an attractive low-complexity method for a first step in the network dimensioning process.

Appendix

Appendix.

A. Satellite Spot Beams and Map Transformations

A.1 Map Projections and Satellite Views

Projections and perspective views are needed to map points on the earth's surface to satellite view angles. It is advantageous to use spherical projection planes as seen from the satellite. For this purpose Fig. A.1 defines Cartesian and polar coordinates for both satellite-centered and earth-centered notation. The geocentric Cartesian coordinate system e_x, e_y, e_z and polar system $e_\lambda, e_\varphi, e_r$ is already known from Sect. 2.1.6. Furthermore, a spherical orthonormal system with base vectors $e_\sigma, e_\vartheta, e_{r'}$ is defined being centered in the satellite. Then, the angles ϑ and σ correspond to the *true view angles* of the satellite antenna. Furthermore, let the base vector $e_{r'}$ coincide with the antenna reference axis pointing to the nadir. The vector e_ϑ describes the off-axis or nadir angle with respect to the antenna reference axis whereas e_σ gives the orientation in the vertical and horizontal direction. The spherical satellite-centered coordinates can be transformed into Cartesian satellite-centered coordinates $e_{x'}, e_{y'}, e_{z'}$ and vice versa through

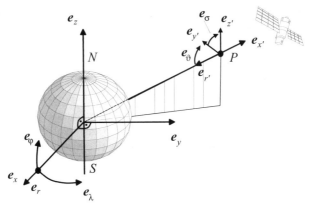

Fig. A.1. Coordinate system for earth–satellite relations

$$
\begin{pmatrix} x' \\ y' \\ z' \end{pmatrix} = \begin{pmatrix} -r' \cos \vartheta \\ r' \sin \vartheta \cos \sigma \\ r' \sin \vartheta \sin \sigma \end{pmatrix} \,, \quad \begin{pmatrix} r' \\ \vartheta \\ \sigma \end{pmatrix} = \begin{pmatrix} \sqrt{x'^2 + y'^2 + z'^2} \\ \arccos \left(\dfrac{-x'}{\sqrt{x'^2+y'^2+z'^2}} \right) \\ \arctan 4 (z'/y') \end{pmatrix} .
$$

$$(\mathrm{A}.1)$$

The Cartesian coordinates can be shifted into geocentric Cartesian coordinates if the satellite is positioned over the point of intersection of the zero-meridian and the equator. Otherwise the earth coordinate system has to be rotated first.

Rotation of the Earth Coordinate System. The following transformation matrices convert a Cartesian coordinate system (x, y, z) into a rotated one (x', y', z') such that a point (λ, φ) is now at zero longitude and latitude. The north–south and east–west directions remain invariant. The total transformation

$$
\begin{pmatrix} x' \\ y' \\ z' \end{pmatrix} = \mathbf{A} \begin{pmatrix} x \\ y \\ z \end{pmatrix} \tag{A.2}
$$

is split into two sub-transformations $\mathbf{B}$ and $\mathbf{C}$, with $\mathbf{A} = \mathbf{C} \cdot \mathbf{B}$. First, the rotation around the z-axis by angle $-\lambda$ is

$$
\mathbf{B} = \begin{pmatrix} \cos \lambda & \sin \lambda & 0 \\ -\sin \lambda & \cos \lambda & 0 \\ 0 & 0 & 1 \end{pmatrix} . \tag{A.3}
$$

Then, the rotation around the (new) y-axis by φ can be performed through

$$
\mathbf{C} = \begin{pmatrix} \cos \varphi & 0 & \sin \varphi \\ 0 & 1 & 0 \\ -\sin \varphi & 0 & \cos \varphi \end{pmatrix} . \tag{A.4}
$$

The reciprocal transformation $(0°, 0°) \rightarrow (\lambda, \varphi)$ can be obtained through the transposed matrix $\mathbf{A}'$.

Transformation of Earth Coordinates into Satellite View Angles. The transformation assumes the satellite is at zero longitude and latitude. Equivalently, its Cartesian coordinates are $(r, 0, 0)$. In all other cases the earth coordinate system must be rotated first (cf. Eq. (A.2)). Figure A.2 displays the geometry for the transformation of the earth coordinates into satellite view angles. A point P of the earth coordinate system is transformed into the spherical satellite coordinate system $e_{r'}, e_\vartheta, e_\sigma$ through two steps:

1. P is translated into the satellite-centered Cartesian coordinates $e_{x'}, e_{y'}$, and $e_{z'}$ by

$$
\mathbf{x}' = \mathbf{x} - \begin{pmatrix} r \\ 0 \\ 0 \end{pmatrix} . \tag{A.5}
$$

2. Using Eq. (A.1) the coordinates are now converted into satellite view angles.

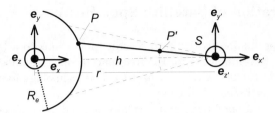

Fig. A.2. Transformation of Cartesian coordinates in a perspective satellite view

Transformation of Satellite View Angles into Earth Coordinates. This transformation is used for example to determine the spot beam projection of a satellite antenna pattern on earth. Usually, such patterns are given in true view angles. An auxiliary pattern sphere around the satellite is introduced at a distance r' from the satellite. Again the transformation assumes the satellite is at zero longitude and latitude. The following steps will carry out the transformation:

1. The view angles ϑ, σ of a point P' can be transformed into Cartesian satellite-centered coordinates $P' = (x'_{P'}, y'_{P'}, z'_{P'})$, cf. Eq. (A.1).
2. Through the translation

$$\boldsymbol{x}_{P'} = \boldsymbol{x}'_{P'} + \begin{pmatrix} r \\ 0 \\ 0 \end{pmatrix} \tag{A.6}$$

   the satellite-centered coordinates are changed into geocentric coordinates.
3. Let P denote the projection of P' onto the earth's surface. P can be determined through the intersection of the line $\overline{P'S}$

$$\boldsymbol{x} = \begin{pmatrix} r \\ 0 \\ 0 \end{pmatrix} + k \cdot \begin{pmatrix} x_{P'} - r \\ y_{P'} \\ z_{P'} \end{pmatrix} \tag{A.7}$$

with the earth's surface $x^2 + y^2 + z^2 = R_e^2$. The resulting quadratic equation for the parameter k is solved with

$$\begin{aligned}
k_{1,2} &= \frac{-b \pm \sqrt{b^2 - 4ac}}{2a} \quad \text{where} \\
a &= (x_{P'} - r)^2 + y_{P'}^2 + z_{P'}^2 \\
b &= 2 x_{P'} r - 2r^2 \\
c &= r^2 - R_e^2 .
\end{aligned} \tag{A.8}$$

The projected point P can be obtained from the insertion of k into Eq. (A.7). In order to resolve the ambiguity of the intersection the smaller value of $k_{1,2}$ must be used to obtain the earth intersection on the side closer to the satellite.

A.2 Generation of Satellite Spot Beams

Satellite footprints are divided into several spot beams to achieve (i) higher antenna gains to compensate for the free space propagation loss (Sect. 3.1.4), and (ii) a better frequency reuse (Chap. 6). The technology for the generation of spot beams is discussed in Chap. 8.

Footprints and spot beams are usually defined by a given value of the signal power on earth that must be exceeded. The signal power decreases according to the antenna characteristics if the user moves away from the main lobe center of the spot beam antenna. Thus, the antenna pattern $G(\vartheta, \sigma)$ as a function of the true view angles ϑ and σ determines the spot beam's border (usually a gain decrease of 3 dB is taken to define the border). For complete coverage the antenna profiles must be arranged in such a way that the whole footprint is filled with cells up to the wanted minimum elevation angle $\varepsilon_{\min}$.

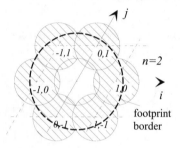

Fig. A.3. Cell arrangement using tier structures

Theoretically, hexagonal cells could be used to achieve coverage without overlapping. However, circular cells must usually be taken for technological reasons. The circular cells circumscribe the hexagons, thus resulting in cell overlapping. Regular patterns can be arranged using grids of equidistant lines that intersect at angles of 60°, see Fig. A.3. Then, the cell centers lie on the intersections denoted by the coordinates (i, j). Circular footprints can be filled with cells by arranging n cell tiers, cf. Sect. 6.1.1. The coordinates of the cell centers in the nth tier are a subset of the possible combinations of i and j:

$$\forall\{i,j\}: \quad \begin{cases} |i| = n-1 & \cap \quad j = -\text{sign}(i) \cdot (1 \ldots n-1) \\ |j| = n-1 & \cap \quad i = -\text{sign}(j) \cdot (0 \ldots n-2) \\ |i| = 1 \ldots n-1 & \cap \quad j = \text{sign}(i) \cdot (n-1-|i|) \;. \end{cases} \quad (A.9)$$

Examples. For a single cell ($n = 1$), the cell center is $(0, 0)$. For a footprint with two tiers ($n = 2$), the cell centers of the outer tier are $\{(1,0),(0,1),(-1,1),(-1,0),(0,-1),(1,-1)\}$, see Fig. A.3. For three tiers ($n = 3$), the cell centers of the second tier are $\{(2,0),(1,1),(0,2),(-1,2),(-2,2),(-2,1),(-2,0), (-1,-1),(0,-2),(1,-2),(2,-2),(2,-1)\}$ are obtained.

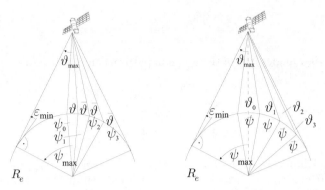

Fig. A.4. Spot beams with equal beamwidth (left) and equal spot area (right)

To transform these regular cell patterns into true view angles ϑ, σ of a satellite antenna, two concepts can be adopted: (i) the concept of equal beamwidth, or (ii) the concept of equal spot size [MDER91], see Fig. A.4. For patterns with equal beamwidth the cell arrangement in Fig. A.3 is interpreted as the true view angles of the spot beams in the spherical, satellite-centered coordinate system. Each spot beam has a constant beamwidth $\bar{\vartheta}$. The corresponding earth central angles ψ_i (see Eq. (2.23)) will increase in the outer tiers of the pattern. For patterns with equal spot size the cell arrangement in Fig. A.3 is interpreted in radio cells in spherical earth-centered coordinates. The earth central angles $\bar{\psi}$ of the cells are constant whereas the beamwidths ϑ_i are becoming smaller in the outer tiers.

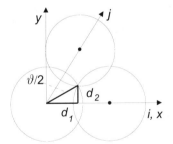

Fig. A.5. Calculation of cell centers

Patterns with Equal Beamwidth. The cell centers in Fig. A.3 are uniquely identified by their indices i, j. Figure A.5 defines the equivalent Cartesian coordinates x, y

$$\begin{aligned}
x &= 2id_1 + jd_1 \\
y &= 3j\,d_2
\end{aligned} \qquad (A.10)$$

with angular distances

$$d_1 = \frac{\sqrt{3}}{4}\bar{\vartheta} \quad \text{and} \quad d_2 = \frac{\bar{\vartheta}}{4} \ . \tag{A.11}$$

The true view angles ϑ, σ of the antenna pattern can now be obtained through the transformation

$$\vartheta = \sqrt{x^2 + y^2}$$
$$\sigma = \text{arctan4}\left(\frac{y}{x}\right) \tag{A.12}$$

using the definition in Eq. (2.18) for the arc tangent for four quadrants.

Gap-free coverage is provided by the cell pattern until the outer intersection of the outermost cells. These intersections define the nadir angle $\vartheta_{\max}$ of the satellite footprint. The nadir angle $\vartheta_{\max}$ can be calculated from $\bar{\vartheta}$ for an n-tier pattern by

$$\frac{\bar{\vartheta}}{2} = \begin{cases} \vartheta_{\max} & \text{for } n = 1 \\ \vartheta_{\max}/(\frac{3}{2}n - 1) & \text{for } n \text{ even} \\ \vartheta_{\max}/\sqrt{(\frac{3}{2}n - 1)^2 + \frac{1}{4}} & \text{for } n \neq 1 \text{ and odd.} \end{cases} \tag{A.13}$$

Cells with Equal Spot Size. The calculation is done similarly as for cells with equal beamwidth. However, the earth central angles $\psi, \bar{\psi}$ replace the nadir angle $\vartheta, \bar{\vartheta}$ in Eq. (A.11) and Eq. (A.12), respectively. Furthermore, the coverage angle $\psi_{\max}$ is determined by the earth central angles $\bar{\psi}$ of the cells for an n-tier pattern through

$$\frac{\bar{\psi}}{2} = \begin{cases} \psi_{\max} & \text{for } n = 1 \\ \psi_{\max}/(\frac{3}{2}n - 1) & \text{for } n \text{ even} \\ \psi_{\max}/\sqrt{(\frac{3}{2}n - 1)^2 + \frac{1}{4}} & \text{for } n \neq 1 \text{ and odd.} \end{cases} \tag{A.14}$$

The nadir angles ϑ_i can now be calculated using Eq. (2.24).

Example. Two examples for the generation of spot beam patterns with $n = 4$ tiers are shown in Fig. A.6, one antenna with equal beamwidth and the other with equal cell size. The antenna patterns were designed to provide coverage for a MEO satellite ($h = 10\,355$ km) with a minimum elevation of $\varepsilon = 20°$. Then, the nadir angle at the edge of coverage is $\vartheta_{\max} = 22.4°$.

Furthermore, the relations from Sect. A.1 have been applied to project the antenna patterns on to the earth's surface (lower figure). Spill-over effects of cells at the horizon are clearly visible for the pattern with equal beamwidth.

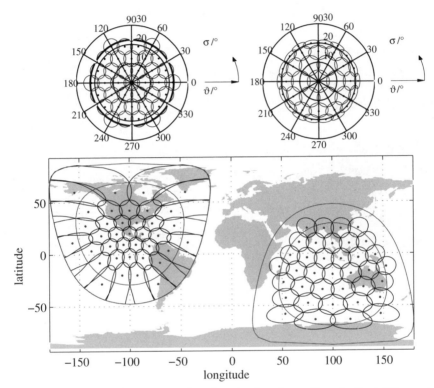

Fig. A.6. Generation (upper figure) and projection (lower figure) of satellite spot beam antennas with 37 cells of equal beamwidth (left) and equal cell size (right). The patterns provide coverage up to a minimum elevation of 20° for a MEO satellite at 10 355 km altitude. The projection centers have been chosen at ±100° east, ∓20° north. The edge of coverage at 20° elevation and the earth horizon are indicated as well

B. Parameters of the Land Mobile Satellite Channel

This appendix provides tables with parameters for the narrowband and wideband models of the land mobile satellite channel as discussed in Chap. 3. The data are given for several frequency bands and operational scenarios.

B.1 Narrowband Two-State Model at L Band

The following tables B.1, B.2, B.3, B.4, B.5, and B.6 provide the parameters of the two-state model for the land mobile satellite channel at L band after [LCD+91] for reception with handheld phones and vehicular terminals in several environments.

Table B.1. Parameters of the two-state model for several environments after [LCD+91], vehicular terminal with roof-mounted antennas, L band

Elevation	Time-share of shadowing A	Rice factor c /dB	μ_{dB} /dB	σ_{dB} /dB
Environment: highway				
13°	0.24	10.2	-8.9	5.1
21°	0.03	16.1	-7.1	5.5
24°	0.25	11.9	-7.7	6.0
34°	0.008	11.7	-8.8	3.8
43°	0.002	14.8	-12.0	2.9
13°				
Environment: urban				
13°	0.89	3.9	-11.5	2.0
18°	0.80	6.4	-11.8	4.0
24°	0.66	6.0	-10.8	2.8
34°	0.58	6.0	-10.6	2.6
43°	0.54	5.5	-13.6	3.8

Table B.2. Parameters of the two-state model for the suburban environment, handheld telephone, L band

Elevation	Time-share of shadowing A	Rice factor c /dB	μ_{dB} /dB	σ_{dB} /dB
$15°$	0.77	5.4	-8.4	4.4
$25°$	0.59	4.8	-6.5	2.5
$35°$	0.54	6.0	-5.8	2.6
$45°$	0.76	7.0	-5.1	2.1
$55°$	0.35	7.0	-4.3	1.1
$65°$	0	7.8	-	-
$75°$	0.10	8.5	-4.7	1.0

Table B.3. Parameters of the two-state model for the suburban environment, vehicular terminal with roof-mounted antenna during drive, L band

Elevation	Time-share of shadowing A	Rice factor c /dB	μ_{dB} /dB	σ_{dB} /dB
$15°$	0.54	4.8	-8.4	1.8
$25°$	0.14	7.0	-9.2	1.3
$35°$	0.34	7.8	-7.6	1.0
$45°$	0.32	7.8	-	-
$55°$	0.28	6.0	-	-
$65°$	0.32	4.8	-	-
$75°$	0.11	6.0	-	-

Table B.4. Parameters of the two-state model for the urban environment, handheld telephone, L band

Elevation	Time-share of shadowing A	Rice factor c /dB	μ_{dB} /dB	σ_{dB} /dB
$15°$	0.97	4.8	-15.6	5.3
$25°$	0.79	4.8	-9.9	5.4
$35°$	0.60	5.4	-7.6	3.7
$45°$	0.56	6.0	-4.5	1.5
$55°$	0.30	7.0	-3.6	0.3
$65°$	0	9.0	-	-
$75°$	0	4.8	-	-

Table B.5. Parameters of the two-state model for the urban environment, vehicular terminal with roof-mounted antenna during drive, L band

Elevation	Time-share of shadowing A	Rice factor c /dB	μ_{dB} /dB	σ_{dB} /dB
15°	0.55	7.0	-12.7	4.8
25°	0.54	6.0	-17.9	4.3
35°	0.36	8.5	-13.8	4.3
45°	0.01	10.8	-	-
55°	0	6.0	-	-
65°	0	7.0	-	-
75°	0	8.5	-	-

Table B.6. Parameters of the two-state model for the highway environment, vehicular terminal with roof-mounted antenna during drive, L band

Elevation	Time-share of shadowing A	Rice factor c /dB	μ_{dB} /dB	σ_{dB} /dB
15°	0.24	10.2	-8.9	5.1
25°	0.19	11.9	-7.7	6.0
35°	0.09	11.7	-8.8	3.8
45°	0.07	14.8	-12.0	2.9
55°	0	-	-	-
65°	0	-	-	-
75°	-	-	-	-

B.2 Narrowband Two-State Model at EHF Band

The following tables B.7 and B.8 provide parameters of the two-state model
after [LCD$^+$91] for the land mobile satellite channel at EHF band for reception with a vehicular terminal in the rural and urban environment.

Table B.7. Parameters of the two-state model for a rural tree-lined road, vehicular
terminal with steered high-gain antenna, EHF band. Note: The increase of the
shadowing factor with elevation is caused by trees with overhanging branches

Elevation	Time-share of shadowing A	Rice factor c /dB	μ_{dB} /dB	σ_{dB} /dB
25°	0.21	21.5	-15.4	2.2
35°	0.40	22.3	-17.1	4.4
45°	0.81	20.6	-25.3	7.2

Table B.8. Parameters of the two-state model for the urban environment, vehicular
terminal with steered high-gain antenna, EHF band

Elevation	Time-share of shadowing A	Rice factor c /dB	μ_{dB} /dB	σ_{dB} /dB
15°	0.71	13.2	-21.5	2.1
25°	0.17	16.8	-26.1	3.3
35°	0.24	15.7	-27.2	1.6
45°	0.08	14.3	-17.50	1.2

B.3 Wideband Model at L Band

The following tables B.9, B.10, and B.11 provide parameters of the wideband tapped delay-line model for the land mobile satellite channel at L band after [BJL96a] for reception with a handheld terminals (except for the highway environment, where a vehicular terminal with a roof-mounted antenna was used) in several environments.

Table B.9. Parameters for the direct path signal of the wideband mobile satellite channel model for handheld phones at L band. Note: The increase of the shadowing factor with elevation in the rural environment is caused by trees with overhanging branches

Parameter:		Time-share of shadowing	Amplitude PDFs$|h_1(t)|$		
			LOS: Rice	nLOS: Rayleigh lognormal	
Environment	Elev.	A	c /dB	μ_{dB} /dB	σ_{dB} /dB
Open	15°	0	6.0	-	-
	25°	0	10.3	-	-
	35°	0	12.0	-	-
	45°	0	10.4	-	-
	55°	0	9.0	-	-
Rural road	15°	0.99	7.0	-11.4	1.1
	25°	0.96	10.8	-9.9	3.3
	35°	0.83	4.8	-6.2	3.9
	45°	0.79	4.7	-5.4	2.3
	55°	0.93	4.8	-7.2	3.9
Suburban	15°	0.77	-	-12.6	4.8
	25°	0.59	4.7	-6.0	3.5
	35°	0.54	10.7	-7.6	3.2
	45°	0.43	4.0	-7.2	3.2
	55°	0.35	11.8	-7.7	2.6
Urban	15°	0.97	-	-15.2	5.2
	25°	0.79	3.2	-12.1	6.3
	35°	0.6	4.8	-4.4	5.1
	45°	0.56	8.5	-3.0	2.7
	55°	0.3	6.0	-	-
Highway	15°	0.24	9.5	-9.3	5.6
	25°	0.19	8.4	-5.8	1.7
	35°	0.01	8.5	-5.0	3.3
	45°	0	7.8	-1.8	1.2
	55°	0	9.0	-	-

Table B.10. Parameters for the near-range echoes of the wideband mobile satellite channel model for handheld phones at L band

Parameter:		$N^{(n)}$ Poisson PDF λ	Max. delay τ_e/ns	$f_{\exp}(\varDelta\tau_i)$ exp. PDF $b/\mu s$	$P_h(\tau)$ exp. decay	
Environ.	Elev.				$P_{h,0}$/dB	$d/\mu s$ dB
Open	15°	1.6	400	0.033	-28.5	3.0
	25°	1.2	400	0.03	-28.6	1.0
	35°	1.2	400	0.027	-25.7	9.5
	45°	0.5	400	0.027	-29.0	1.1
	55°	-	400	-	-	-
Rural road	15°	1.8	400	0.061	-25.9	10.7
	25°	1.5	400	0.055	-24.9	19.2
	35°	1.6	400	0.043	-25.3	14.1
	45°	1.8	400	0.051	-24.5	13.4
	55°	1.6	400	0.047	-21.7	36.8
Suburban	15°	1.2	400	0.037	-22.6	-21.9
	25°	1.4	400	0.038	-23.8	23.7
	35°	1.2	400	0.039	-24.9	19.4
	45°	1.5	400	0.027	-24.4	23.0
	55°	1.6	400	0.033	-24.7	18.7
Urban	15°	1.2	600	0.118	-16.5	11.0
	25°	4.0	600	0.063	-17.0	26.2
	35°	3.5	600	0.069	-23.6	6.5
	45°	3.6	600	0.081	-23.5	8.5
	55°	3.8	600	0.079	-26.1	6.3
Highway	15°	1.2	600	0.072	-27.0	6.4
	25°	2.2	600	0.077	-25.8	7.3
	35°	2.8	600	0.091	-26.8	30.6
	45°	1.8	600	0.043	-27.1	29.5
	55°	-	-	-	-	-

Table B.11. Parameters for the far-range echoes of the wideband mobile satellite channel model for handheld phones at L band

Background	Elev.	$N^{(f)}$: Poisson λ	$h_i(t)$: Rayleigh $2\sigma^2/\mathrm{dB}$	Max. delay $\tau_{\max}/\mu\mathrm{s}$
Flat	15°			15
	25°	0.3	-26.4	15
	35°			15
	45°			15
	55°			15
Rural	15°			5
	25°	0.8	- 28.2	5
	35°			5
	45°			5
	55°			5
Hilly	15°	-	-	10
	25°	1.2	-29.0	10
	35°	-	-	10
	45°	-	-	10
	55°	-	-	10
Mountainous	15°	0.9	-29.0	15
	25°	1.8	-28.5	15
	35°	4.4	-23.5	15
	45°	4.0	-21.7	15
	55°	-	-	15

Table B.12 provides the parameters of the wideband ITU model at L band after [BJL96c] for reception with a handheld terminal. The model uses a fixed number of echoes with given delay. The environments urban, rural, and suburban correspond to the 90%, 50%, and 10% percentiles of the delay spread.

Table B.12. Parameters of the wideband ITU model for handheld reception at L band

90% percentile of delay spread: urban environment

Echo No.	Delay $\Delta\tau_i$ / ns	Amplitude PDF	Parameter of PDF	Value / dB
1	0	LOS: Rice	c	5.2
		nLOS: Rayleigh	$2\sigma^2$	-12.1
2	60	Rayleigh	$2\sigma^2$	-17.0
3	100	Rayleigh	$2\sigma^2$	-18.3
4	130	Rayleigh	$2\sigma^2$	-19.1
5	250	Rayleigh	$2\sigma^2$	-22.1

50% percentile of delay spread: rural environment

Echo No.	Delay $\Delta\tau_i$ / ns	Amplitude PDF	Parameter of PDF	Value / dB
1	0	LOS: Rice	c	6.3
		nLOS: Rayleigh	$2\sigma^2$	-9.5
2	100	Rayleigh	$2\sigma^2$	-24.1
3	250	Rayleigh	$2\sigma^2$	-25.2

10% percentile of delay spread: suburban environment

Echo No.	Delay $\Delta\tau_i$ / ns	Amplitude PDF	Parameter of PDF	Value / dB
1	0	LOS: Rice	c	9.7
		nLOS: Rayleigh	$2\sigma^2$	-7.3
2	100	Rayleigh	$2\sigma^2$	-23.6
3	180	Rayleigh	$2\sigma^2$	-28.1

C. Existing and Planned Satellite Systems

C.1 Survey of Satellite Systems

Table C.1. Operational geostationary satellite systems for speech and data transmission in the L band

System	Service	Region	Satellites	Start of service
Inmarsat-A	speech	global	4 GEOs	1982
Inmarsat-B	speech	global	4 GEOs	1993
Inmarsat-M	speech	global	4 GEOs	1993
Inmarsat-3	speech	global	4 GEOs (3rd gen.)	1996
Inmarsat GAN	mobile ISDN mobile IP	global	4 GEOs (3rd gen.)	1999
Inmarsat Aero-H	speech	global	4 GEOs	1990
Inmarsat Aero-I	speech	global	4 GEOs (3rd gen.)	1997
Inmarsat-C	data	global	4 GEOs	1991
Inmarsat-D/D+	paging	global	4 GEOs (3rd gen.)	1996
Inmarsat-E	emergency	global	4 GEOs	1997
Omnitracs	fleet management	USA	2 GEOs	1989
Euteltracs	fleet management	Europe	2 GEOs	1991
MobileSat	speech	Australia	2 GEOs	1994
Movisat	speech	Mexico	3 GEOs	1994
MSAT	speech	USA+Can.	1 GEO	1995
ACTEL	speech	Africa	2 GEOs	1998
Marathon	speech	CIS/USA	5 GEOs + 4 HEOs	?
N-Star (NTT)	speech	Japan	2 GEOs	1996

Table C.2. Planned geostationary satellite systems for speech and data transmission in the L band

System	Region	Satellites	Beams per satellite	Terminals	Start of service
ACeS	Asia	1 GEO	140	handheld	2000
APMT	China	1 GEO		handheld	2000
ASC/Agrani	India	2 GEOs	250	handheld	2001/2002
EAST	Africa	2 GEOs	150	handheld	2002
Inmarsat-4	USA, Asia, Europe	2 GEOs		handheld, laptop	2004
LLM (ESA)	Europe	1 GEO (Artemis)	3 spots + 1 Euro	20×40 cm	2000
N-Star C (S band)	Japan	1 GEO		mobile	2002
Satphone	Mediterr.	3 GEOs	150	handheld	1999
Thuraya	Middle East, etc.	1 GEO	256	handheld	2000/2001

Table C.3. Planned and existing LEO/MEO satellite systems for worldwide mobile telephony in the L and S band (big-LEOs)

System	Satellites	Beams per satellite	Access scheme	Start of service
Iridium	66 LEOs at 780 km polar orbits	48	TDMA	Nov. 1998
Globalstar	48 LEOs at 1414 km inclined orbits	16	CDMA	1999/2000
ICO	10 MEOs at 10 390 km inclined orbits	163	TDMA	2001
Ellipso	10 HEOs, inclined + 6 equatorial MEOs	61	CDMA	2003
Constellation	46 LEOs, equatorial and inclined	24 32	CDMA	2001 2003
Courier	72 LEOs at 800 km inclined orbits	37	CDMA	2001
Gonets	45 LEOs at 1400 km inclined orbits		TDMA	pre-operat-ional 1995

Table C.4. Satellite systems for IMT-2000/UMTS, at 2 GHz

System	Satellites	Bit rate	Status
Boeing	16 MEOs at 20 180 km	communication and navigation	service begins 2002
Celsat (Echostar)	3 GEOs North America		FCC filing Sept. 1997
GS-2 (Globalstar)	64 LEOs at 1420 km + 4 GEOs	144 kb/s	FCC filing Sept. 1997
Horizons (Inmarsat)	3 GEOs for continents	144 kb/s (portable unit)	service begins 2002
Iridium Next Generation	LEOs or MEOs	144 kb/s	ITU proposal Sept. 1998
Macrocell (Motorola)	96 LEOs at 850 km polar orbits	64 kb/s	service begins 2003
S. Korea TTA	49 LEOs at 2000 km inclined orbits	9.6/64/144 kb/s	ITU proposal June 1998
ESA SW-CDMA	for LEO/MEO constellations	32 kb/s handheld 144 kb/s portable	ITU proposal June 1998
ESA SW-C/TDMA	for regional GEO/HEOs	183 kb/s	ITU proposal June 1998

Table C.5. LEO satellite systems for worldwide low-rate data communication (little-LEOs)

System	Satellites	Frequency band	Start of service
Orbcomm	28 LEOs at 825 km inclined orbits	up: 150 MHz down: 140/400 MHz	1996 USA 1998 global
Vita	3 LEOs at 1000 km polar orbits	up: 150 MHz down: 400 MHz	pre-operational
E-Sat	6 LEOs at 1261 km polar orbits	up: 150 MHz down: 140 MHz	2001/2002
Final Analysis	26 LEOs at 1000 km inclined orbits	up: 150 MHz down: 400 MHz	2002/2003
Leo One	48 LEOs at 950 km inclined orbits	up: 150 MHz down: 140 & 400 MHz	2002/2003
Safir (OHB Bremen)	6 LEOs at 690 km polar orbits	VHF/UHF	1994

Table C.6. Existing GEO satellite systems dedicated to the provision of fixed Internet services

System	Coverage region	Satellite(s)	Satellite download link	Request link	Start of service
Astra ARCS	Europe, West Asia	Astra 1H, Astra 1K	max. 38 Mb/s (Ku band)	max. 2 Mb/s (satellite, Ka band)	2000
Gilat	North America		40 Mb/s	154 kb/s (satellite)	2000
Hughes DirecPC	Europe USA	Eutelsat Hotbird 3, Galaxy IV	typ. 400 kb/s max. 6 Mb/s (Ku band)	telephone	1997 2000
Web-Sat	Europe	Eutelsat W3	typ. 350 kb/s max. 4 Mb/s (Ku band)	16 kb/s (satellite)	1999

Terminal antenna diameters range from 40 to 100 cm.

Table C.7. Broadband satellite systems in the Ka band (also Ku band). Also, the organizations Eutelsat, Intelsat, Orion, and PanAmSat intend to expand their existing satellite fleets into the Ka band

System	Satellites	Coverage region	Satellite capacity	System setup
Astrolink	9 GEOs	global	6.5 Gb/s	2002–2003
Contact	16 MEOs	global		2006
Cyberstar	3 GEOs	multiregional	4.9 Gb/s	1998–2001
EchoStar	2 GEOs	USA	5.8 Gb/s	
EuroSkyWay	5 GEOs	Europe	8 Gb/s	2003
GE Star	9 GEOs	multiregional	4.7 Gb/s	2002
HALO	aircraft	cities	10 Gb/s	2000–2001
HughesLINK (Ku Band)	22 MEOs			FCC filing Jan. 1999
HughesNET (Ku Band)	70 LEOs			FCC filing Jan. 1999
iSky (Ka-Star)	2 GEOs	USA	7.5 Gb/s	2001–2002
Morning Star	4 GEOs	multiregional	0.5 Gb/s	FCC-licensed
NetSat 28	1 GEO	USA	40 Gb/s	2002
SkyBridge (Ku Band)	80 LEOs at 1457 km	global	1 Gb/s per beam	2001–2002
Spaceway	8 GEOs	global	4.4 Gb/s	2002–2003
Teledesic	288 LEOs	global	10 Gb/s	2004
VIRGO (Ku Band)	15 ellipt. MEOs			2004–2005
Vision Star	1 GEO	CONUS	1.9 Gb/s	FCC-licensed
WEST (Matra Marconi)	2 GEOs + 9 MEOs	Europe global	9 Gb/s	2002–2005

Table C.8. Broadband satellite systems in the V band (EHF band). FCC filing for all systems: Sept. 1997

System	Satellite constellation	Data rate (small terminal)	Start of service
Aster	25 GEOs	51 Mb/s	2002
Cyberpath (Loral)	4 ... 10 GEOs	3 Mb/s up 90 Mb/s down	
Expressway (Hughes)	14 GEOs	155 Mb/s	2003
GE Star Plus	11 GEOs	1.5 ... 155 Mb/s	2003
GESN (TRW)	4 GEOs + 15 MEOs	155 Mb/s	2002
GS-40 (Globalstar)	80 inclined LEOs at 1440 km	1.2 Mb/s	≈ 2004
Q/V Band (Lockheed Martin)	9 GEOs	9 Mb/s	
Orblink (Orbital Sciences)	7 equator. MEOs at 9000 km	10 Mb/s	2002
Sky Station	250 blimps at 21 km	2 Mb/s up 10 Mb/s down	2001
SpaceCast (Hughes)	6 GEOs	155 Mb/s	
StarLynx (Hughes)	4 GEOs + 20 MEOs	2 Mb/s	≈ 2005
VBS (Teledesic)	72 inclined LEOs at 1375 km	10 ... 100 Mb/s	≈ 2006
V-Stream (PanAmSat)	12 GEOs	1.5 Mb/s	

C.2 ACeS (Asia Cellular Satellite)

ACeS, owned by ACeS International Ltd, is a regional GEO system for mobile telephony in South East Asia, India, and China. Financial partners are

- PT Pasifik Satelit Nusantara
- Philippine Long Distance Telephone Company
- Jasmine International, Thailand
- Lockheed Martin

in roughly equal parts. In Oct. 1996, ACeS was fully financed. Industrial partners are:

- Lockheed Martin Telecommunications, holding a US-$ 650 million contract for delivery of one satellite, two control centers, and GWs
- Spar Aerospace, providing satellite antennas
- Alcatel CIT, Comsat RSI, and IEX Corp., contributing to the ground segment.

The following service providers are envisaged:

- PT Pasifik Satelit Nusantara
- Philippine Long Distance Telephone Company
- Jasmine International.

MACH (Multinational Automated Clearing Houses S.A.) will be the data clearing house for ACeS, the service providers, and network operators.

ACeS will provide personal communications via voice (6 kb/s incl. FEC), fax (group 3), data (2.4 kb/s), and paging services. For broadcasting messages and paging a high-penetration signal with 30 dB link margin will be used. Mobile-to-mobile connections will be provided via single hops. The main market perspective of ACeS is to provide fill-in services for terrestrial cellular systems.

ACeS uses the "Garuda" GEO satellite (see Fig. 8.2) positioned at 123° east. Garuda is characterized by the following features:

- A2100 AX bus, Lockheed Martin Astro Space
- launch mass 4500 kg
- life cycle 12 years
- DC peak power 12 kW EOL
- two L band antennas with 12 m diameter (mobile link)
- 140 mobile link spot beams, 400 ... 500 km diameter, cluster size 7
- one feeder link beam
- $G/T = 15.3$ dB/K
- satellite RF power (mobile link): 88 SSPAs @ 20 W
- transparent transponder
- on-board processing.

ACeS is based on GSM protocols, provides a satellite capacity of 11 000 duplex channels, and is dimensioned for 2 million users. 20% of the satellite capacity can be concentrated in a single spot beam (power-limited). No satellite/terrestrial handover is possible.

The L band mobile link uses 1626.5–1660.5 MHz for the uplink (return link) and 1525–1559 MHz for the downlink (forward link). The downlink burst format (156.25 bits in 0.577 ms) is similar to the GSM format. The uplink is organized in MF-TDMA; the uplink burst rate is a fraction of the downlink burst rate. Hence, several uplink carriers correspond to one downlink carrier. The uplink provides a link margin of 10 dB (7 dB for mobile-to-mobile).

The feeder link uses the C band at 6425–6725 MHz for the uplink and 3400–3700 MHz for the downlink.

The ground segment includes a satellite and network control center located in Indonesia, having an antenna diameter of 15.5 m. Gateways are located in Indonesia (Jakarta), the Philippines, and Thailand.

The terminals are handheld with SIM cards, providing a mean transmit power of 0.25 W and exhibiting an antenna gain of 2 dBi. Dual-mode terminals (ACeS/GSM or ACeS/AMPS) will be manufactured by Ericsson. Additionally, mobile terminals and satellite telephone boots can be used.

The development of the ACeS system can be characterized by the following steps:

1993:	ITU filing
1995:	contract to Lockheed Martin
Oct. 1996:	ACeS is fully financed
1997:	preliminary rural satellite telephone service "xpress connection"
Feb. 2000:	Garuda I satellite launched by Proton
April 2000:	start of commercial service (as of Oct. 1999).

The cost of the ACeS system amounts to US-$ 900 million; the service charge will be around US-$ 0.5 per minute.

More detailed information on the ACeS system can be found in [NBA97] and at http://www.aces.co.id.

C.3 Astrolink

Astrolink, owned by Astrolink International LLC, is a global broadband GEO Ka band system for fixed on demand services. In May 1999, an initial equity of US-\$ 900 million was provided by the financial partners:

- Lockheed Martin Global Telecommunications (US-\$ 400 million)
- Telespazio/Telecom Italia (US-\$ 250 million)
- TRW (US-\$ 250 million)
- In Oct. 1999, Liberty Media invested another US-\$ 425 million, increasing the total investment to US-\$ 1.325 billion.

Overall, the system cost is planned to be covered by 40% equity and 60% debt. The investors also act as industrial partners:

- Lockheed Martin Global Telecommunications will deliver 4 satellites and provide the system integration and the satellite launches.
- Telespazio is responsible for the gateway stations, the network control center, and the satellite control center.
- TRW will manufacture the satellite payload.
- In addition, Scientific-Atlanta will build the gateway stations.

Service providers will be Lockheed Martin Global Telecommunications, Telecom Italia, and TRW.

Astrolink will concentrate on the provision of high-speed Internet services and flexible, high-quality multimedia services (bandwidth-on-demand) for the following application scenarios:

- multinational companies
- SOHO (small office, home office)
- e-commerce
- desktop video conferencing
- business TV
- LAN-to-LAN intranets
- virtual private corporate networks
- business multicasting
- distance learning
- telemedicine
- email
- telephony.

Astrolink offers an average availability of 99.8% and provides ATM-based services with a cell loss ratio of 10^{-6}.

9 GEO satellites will be launched into 5 orbital positions, cf. Fig. C.1:

- 2 satellites at 97°W, for the Americas
- 2 satellites at 21.5°W, for the Atlantic area
- 2 satellites at 2°E, for Europe, Africa, and West Asia

- 2 satellites at 130°E, for East Asia and Australia
- 1 satellite at 175.25°E, for Oceania.

In a first phase, 4 satellites will be launched, in a second phase, the remaining 5 satellites will be added.

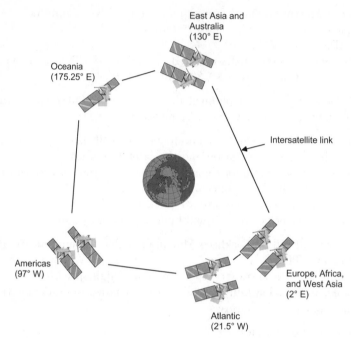

Fig. C.1. Astrolink geostationary satellite constellation

The satellites are characterized by the following features:

- A2100 bus, Lockheed Martin
- satellite dry mass: 2185 kg
- satellite life: 12 ... 15 years
- DC peak power: 10.5 kW
- 4 uplink and 4 downlink multibeam reflector antennas
- 24 ... 88 uplink/downlink beams, beamwidth 0.8°.
- frequency reuse cluster size 4
- satellite RF power 56 W per carrier, generated by a TWTA
- EIRP: 56 dBW
- on-board processing with ATM-based fast packet switching in the baseband, comprising the following steps: resource control, channelization, demodulation, decoding, ATM fast packet switching, encoding, modulation
- satellite throughput 6.5 Gb/s
- optical ISLs

− 14 GW beams per satellite.

Overall, the satellites will provide (quasi-)global coverage with a minimum elevation of 20°.

The Astrolink protocol architecture is based on ATM. The Ka band user links have the following features:

− uplink:
  − 29.5–30 GHz
  − QPSK
  − MF-TDMA, one ATM cell per time slot
− downlink:
  − 19.7–20.2 GHz
  − QPSK
  − adaptive TDM, 4 ATM cells per packet.

Three types of user terminals are envisaged, according to Table C.9. They provide a download speed of up to 110 Mb/s.

Table C.9. Astrolink user terminals

Terminal:	Class A	Class B	Class C
Application	SOHO	medium enterprise	major enterprise
Antenna size	90 cm	90 cm	
Transmit power	2 W	10 W	
Data rate	16 ... 384 kb/s	16 kb/s ... 2.3 Mb/s	16 kb/s ... 20 Mb/s
Price	US-$ 1000		US-$ 8000

Also the feeder links use the Ka band: 28.35–28.6 and 29.25–29.5 GHz for the uplink and 18.3–18.8 GHz for the downlink. Astrolink will include 30 ... 50 gateway earth stations with 3 ... 5.5 m antenna diameter, providing protocol conversions for T1/E1, TCP/IP, frame relay, X.25, and others.

The development of the Astrolink system can be characterized by the following steps:

Sept. 1995:	FCC filing for space segment
May 1997:	FCC license for space segment received
July 1999:	regulatory approval received from the EC
2002:	first satellite launch (as of May 1999)
2003:	start of operation in Europe and America (as of May 1999)
2004:	full deployment (as of Nov. 1999).

As of May 1999, the cost for the first phase of the Astrolink system with 4 satellites is estimated at US-$ 3.6 billion. The service charge will be usage-based and will range between US-$ 0.05 and US-$ 0.5 per Mbyte of transmitted data. More detailed information on the Astrolink system can be found in [EGMG97] and at http://www.astrolink.com.

C.4 EuroSkyWay

EuroSkyWay is a European Ka band GEO system for fixed and mobile multimedia communications. The following services are envisaged:

- (mobile) Internet, WWW
- multimedia (video, audio, data)
- private networks
- telephony
- enhanced email.

Typical applications are:

- teleworking
- telemedicine
- tele education
- entertainment.

Customers can be connectivity customers, such as telecom operators and Internet service providers, companies, and end users.

EuroSkyWay will be based on 5 GEO satellites: In a first phase, 2 co-located satellites will be installed at 5° east, and in a second phase, 3 additional satellites will be placed at 30° east. The satellites will be equipped with on-board processing, including a T-stage switch. They will provide coverage to extended Europe (32 spot beams), see Fig. 12.3, and to the North Atlantic area (46 spot beams). In the uplink, a cluster size of 4 will be used; in the downlink the cluster size is 3.

Each satellite provides a gross throughput of 9 Gb/s, giving a total system throughput of 45 Gb/s. The EuroSkyWay satellites are interconnected by ISLs at 56–64 GHz; other satellite systems can be connected via interlink stations.

EuroSkyWay can act as an access network and as the core network of an integrated broadband communication (IBC) system, and supports several protocols, such as TCP/IP, ATM, X.25, frame relay, and DVB. The most time-critical functions, such as cell commutation and traffic management, are hosted on-board to minimize response times. Connection management and mobility management are hosted on-ground, e.g. in the master control station (MCS), Fig. C.2.

The user links use the Ka band at 27.5–30.0 GHz for the uplink and at 17.7–20.2 GHz for the downlink. The uplink is organized in synchronous MF-TDMA (no guard time, no preamble). A satellite data packet contains a payload of 53 bytes (the size of a full ATM cell) plus 11 bytes of signaling overhead. These uplink data packets are RS-coded into 80 bytes. In the downlink a concatenation of RS(204,188,16) and rate-3/4 convolutional coding is applied.

Different classes of user terminals are foreseen, Fig. C.3:

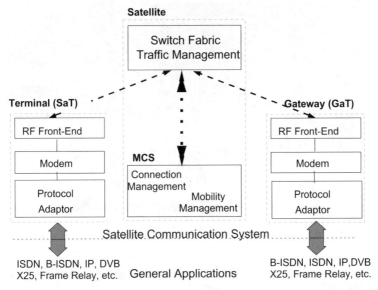

Fig. C.2. Distributed switching architecture of the EuroSkyWay system [PFV⁺99]
©1999 IEEE

- portable laptop-sized Sat-A terminals provide 160 kb/s transmit rate and 2 Mb/s receive rate
- standard PC-sized Sat-B terminals allow 512 kb/s transmit rate and 2 Mb/s receive rate
- high-capacity PC-sized Sat-C terminals provide a rate of 2 Mb/s in transmit and receive directions.

In addition, land mobile and aeronautical terminals will be available, as well as extension boards to PCs.

The development of the EuroSkyWay system can be characterized by the following steps:

Sept. 1995:	ITU filing for 7 GEO slots
Apr. 1996:	preliminary authorization by the Italian Ministry of PTT to construct, launch, and operate the network at orbital slots at 5° east and 30° east
Jan. 1997:	creation of the EuroSkyWay company
Sept. 1997:	official publication of the filed slots at the ITU, for frequency coordination
1998:	system financing
1998:	pre-operational service in the frame of the ACTS project ISIS
2001–2002:	satellite launches
2003:	start of operation with 2 satellites (as of July 1999).

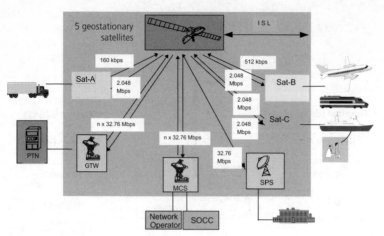

Fig. C.3. System architecture of the EuroSkyWay system

More detailed information on the EuroSkyWay system can be found in [PFV$^+$99] and at http://194.20.199.3.

C.5 Globalstar

Globalstar is a simple and relatively cheap LEO system for extending terrestrial cellular networks. The Globalstar Ltd Partnership comprises the following companies:

- Loral Space & Communications Ltd is the managing partner (with a 42% stake in Globalstar).
- Space Systems/Loral is the prime contractor for the space segment, designs the satellites, and operates the satellites and ground stations.
- Qualcomm Inc. is responsible for the ground segment and terminals, develops gateways, and is the provider of the CDMA technology used in Globalstar.
- Alcatel Espace develops the satellite payload and the internetworking.
- Alenia Spazio performs satellite integration and test.
- DASA delivers the satellite bus systems, solar generators, orbit and attitude control modules, and propulsion modules.
- Vodafone acts as network operator.
- Hyundai delivers satellite subsystems.
- Airtouch, DACOM, Elsag Bailey, and France Telecom are financial partners.

Globalstar has raised US-$ 3.8 billion and is fully financed. Among the Globalstar service providers, AirTouch and Vodafone are responsible for the USA and some other countries. Elsacom is the service provider for several European countries including Germany and Italy. Also TESAM, a joint venture of France Telecom and Alcatel, acts as service provider in Europe. Globalstar provides the following services:

- mobile and fixed telephony with an adaptive net bit rate of 0.6 ... 4.8 kb/s (9.6 kb/s for high-quality speech)
- group 3 fax
- real-time data up to 9.6 kb/s net rate
- paging, messaging (short message service, voice message service)
- position determination using integrated GPS.

The market segments of Globalstar are areas underserved or not served at all by existing wireline and cellular telecommunication systems, particularly in developing countries. The Globalstar satellite constellation comprises 48 active LEO satellites plus 4 in-orbit spares, cf. Fig. 2.14. The active satellites are placed in 8 circular orbits with 1414 km altitude and 52° inclination, constituting a 48/8/1 Walker constellation. The in-orbit spares are parked in a common 900 km parking orbit.

The satellites are characterized by the following features:

- satellite in-orbit wet mass: 450 kg
- payload mass: 152 kg

- life cycle: 7.5 years
- DC power consumption: 650 ... 2260 W, average: 1100 W, EOL
- 16 transparent transponders with 3 ... 16.5 MHz adjustable channel bandwidth
- 16 isoflux active phased-array antennas for the mobile link, with 61 elements for the L band receive antenna and 91 elements for the S band transmit antenna, cf. Fig. 8.4
- maximum transmit power: 91 × 4.2 W = 380 W
- global antennas for feeder link, 105° view angle
- satellite position determination with GPS (20 m accuracy)
- launches with Delta II, NPO Yuzhnoye Zenith-2, and Soyuz.

The satellite coverage area extends from 70° north to 70° south, cf. Fig. 2.20. The diameter of a satellite footprint is 5760 km at a minimum elevation of 10°. Globalstar provides 100% double coverage within CONUS (Continental USA). In order to obtain Globalstar service, a user must also be within the service area of a gateway, having a diameter of 3000 km. No full coverage of the Atlantic and Pacific Oceans is foreseen.

Globalstar can use the following frequency bands:

- terminal to satellite: 1610–1626.5 MHz (L band)
- satellite to terminal: 2483.5–2500 MHz (S band)
- gateway to satellite: 5091–5250 MHz (C band)
- satellite to gateway: 6875–7055 MHz (C band).

In the feeder link, the radio channels for the 16 spot beams are separated in frequency and in polarization. All spot beams in the user link use the same frequency band, Fig. C.4.

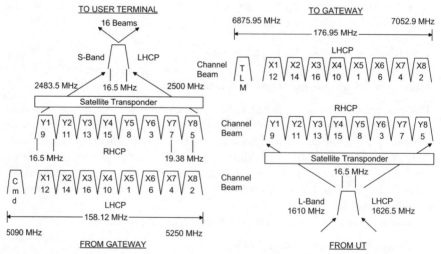

Fig. C.4. Frequency plan for the Globalstar forward and return link [Glo95]

Each Globalstar satellite can provide 2500 communication channels (this number is limited by the transmit power) with frequency reuse in every spot beam. Up to 25% of the satellite capacity can be concentrated in a single spot beam. The Globalstar CDMA multiple access scheme is based upon the IS-95 CDMA standard; it allows soft beam and satellite handovers, as well as satellite diversity.

In the S band mobile downlink, the following transmission scheme is used:

- convolutional coding with rate 1/2 and $K = 9$
- scrambling with a long PN sequence, the offset (code phase) representing the user address
- multiplication of each 9600 b/s transmission symbol with a Walsh function of length 128 giving rise to a chip rate of $128 \times 9600 = 1.23$ Mch/s (the Walsh function is determined by the channel allocated to the user)
- QPSK modulation
- max. 13 CDMA carriers with a bandwidth of 1.23 MHz in FDM, resulting in a total bandwidth of 16.5 MHz (MF-CDMA). Actually, less than 13 carriers are used because of the FCC L band separation, Fig. 9.2.
- a forward synchronous CDMA carrier contains 1 pilot channel for timing phase and signal power, 1 synchronization channel for time synchronization, up to 7 paging channels, and 119 traffic channels.

In the L band mobile uplink, the following transmission scheme is used:

- convolutional coding with rate 1/3 and $K = 9$
- grouping of the coded sequence into groups of 6 bits
- each group determines one out of 64 orthogonal Walsh functions (the Walsh functions are thus determined by the transmitted information, representing a 64-level orthogonal modulation)
- combining with a long PN sequence with length $2^{42} - 1$ (the offset of the long PN sequence corresponds to the user address)
- O-QPSK modulation with two orthogonal PN sequences with 1.23 Mch/s
- a return CDMA carrier contains 9 random access channels and 55 traffic channels
- the orthogonally modulated signals are non-coherently demodulated.

In Globalstar, dual-mode and triple-mode handheld terminals (Globalstar/IS-95, Globalstar/GSM, Globalstar/AMPS, Globalstar/AMPS/IS-95) are used. They comprise a 3-path Rake receiver for dual-path diversity plus one search finger. The quadrifilar helix terminal antenna is roughly hemispherical with an antenna gain of 2.6 dBi. The antenna characteristic is more pronounced at low elevation angles. The L band receive and S band transmit antennas are stacked upon each other. The nominal transmit power ranges between 50 and 400 mW. Terminals will be produced by Qualcomm, Telit and Ericsson. The Qualcomm tri-mode phone weighs 370 grams and provides 3.5 h talktime and 9 h standby. In addition to handheld terminals, mobile and fixed terminals will be available.

Globalstar plans for 50 gateways with up to 4 antennas of $\geq$ 5.5 m diameter. In the gateways enhanced Rake receivers for up to 5 paths are used, allowing multiple satellite diversity for the return link.

Two ground operation control centers control the system and allocate resources to satellites and gateways. Two satellite operation control centers monitor the satellite constellation and keep the satellites within their orbits.

The development of the Globalstar system can be characterized by the following steps:

June 1991:   FCC filing for construction permit and frequency allocation

Jan. 1995:   FCC license received, for construction, launch, and operation in the USA

Feb. 1998:   first satellite launch with Delta II (4 satellites)

Sept. 1998:   loss of 12 satellites during a Zenith launch

Oct. 1999:   start of pre-operational service with 32 satellites and 9 gateways

End of 1999:   40 000 handsets shipped to distributors

March 2000:   start of commercial service with full satellite constellation and 38 gateways (as of Oct. 1999).

The total cost of the Globalstar system amounts to US-\$ 3.3 billion. A gateway costs below US-\$ 10 million. A handheld terminal will be sold for around US-\$ 1000, and the service charge should be around US-\$ 1 ... 2 for domestic and international calls, and up to US-\$ 3 for worldwide calls, including roaming charges and interconnection fees. The end user bill includes a Globalstar satellite component, a non-resident gateway component, a service provider mark-up component, and a PSTN component. In turn, Globalstar bills the service provider for satellite time and for non-resident gateway usage. Globalstar expects a subscriber base of around 500 000 by the end of 2000 and an ultimate subscriber base of 7.5 million (as of Feb. 1999).

More detailed information on the Globalstar system can be found in [Die98] and at http://www.globalstar.com.

C.6 ICO (Intermediate Circular Orbits)

ICO Global Communications is developing a MEO system which will provide personal mobile global communication services, mainly to users with handheld terminals.

ICO is supported by approx. 60 strategic partners. The main investors are Beijing Maritime & Navigation Co. (China), Inmarsat, TRW, Videsh Sanchar Nigam Ltd. (India), Satellite Phone Japan, T-Mobil (Germany), Korea Telecom, Etilasat, and Singapore Telecom. In 1999, ICO failed to raise additional required funding and filed for Chapter 11 bankruptcy protection. Fortunately, C. McCaw (Teledesic) and S. Chandra (Essel, Agrani) agreed on a US-$ 1.2 billion investment, enabling ICO to complete the system deployment.

The main technology partners of ICO provide the following contributions:

- Hughes Space & Communications, USA, received a US-$ 1.4 billion contract for the production of 12 satellites. Hughes will also manage the satellite launches.
- A consortium consisting of NEC, Ericsson, and Hughes Network Systems received a US-$ 616 million contract for the construction of 12 gateway stations.
- Mitsubishi, Samsung, NEC, and others produce ICO phones.

The ICO services will be distributed by national and regional service providers. They include:

- Satellite Phone Japan, for Japan
- a joint venture of all European investors, for Europe
- TRW, for USA.

ICO provides the following services:

- voice at a net bit rate of 3.6 kb/s
- group 3 facsimile
- circuit-switched data up to 9.6 kb/s; high-speed data up to 38.4 kb/s
- short message services.

Value-added services are:

- voicemail
- high-penetration notification with confirmation, to inform users of incoming voice calls or data messages in locations where normal communication is not possible
- three-way calling
- call forwarding.

For mobile voice services, the following business segments are envisaged:

- cellular extension (for domestic and international travelers who are already users of terrestrial cellular systems, to extend their coverage and to overcome incompatible terrestrial standards)

- basic mobile (for people located in areas without mobile terrestrial coverage, e.g. in developing countries)
- specialty mobile (for aeronautical, maritime, trucking, and government applications, including vehicle location and fleet management services)
- basic semi-fixed, incl. high-speed data (for people living in rural and remote areas lacking adequate telecommunications infrastructure).

After the entrance of Teledesic, ICO is going to refocus its efforts on Internet and data services rather than voice telephony. The revised plan is to offer fixed and mobile services at variable data rates up to about 288 kb/s, for accessing computer systems, corporate networks, and the Internet. In this sense, ICO will provide a precursor to a later Teledesic system.

The ICO satellite constellation consists of 10 active MEO satellites at an altitude of 10 390 km. The satellites are placed in 2 orbital planes with 45° inclination and zero phase offset, forming a 10/2/0-Walker constellation. Figure C.5 shows statistics for satellite visibility in terms of minimum and average elevation angle and percentage of multiple satellite visibility.

The ICO satellites (Fig. C.6) are based on the Hughes HS601 platform and provide the following features:

- 2750 kg launch mass
- 26 m span
- 12 year life cycle
- 9 kW DC power, EOL
- 5.5 kW power requirement of the payload
- 163 spot beams with a beamwidth of 3.86°
- two separated S band mobile link antennas of more than 2 m diameter
- direct radiating array transmit antenna with 127 elements and 127 SSPAs
- 500 W RF transmit power
- EIRP: 58.1 ... 55.4 dBi (spot beam ring 1 ... 5)
- G/T: 2.6 ... 1.5 dB/K (spot beam ring 1 ... 5)
- fixed global C band feeder link antennas for transmit and receive
- no intersatellite links
- transparent transponder without demodulation
- digital OBP for channelization, routing, and beam-forming
- 30 MHz processor bandwidth.

For the satellite body and antennas, see Fig. 8.5. The satellites will be launched using the Atlas 2AS, Delta 3, Proton, and the Zenith/Sea Launch launchers.

The coverage area of a satellite will extend over 12 900 km with a minimum elevation of 10°. The nominal cluster size is 4. Each satellite can provide 4500 duplex channels using a total of 750 TDMA carriers per satellite.

The mobile user links use the S band at 1980–2010 MHz for the uplink and 2170–2200 MHz for the downlink. The frequency band is channelized

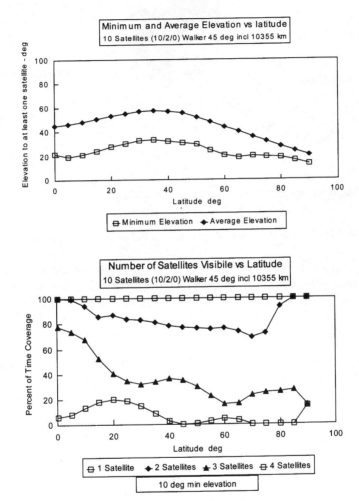

Fig. C.5. Statistics of ICO satellite visibility [Bai99]

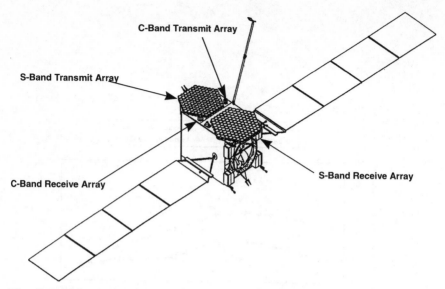

Fig. C.6. ICO satellite (ICO Global Communications)

into 150 kHz segments, with a spot beam being able to use up to 30 MHz bandwidth.

The uplink is organized in MF-TDMA with a frame duration of 40 ms and 6 time slots per frame. The voice signals are IMBE (improved multiband excitation) coded at a net bit rate of 3.6 kb/s. With rate-3/4 convolutional coding ($K = 7$) and other overhead, the gross bit rate amounts to 6 kb/s. The bit error rate requirement on the channel is 4 per cent. The TDMA carrier bit rate results in 36 kb/s.

For data services, rate-1/2 convolutional coding with $K = 7$ is used. Each TDMA slot provides a net bit rate of 2.4 kb/s at a bit error rate of 10^{-5} after decoding. Data rates higher than 2.4 kb/s are realized by use of multiple TDMA slots.

The transmit signal is GMSK-modulated and Nyquist-filtered with a roll-off factor of 0.4, resulting in a signal bandwidth of 25 kHz. A link margin of 8 dB minimum and 10 dB average is provided. Moreover, satellite diversity and voice activation are used.

For the feeder link, the C band at 5 GHz and 7 GHz is used, with 100 MHz bandwidth, uplink and downlink.

The following terminals will be available:

– handheld terminals
– vehicular mobile terminals
– aeronautical terminals
– maritime terminals
– laptop and palmtop terminals for data services and Internet access

– semi-fixed terminals
– fixed terminals (rural phone boots, community phones).

The handheld terminals will be dual-mode (terrestrial/satellite) or triple-mode and will be provided by NEC, Mitsubishi, and Samsung. The antenna will be a 10 cm long and 1.4 cm thick quadrifilar helix, providing 2.5 dBi maximum gain. The average transmit power will be 250 mW, and the G/T will be around -24 dB/K. The terminals will have a data interface and will accept SIM cards. Talk time should be longer than 1 h, standby time more than 24 h, and the weight should be below 300 g. Figure C.7 shows an example of an ICO handheld terminal.

Fig. C.7. ICO handheld terminal. Courtesy: ICO Global Communications

The ICO ground segment is called ICONET and consists of 12 gateways called satellite access nodes (SANs), plus a network management center, Fig. C.8. The SANs are linked by a high-capacity terrestrial backbone network for signaling and call routing.

A SAN will comprise three main elements:

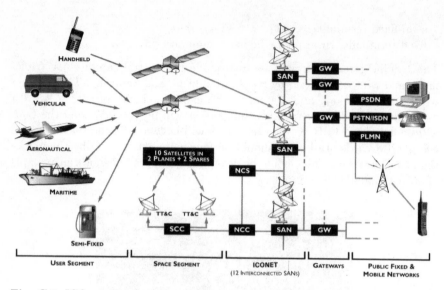

Fig. C.8. ICO satellite access network (SAN)

- 5 antennas of 7.6 m diameter, to communicate with the satellites (land earth station)
- a switch to route traffic within the ICONET and other terrestrial networks (mobile switching center)
- databases to support mobility management. Each SAN is equipped with a visitor location register, VLR.

The SANs are owned by ICO; their operation is contracted to national or regional operators. The SANs are located in Brewster/USA (operated by US Electrodynamics Inc.), Australia (operated by Telstra Australia), Chile, Germany (operated by Deutsche Telekom), India (operated by Indosat), South Africa, Indonesia, Korea (operated by Korea Telecom), UAE, Brazil, Mexico, and China. As of July 1999, the 6 SANs mentioned first are operational.

Six SANs are equipped with TT&C facilities. The network management center is located in Tokyo, and the satellite control center in London.

The development of the ICO system can be characterized by the following steps:

1990–94:	concept development
Jan. 1995:	ICO company founded
May 1995:	FCC filing
Oct. 1995:	contract with Hughes for construction of 12 satellites
Feb. 1997:	contract for construction of SANs
Sept. 1997:	FCC filing for usage of the 2 GHz band
end of 1998:	commencement of ICO global roaming service via T-Mobil and GTE Wireless

summer 1999: failure to raise additional funding
Aug. 1999:   ICO files for chapter 11 bankruptcy protection
Dec. 1999:   C. McCaw (Teledesic, USA) and S. Chandra (Essel, Agrani, India) agree on US-$ 1.2 billion investment in ICO
March 2000:   first satellite launch (as of Feb. 2000)
mid 2001:   start of commercial service (as of Dec. 1999).

Until the start of operation, the system cost of ICO will amount to US-$ 4.7 billion, of which 1.3 billion are required for the satellites, 1.2 billion for the ground infrastructure, 1 billion for launches and insurance, 0.6 billion for pre-operating and marketing, and 0.3 billion for financing costs. The handheld terminals are expected to cost US-$ 1000, and the basic tariff for telephony will be below US-$ 2 per minute. The subscriber bill will include

– an ICO satellite component,
– a PSTN component, and
– a component for the service provider.

In 2005, ICO expects revenues of US-$ 5 billion (as of Feb. 99).

More detailed information on the ICO system can be found in [MS98, GST99, Bai99] and at http://www.ico.com.

C.7 Inmarsat-3/Inmarsat mini-M

The Inmarsat organization (International Maritime Satellite Organization) was founded in 1979 to provide maritime satellite services for communications and emergency applications. Inmarsat now has about 80 signatories from around the world, and became a private company in 1999.

Since 1996, the 3rd generation of Inmarsat satellites has been in commercial use, providing global Inmarsat-3 (also called Inmarsat mini-M) services to laptop-sized portable terminals. The services include:

- mobile telephony at a net rate of 3.6 kb/s
- fax
- email at 2.4 kb/s
- data at 2.4 kb/s and up to 64 kb/s
- paging
- short messaging
- Internet access
- positioning.

Among the technology partners are Lockheed Martin, which acted as prime contractor, and Matra Marconi Space, which built the satellite payload. Comsat is a service provider and owns the service license for the USA.

The Inmarsat 3rd-generation satellite constellation consists of 4 GEO satellites plus one in-orbit spare. The satellites have the following technical features:

- 2066 kg launch mass, leaving 1100 kg BOL in-orbit mass
- 13 year life cycle
- 2800 W DC power, EOL
- 5 (max. 7) spot beams + 1 global beam, resulting in approximately two-fold frequency reuse, cf. Fig. 2.18
- transparent transponder with SAW-filter channelization, crossbar MMIC switching, and RF beam-forming
- 48 dBW EIRP
- 2200 duplex channels per satellite.

The main terminal types are laptop-sized portable terminals and mobile vehicular terminals. Typical features of the portable terminals are:

- 25 cm × 22 cm × 5 cm size
- 2.4 kg weight (including batteries)
- 2.5 h talk time, 50 h standby
- 10 W power consumption at transmit state
- removable antenna, $\pm 15°$ beamwidth
- 11 ... 17 dBW EIRP
- -17 dB/K G/T
- SIM card for common billing of satellite service and terrestrial service.

Fig. C.9. Nera world phone for the Inmarsat mini-M service. Courtesy: Nera

Examples are the Planet-1 terminal from NEC, the Worldphone terminal from Nera (Fig. C.9) and the Capsat mobile telephone from Thrane & Thrane.

The mobile user link uses the L band at 1625–1660.5 MHz for the uplink, and at 1525–1559 MHz for the downlink. The codecs use AMBE (advanced multiband excitation) voice coding at a net bit rate of 3.6 kb/s. Block coding of selected bits results in a data rate of 4.8 kb/s. SCPC with 5.6 kb/s O-QPSK modulation is used for transmission in the up- and downlink.

The Inmarsat-3 satellites were launched in 1996 and 1997, with service commencing in Oct. 1996. Presently, several tens of thousands of terminals are in use, at a cost of US-$ 3000 for the terminal and a service charge of US-$ 3 per minute airtime. The total cost of the Inmarsat-3 system amounts to US-$ 690 million.

Inmarsat GAN. In autumn 1999, Inmarsat launched its Global Area Network (GAN), providing mobile ISDN services (high-quality speech, group IV fax, image file transfer, video conferencing) and mobile IP services (mobile internet and email access, etc.), both at a rate of 64 kb/s.

The system uses the existing Inmarsat-3 satellites. The laptop-sized terminals ($30 \times 30 \times 6$ cm) weigh around 4 kg, include an unfoldable antenna (approx. 40×70 cm), and provide a number of standard interfaces. The terminals are priced from around US-$ 10 000, and the ISDN airtime costs around US-$ 7 per minute. The mobile IP services are charged for the amount of data transmitted.

Inmarsat-4. For the year 2004, Inmarsat plans a new generation of Inmarsat spacecraft, Inmarsat-4, at a cost of US-$ 1.4 billion. Two in-orbit GEO satellites in the Atlantic Ocean West and Indian Ocean regions will provide a range of personal multimedia communication services at data rates of 144 to 432 kb/s for applications, such as e-business, on-line access, voice telephony (IMT-2000), and GMDSS.

More information on Inmarsat systems and services can be found at http://www.inmarsat.org.

C.8 Iridium

Iridium is a pioneering LEO system for global telephony, which employs intersatellite links and on-board switching. The system is owned and operated by Iridium LLC, a subsidiary of Motorola Satellite Communication Inc.

Partners. Iridium is backed by 19 strategic investors. The main investors are:

- Motorola Inc., USA (19.5%)
- Nippon Iridium Ltd, Japan
- Vebacom, Germany
- South Pacific Iridium Holdings Ltd
- Iridium SudAmerica Corp.
- Iridium Middle East Corp.
- Khrunichev State Research and Production Space Center, CIS
- Iridium Italia.

In total, around US-$ 2 billion of equity have been invested in the system. In July 1997, Iridium was fully financed with an amount of US-$ 4.4 billion. However, due to the delayed start of commercial service and the very slow development of subscriber numbers, Iridium got into severe financial trouble and in 1999 filed for Chapter 11 bankruptcy protection.

Iridium has been supported by a large number of industrial partners. The main partners are as follows:

- Motorola Inc. Satellite Communication Division, USA, was the main contractor for the system and in 1992 received a US-$ 3.4 billion contract from Iridium Inc. for:
  - design, production, and launch of the satellites
  - production of the satellite payload including OBP, switch, and ISLs
  - setup of the satellite control system.
  Moreover, Motorola received a US-$ 2.8 billion contract for operating and maintaining the Iridium system for 5 years, starting from 1998.
- Lockheed Missiles & Space Co., USA, received a US-$ 700 million contract for delivery of 125 satellite buses.
- McDonnell Douglas Corp., USA, received a US-$ 400 million contract for 8 Delta II launches with 5 satellites each.
- Raytheon Company, USA, was contracted to deliver phased-array satellite antennas and transceivers for US-$ 273 million.
- Siemens AG, Germany, built EWSD-D900 mobile switching centers for the Iridium gateway stations.

The Iridium ground segment comprises 12 regional gateways, which are operated by 17 companies distributing the Iridium services. Examples are:

- Motorola, Sprint Iridium, and Iridium Canada provide Iridium services in the USA and Canada via the gateway station in Phoenix, USA, which is operated by their jointly-owned company, Iridium North America.

- Khrunichev State Research and Production Space Center uses the Moscow gateway for service distribution within the Russian Federation and neighboring countries.
- Iridium Communications Germany, a subsidiary of o.tel.o, is a service provider in Europe, e.g. in Germany, Austria, Spain, Sweden, the UK, and other countries.
- Iridium Italia cooperates with Iridium Communications Germany in operating the Rome gateway, and provides Iridium services in Italy, France, Greece, Switzerland, and other European countries.
- Nippon Iridium is the service provider in Japan, using the gateway in Nagano.

Market. Iridium addresses the following market segments:

- business users, such as international business travelers, aeronautical users, maritime users, and trucking companies
- governmental users, such as security forces and emergency services
- private users, such as travelers in countries without adequate communications infrastructure.

For these market segments, Iridium offers the following services:

- telephony at 2.4 kb/s net rate and high-penetration incoming call alert
- fax
- modem data at 2.4 kb/s net rate
- paging (simplex paging, acknowledgement-back paging via speech channel) with a mean link margin of 30 dB
- aeronautical services (beginning in 2001).

Also, Iridium offers a number of value-added subscriber features, such as call barring, call forwarding, voice-mail, and subscriber identity confidentiality.

Space Segment. The satellite constellation consists of 66 LEO satellites at an altitude of 780 km. 11 satellites are regularly distributed in each of 6 polar orbit planes with an inclination of 86.4°. The equatorial separation between co-rotating orbits is 31.6°; the equatorial separation between counter-rotating orbits is 22.0°. Six spare satellites are placed in a parking orbit below the operational orbits. The Iridium satellites (Fig. C.10) are characterized by the following features:

- satellite in-orbit mass 689 kg (including 115 kg fuel)
- payload mass 140 kg
- 8.25 m span, 4.6 m height
- three-axis stabilization
- launches by Delta 2 (Boeing), Long March (China Great Wall Ind.), and Proton (Khrunichev Enterprises)
- life cycle: business plan, 5 years; station-keeping (fuel), 8 years
- de-orbiting into atmosphere

- solar cells, 1200 W; battery, 50 Ah; RF power, approx. 400 W
- regenerative transponder with baseband switching
- 3 phased-array antennas for the mobile link, each with 106 microstrip patch elements
- spot beam antenna gain approx. 24 dBi EOC
- 4 steerable feeder link antennas of approx. 30 cm diameter
- 4 intersatellite links per satellite.

Fig. C.10. Iridium satellite

The Iridium satellite constellation provides global coverage. Each satellite covers an area with a diameter of 4650 km, at a minimum satellite elevation of 8.2°. Due to the high satellite velocity (6.6 km/s relative to the ground), satellite visibility is limited to 9 minutes. Each satellite provides 48 spot beams with 750 ... 950 km cell diameter at the earth's surface, Fig. C.11. Thus, the whole constellation would provide 3168 radio cells. However, at high latitudes, overlapping cells are switched off, leaving 2150 active cells in the system. A cluster size of 12 is used, resulting in a global frequency reuse of 179.

Communication Links. For the mobile user uplink and downlink, Iridium can use L band frequencies from 1616 MHz to 1626.5 MHz. Due to FCC rules, the usage is restricted to the band 1621.35–1626.5 MHz. The available bandwidth is divided into channels of 41.67 kHz, giving approx. 120 channels. With a cluster size of 12, 10 channels can be used in each spot beam. These channels are used with a 50 kb/s QPSK modulation and a roll-off factor of 0.26, resulting in a 31.5 kHz channel bandwidth and approx. 10.2 kHz guard

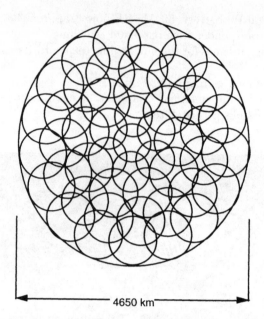

4650 km

Fig. C.11. Iridium spot beam pattern

band. The frequency channels are organized in TDD/TDMA with 90 ms time frames and carry 4 duplex traffic channels, cf. Fig. 5.1. Since the number of traffic channels per satellite is power-limited to 1100, the system capacity is 72 600 duplex channels.

The voice coder uses advanced multiband excitation (AMBE) at a net rate of 2.4 kb/s. For error control, rate-3/4 convolutional coding with constraint length 7 plus interleaving is applied, resulting in a channel data rate of 4.6 kb/s including overhead; 414 bits are transmitted in each TDMA burst. The link margin for voice is 16 dB, and the link margin for paging is 32.5 dB.

Up to 4 intersatellite links (ISLs) are used per satellite, cf. Fig. 2.22. No ISLs exist at the "seam" of the constellation. The ISLs use the Ka band at 23.18–23.38 GHz with a bandwidth of 200 MHz. They are organized in TDD with a frame duration of 9 ms, a transmission rate of 25 Mb/s, rate-1/2 channel coding, and QPSK modulation.

ISL routing is performed with the help of routing tables, taking into account the time of day and the satellites' position. The routing tables allow 8 different routes; the tables in the satellites are updated every day. All services are packet-switched. Speech packets take the shortest route, management packets use detours with less traffic. The gateway stations reorder the incoming packets of a connection with the help of a buffer.

The feeder links are located in the Ka band at 29.1–29.3 GHz for the uplink and at 19.4–19.6 GHz for the downlink, each with a bandwidth of

100 MHz and a transmission rate of 3.125 Mb/s, including rate-1/2 channel coding.

Ground Segment. The Iridium ground segment comprises 12 gateway stations. Each gateway is equipped with 3 antennas of 3 m diameter. To provide site diversity, additional GW antennas are installed at a distance of up to 50 km. The gateway stations contain user databases, monitor the calls for billing purposes, and are equipped with connections to international switching centers. Iridium gateways exist in the following locations: Beijing (China), Seoul (Korea), Nagano (Japan), Taipei (Taiwan), Bangkok (Thailand), Tempe (Arizona, USA), Hawaii (for the US government), Rio de Janeiro (Brazil), Rome (Italy), Moscow (Russia), Jeddah, and Mumbai (India).

In order to provide service only in countries in which Iridium owns a service license, location determination of subscribers is mandatory. This is performed by measurements of signal delay and frequency offset of the user signal and achieves an accuracy of 5 km.

The master control station is located in Virginia, USA; a backup control station exists in Italy. The main tasks of the control station are:

- network management
- frequency planning and generation of routing tables for satellites and gateways
- loading of frequency plans and routing tables into the satellites
- assignment of ground stations to satellites
- satellite control.

Terminals. Iridium handheld terminals are being produced by Motorola and Kyocera, Fig. C.12. Dual-mode and satellite-only terminals, as well as pagers, are available. Typical parameters of terminals are:

- mass 400 g
- antenna length 15 cm
- talk time 2 h
- standby time 20 h
- peak transmit power 7 W
- mean transmit power 0.6 W
- antenna gain 2 dBi
- G/T -23 dB/K.

Iridium subscribers are identified by several numbers:

- The mobile subscriber ISDN number (MSISDN) is the number which is dialed by the caller. The MSISDN contains the Iridium country code (8816 or 8817), the Iridium numbering plan area (INPA, 3 digits) determining the home gateway of the callee, the service provider number (2 digits), and the subscriber number (7 digits).

Fig. C.12. Iridium terminals

- The temporary mobile subscriber ID (TMSI) is the number that is transferred via the radio interface during call setup. This number is changed periodically to protect subscriber confidentiality.
- The Iridium mobile subscriber ID (IMSI) is a permanent number stored on the SIM card of the subscriber.

Calls are recorded in Iridium's clearing house. There are five components to the per-minute charge (Satellite Communications, Feb. 99):

- Iridium LLC satellite usage component (includes operating costs and profit)
- non-resident gateway component (if used)
- PSTN component
- resident gateway component
- service provider mark-up component.

History. The development of the Iridium system can be characterized by the following steps:

1987:	start of concept design
Dec. 1990:	FCC filing for construction permit and frequency allocation
1991:	founding of the Iridium company

Jan. 1995:	FCC license received (for construction, launch, and operation of Iridium)
May 1997:	first satellite launch (5 satellites by Delta 2)
May 1998:	full satellite constellation in orbit
Nov. 1998:	start of operation (telephony, paging, and messaging)
end of 1998:	problems with 12 satellites of the constellation
Aug. 1999:	Iridium files for Chapter 11 bankruptcy protection, having a debt level in excess of US-$ 4 billion.
Dec. 1999:	Iridium has 50 000 subscribers.

The total system cost is estimated at US-$ 5.7 billion, including satellite construction, launches, control centers, and one year of operation. Moreover, for every year of operation, a cost of US-$ 1 billion is needed for satellite replenishment and interest payments.

The handheld terminals used to sell for US-$ 3000, but after the severe financial problems in June 1999, they are sold for US-$ 1500. Similarly, the airtime charge dropped from US-$ 2 ... 7 per minute to US-$ 1.5 ... 3 per minute.

More detailed information on the Iridium system can be found in [PRFT99] and at http://www.iridium.com.

C.9 Orbcomm

Orbcomm is a "little-LEO" system for near real-time transmission of data and messages. Orbcomm Global L.P., the Orbcomm system owner and operator, is a partnership owned by Orbital Sciences Corp. (USA) and Teleglobe Inc. (Canada). In 1995, these partners invested US-$ 160 million in the Orbcomm system. Meanwhile, the system cost of approx. US-$ 500 million is fully financed.

Technology partners of Orbcomm are Fairchild Space and Defense Corp. and Orbital Sciences Corp. In 1995, the latter company received a contract for producing and launching 34 satellites.

Typical applications of Orbcomm are:

- tracking of mobile property (e.g. for land mobile and maritime transport)
- remote monitoring & control (e.g. for the oil and gas industries)
- messaging for personal users.

The following basic services are provided:

- two-way real-time and store-and-forward transmission of alphanumeric data packets at 2.4 kb/s uplink burst rate and 4.8 kb/s downlink rate; typical message lengths are up to 250 characters.
- email
- fax
- position determination by means of terminals with integrated GPS or by repeated Doppler measurements.

The services are available for a one-time fee of US-$ 50, a monthly fee of US-$ 30, and a service charge of US-$ 0.25 ... 1 per 100 bytes of transmitted data.

The satellite constellation contains 28 active satellites:

- 4 satellites at 740 km, in two orbit planes with 70° and 108° inclination
- 24 satellites at 825 km, in 3 orbit planes with 45° inclination.

8 more satellites in an equatorial orbit at 825 km are planned. The satellites are "Microstar" satellites, Fig. C.13, with

- 42 kg launch mass
- 1 m diameter and 17 cm height of the satellite body
- life cycle 4 years
- 70 W average DC power, EOL
- 3 transmitters, one for the VHF subscriber link, one for the VHF feeder link, and one for a UHF beacon
- 7 receivers for the subscriber uplink, including a scanning DCAAS receiver
- 1 receiver for the VHF feeder link
- GPS on board the satellites.

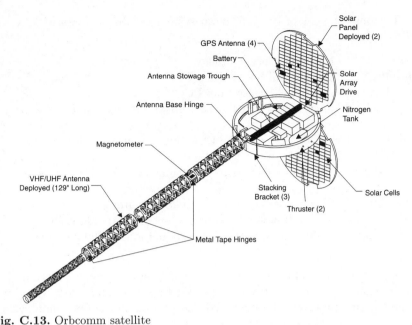

Solar
Panel
Deployed (2)

GPS Antenna (4)

Battery

Antenna Stowage Trough

Solar
Array
Drive

Antenna Base Hinge

Nitrogen
Tank

Magnetometer

VHF/UHF Antenna
Deployed (129" Long)

Stacking
Bracket (3)

Solar Cells

Thruster (2)

Metal Tape Hinges

Fig. C.13. Orbcomm satellite

Eight satellites at a time can be launched by the Pegasus XL launcher.

The subscriber links are organized in FDMA. The 7 VHF uplink channels use the frequency band 148–150 MHz. The burst rate in a channel is 2.4 kb/s and the bandwidth is 10 kHz. Because the VHF band is also used by terrestrial systems, the usage of the satellite uplink channels must be adapted via the dynamic channel activity assignment system (DCAAS): A satellite measures the interference power in the VHF channel bands using its scanning receiver and dynamically selects the 6 channels exhibiting the least interference. This choice is broadcast to the subscriber terminals via signaling packets [Gif96]. In this way, DCAAS avoids interference to terrestrial systems and guarantees the quality of satellite transmissions.

Moreover, the uplink channels are divided into random access channels and reservation/messaging channels.

The downlink subscriber link uses the frequency band 137–138 MHz. Each carrier provides 4.8 kb/s, with a typical transmit power of 12 W per FDMA carrier.

Fixed and mobile terminals are available. The handheld Orbcomm Communicator weighs 500 ... 1000 grams, provides a transmit power of 5 W, and is equipped with a $\lambda/2$ whip antenna. Optionally, the terminals have integrated GPS. Terminal providers are Panasonic, Magellan, and others. The Magellan GSC 100 handheld unit sells for US-\$ 1000. The Magellan OM 200 satellite modem card combines GPS location determination with an Orbcomm receiver/transmitter and is available for US-\$ 200.

The development of the Orbcomm system can be characterized by the following steps:

1990:	Orbcomm concept design
May 1992:	experimental license from FCC
Oct. 1994:	operational license from FCC
April 1995:	launch of the first two satellites by Pegasus and preliminary service
Feb. 1996:	start of service in the USA (via two satellites)
Dec. 1997:	launch of 8 more satellites
April 1998:	FCC grants additional spectrum and license for 12 additional satellites
Nov. 1998:	full commercial service with 28 satellites
1999:	constellation is planned to be extended to 36 satellites
Dec. 1999:	30 000 terminals are sold and another 200 000 on order.

After five years, a final subscriber base of several million is expected. The service charge amounts to US-$ 4.5 per kbyte of data and a few cents for a radio location message. More detailed information on the Orbcomm system can be found in [Gif96, Maz99] and at http://www.orbcomm.net.

C.10 SkyBridge

SkyBridge is a LEO system for broadband access which intends to remove the bottleneck in the local loop.

System proponent SkyBridge Ltd Partnership is responsible for the space segment and is supported by Alcatel Alsthom with regard to system development. Alcatel Espace will provide the satellite payload, and Alenia Aerospazio will integrate the satellites.

Financially, the system is supported by Alcatel as general partner, Loral Space & Communications, CNES, Toshiba, Mitsubishi, SPAR, and others. It is planned to market SkyBridge together with the Cyberstar system.

The market segments of SkyBridge are the residential consumer market as well as business markets such as enterprises, public service agencies, telecommuters, and the SOHO environment. The following high-speed local loop and interactive real-time applications are foreseen:

- high-speed access to the Internet and other on-line services
- e-commerce
- telecommuting (access to company servers and LANs, email, file transfer)
- corporate networking (LAN interconnection, WANs)
- video conference and video telephony
- telemedicine
- interactive video-on-demand, games, etc.

In addition, narrowband speech, data, and video communications will be provided.

SkyBridge uses a constellation of 80 satellites at 1469 km, Fig. C.14, consisting of two (40/10/4)-Walker subconstellations, each one containing 40 satellites in 10 orbit planes. The orbit planes have an inclination of 55°.

The satellites are based on the Proteus platform from Aerospatiale and are characterized by the following features:

- launch mass 1250 kg, payload 300 kg
- span 15.5 m
- life cycle 8 years
- DC power 3 kW EOL, battery 150 Ah
- 18 steerable (earth-fixed) spot beams per satellite
- around 1 Gb/s maximum capacity per spot beam
- one transparent transponder (bent-pipe) per spot beam
- phased-array antennas
- no ISLs.

The coverage of the satellite constellation extends up to ±68° latitude, with a minimum satellite elevation of 10°. The footprint diameter of a satellite is 6000 km. The diameter of a spot beam radio cell is 700 km. The frequency band is reused in every radio cell. Each radio cell corresponds to the service

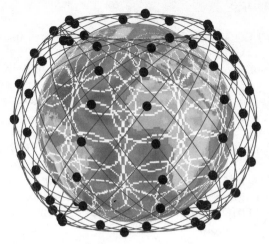

Fig. C.14. SkyBridge satellite constellation

area of a gateway station, Fig. C.15. 200 gateways of 3 Gb/s capacity are foreseen for the whole SkyBridge system.

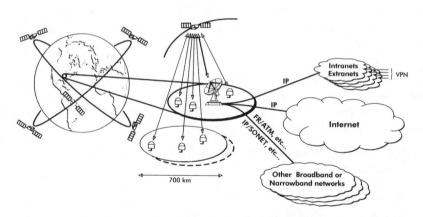

Fig. C.15. SkyBridge system architecture

SkyBridge has been designed to avoid interference to GEO systems working in the same frequency band: No SkyBridge satellite transmits within an angular spacing of 10° around the geostationary arc, as seen from a user. If a satellite is approaching this "non-operating zone", the traffic of the respective spot beam is switched to another satellite that is well clear of it.

User links and gateway links both use the Ku band at 12.75–18.1 GHz for the uplink and 10.7–12.75 GHz for the downlink. The links are organized in a combined FDMA/TDMA/CDMA scheme based on the ATM format.

Residential and professional/collective terminals are planned with features according to Table C.10.

Table C.10. SkyBridge user terminals

Terminal:	Residential	Professional/collective
Antenna size	50 cm	80 cm
Antenna tracking	Luneberg lens, Fig. C.16	2 separate antennas
Receive data rate	20 Mb/s	100 Mb/s
Transmit data rate	2.5 Mb/s	10 Mb/s
Terminal cost	US-$ 700	US-$ 2000

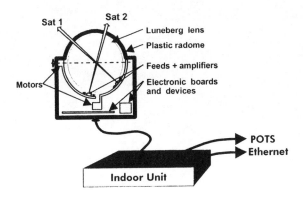

Fig. C.16. SkyBridge residential terminal with Luneberg lens

The cost of the whole system amounts to US-$ 4.6 billion, including one year of operation. Additional US-$ 2 billion are envisaged for the 200 gateways. The user charges include a monthly fee of US-$ 30 and US-$ 0.03 per Mbyte of transmitted data, corresponding to US-$ 0.04 for a 3 minute call at 64 kb/s or US-$ 0.5 for a ten minute video conference at 256 kb/s.

The development of the SkyBridge system can be characterized by the following steps:

1993:	start of system development
Feb. 1997:	FCC filing for space segment
WRC'97:	frequency allocation "subject to non-interference with existing GEO systems"

June 1999:   final international regulatory approval received from ITU
2001:   first satellite launches
end of 2001:   start of operation with 40 satellites
end of 2002:   full service with 80 satellites (as of Jan. 1999).

More detailed information on the SkyBridge system can be found at
http://www.skybridgesatellite.com.

C.11 Sky Station

Sky Station is a non-satellite system concept which is based on steerable blimps serving as telecommunication platforms positioned above large cities, e.g. With this approach, Sky Station combines the advantages of geostationary and low-altitude satellites.

The Sky Station idea is pursued by Sky Station International Inc. supported by the following industrial partners:

- Alenia Aerospazio delivers the communications payload.
- Thomson-CSF Communications is responsible for the ground earth stations and deals with ground electronics and user terminals.
- Dornier Satellitensysteme delivers subsystems.
- Comsat Labs integrates the communications system.
- United Solar is the supplier of photovoltaic modules.

Providing up to 2 Mb/s in the uplink and up to 10 Mb/s in the downlink of a blimp, Sky Station is suitable for the following applications:

- mobile/portable telephony, video telephony, and video conferencing
- mobile/portable high-speed Internet access and WWW
- WLL
- video on demand.

The terminals comprise handheld terminals, palmtop terminals, laptops, and desktops. The blimp platforms are characterized by the following features, Fig. C.17:

- 250 blimp platforms above large cities or temporary events
- 10.5 t blimp mass, 62 m diameter, 157 m length
- 21 km flight altitude
- 520 kW DC power generation
- propulsion by electrically driven propellers
- GPS controlled, with an accuracy of a few hundred meters
- life cycle (flight duration) 5 ... 10 years
- 1 t payload mass, 20 kW power consumption
- 11 Gb/s ATM switch
- up to 130 receive antennas and 130 transmit antennas for the user links, each one providing up to 7 spot beams
- 15 gateway beams for feeder links to up to 15 gateways.

The Sky Station system may operate in two different frequency ranges, Table C.11:

- In the 47 GHz range, Sky Station can provide fixed users and portable terminals with personal T1/E1 broadband links for high-speed Internet access and other broadband services such as video conferencing. Also, telephone services can be brought to developing countries.

Fig. C.17. Sky Station platform

– In the 2 GHz range allocated to IMT-2000, Sky Station can be used for the rapid deployment of 3rd generation mobile services.

The development of the Sky Station system can be characterized by the following steps:

1996:	FCC filing
July 1998:	the FCC approves the use of stratospheric platforms as telecommunications stations
Sept. 1998:	national filings for 50 platforms
2000:	first launches (as of June 1998)
2002:	start of service (as of June 1999).

The system cost of Sky Stations is estimated at US-$ 800 million, and the service charge should be around US-$ 0.1 per minute for broadband services.

More detailed information on the Sky Station system can be found at http://www.skystation.com.

Table C.11. Sky Station service characteristics

	47 GHz broadband service	2 GHz mobile service
Platforms	250 platforms worldwide, optionally interconnected	one platform per metropolitan area
Coverage area	150 km diameter	1000 km diameter
Minimum elevation angle	15°	-
Number of spot beams	700 per platform	1000 per platform
Capacity	11 Gb/s	2 Gb/s
Frequency band	47.9–48.2 GHz uplink 47.2–47.5 GHz downlink	1885–1980 MHz 2010–2025 MHz 2110–2160 MHz
Bandwidth per direction	100 MHz	5 MHz
Modulation scheme	QPSK (subscriber) 64QAM (ground station)	QPSK
Multiplexing	MF-TDMA uplink TDM downlink	wideband CDMA
Data rates	2 Mb/s uplink 10 Mb/s downlink	8 ... 16 kb/s for voice 384 kb/s for data
Terminal transmit power	100 ... 250 mW	25 mW

C.12 Spaceway

In the final constellation, Spaceway is planned as a broadband quasi-global hybrid GSO/NGSO Ka band system. This goal is pursued by Hughes, who gave the development a new momentum by investing US-$ 1.4 billion in March 1999. In June 1999, America Online invested another US-$ 1.5 billion. Thus, Spaceway is now fully financed.

Industrial partners are Hughes Space and Communications Co. which will produce the satellites, and Hughes Network Systems which will design the ground earth stations and the user terminals.

Typical applications for consumers and business customers are high-speed Internet access, corporate intranets, virtual private networks, multimedia broadcasting, and high-speed data delivery. Uplink data rates will vary between 16 kb/s and 6 Mb/s.

Initially, the satellite constellation will consist of 2 GEO satellites above North America, located in orbital slots assigned at 101°W and 90°W. Six more satellites are planned to be positioned in internationally assigned slots at 49°W, 25°E, 54°E, 101°E, 111°E, and 164°E above 4 geographic regions. Later on, an additional MEO or LEO constellation may be added. The GEOs would be used for high-speed data transport; the NGSOs would be suited for interactive multimedia services.

The GEO satellites are based on the Hughes HS 702 platform, and are designed for a life cycle of 15 years. They will use OBP and on-board packet switching to provide a throughput of 4.4 Gb/s per satellite. Moreover, they use spot beam technology and may be interconnected by ISLs.

The Ka band user uplink is organized in MF-TDMA and uses the frequency bands 28.35–28.6 and 29.25–30.0 GHz. For the downlink, the frequency bands 17.7–18.8 and 19.7–20.2 GHz can be used.

Standard USAT terminals with a diameter of 0.66 m provide an uplink data rate of 384 kb/s and a downlink rate around 100 Mb/s. Such terminals should be available for approx. US-$ 1000. Larger broadband terminals can provide 6 Mb/s in the uplink.

As of March 1999, a cost of US-$ 4 billion is estimated for the GEO system. The development of the Spaceway system can be characterized by the following steps:

1994:	FCC filing for 2 GEO satellites for the USA
1997:	FCC license received
1997:	FCC filing for 15 orbital slots for 15 GEO satellites
2002:	launch of 2 GEO satellites for the USA (as of March 1999)
end of 2003:	launch of 6 more GEO satellites (as of March 1999)
later:	optional MEO or LEO constellation.

More detailed information on the Spaceway system can be found at http://www.hns.com/spaceway.

C.13 Teledesic

Teledesic is a concept for a global broadband LEO satellite network, which can be seen as "Internet in the Sky" or as a "fiber network" in the sky.

The Seattle-based Teledesic Corp. is financially backed by the following main investors:

- Motorola, investing US-$ 750 million in May 1998 and holding a stake of 26%. A system agreement was signed in July 1999.
- Craig McCaw, Chairman of Teledesic Corp., holding a private investment.
- Bill Gates, another private investor.
- Prince Alwaleed Bin Talal investing US-$ 200 million in April 1998 and holding a stake of 14%.
- Boeing, holding a stake of 5% for an investment of US-$ 50 million.

As of July 1999, more than US-$ 1.5 billion of equity had been raised.

As prime contractor, Motorola would be responsible for engineering and construction of the system, for the coordination of satellite launches, and for the operation of the network control center. Hughes may provide the satellite payload, and in July 1999 Lockheed Martin received a launch contract for Proton M and Atlas V.

Customers of Teledesic could be ISPs, telecom operators, corporate users, fixed end users (in a SOHO environment), and mobile end users on-board ships and aircraft. The ambition of Teledesic to provide a fiber-like network in the sky means:

- low bit error rate
- low latency (therefore, a LEO constellation was chosen)
- high bandwidth (therefore, the Ka band will be used)
- high availability (made possible by a high satellite elevation)
- compatibility with terrestrial fiber networks.

Based on these characteristics, the following applications and services are foreseen:

- computer networking, intranets, virtual private networks, LAN interconnection
- high-speed Internet access, intranet access
- interactive multimedia, video conferencing, high-quality voice
- wireless backhaul.

The Teledesic satellite constellation as seen in 1999 comprises 288 LEO satellites at an altitude of 1375 km. 24 active satellites are placed in each of 12 polar orbital planes with an inclination of 98.1°. Teledesic uses an asynchronous polar constellation, which means that the phase shifts between different orbital planes are not controlled. The high number of satellites provides a high system capacity and a high satellite elevation enhancing the link availability and reducing rain fading.

Up to now, the satellites have been characterized by the following parameters:

- 795 kg launch mass
- 144 kg payload mass
- 6.6 kW DC power EOL
- active phased-array antennas of 12 m diameter, providing electronic beam steering
- transponder with OBP and fast packet switching
- 10 Gb/s data throughput
- 6 intersatellite links, each one providing a data rate of 2 Gb/s.

In the light of reducing the number of satellites from initially 840 to presently 288, the number of satellites may be further reduced to around 100. Their technical parameters could accordingly change during the further development of the system.

Teledesic intends to provide global coverage with earth-fixed cells, Fig. C.18, served by scanning satellite beams. Therefore, cell handover is required only in exceptional cases. The minimum satellite elevation of 40° mitigates the influence of signal blockage and rain fading. Thus, Teledesic can provide 99.9% availability. Within a satellite, frequencies are allocated dynamically, and are reused up to 100 times. The traffic capacity within a circular area of 200 km diameter amounts to 500 Mb/s.

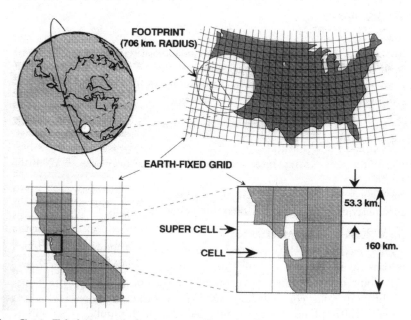

Fig. C.18. Teledesic earth-fixed super cells and cells

The Teledesic satellite network is based on fast packet switching and connectionless transport of fixed-length packets through the ISL network. A distributed adaptive routing scheme with destination-based packet addressing is used, providing low delay and low delay variation. Packets of the same session may follow different paths through the network. The terminal at the destination buffers and if necessary reorders the received packets to eliminate the effect of timing variations.

For the communication between terminals and satellites the Ka band is used, at 28.6–29.1 GHz for the uplink, and at 18.8–19.3 GHz for the downlink. The uplink is organized in MF-TDMA, whereas in the downlink asynchronous TDMA is used: All waiting downlink packets in a radio cell are sent one after another, and the terminals pick up their respective packets, see Fig. C.19.

Most users will resort to a standard terminal of approx. 30 cm antenna diameter, providing 16 kb/s ... 2.048 Mb/s on the uplink and 16 kb/s ... 64 Mb/s on the downlink. Broadband terminals can provide a two-way capacity of 64 Mb/s or more. The terminals perform the translation between the Teledesic network's internal protocols and the standard protocols of the terrestrial world.

The development of the Teledesic system can be characterized by the following steps:

1990:   founding of the Teledesic Corp.
March 1994:   FCC filing
March 1997:   FCC license to build, launch, and operate the system
Nov. 1997:   allocation of international radio spectrum to non-geostationary fixed satellite services
Dec. 1999:   Teledesic invests in the ICO system and aims to use ICO as a precursor to Teledesic
2001:   first satellite launches
2004:   commencement of commercial operation (as of Dec. 1999).

As of 1998, a cost of US-$ 9 billion is estimated for the Teledesic system.

More detailed information on the Teledesic system can be found at http://www.teledesic.com.

Cell Scan Pattern

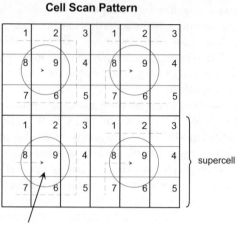

cell 9 illuminated in all supercells

Cell Scan Cycle

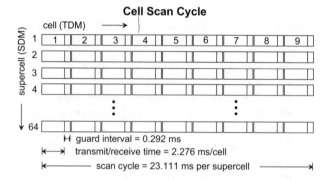

guard interval = 0.292 ms
transmit/receive time = 2.276 ms/cell
scan cycle = 23.111 ms per supercell

Channel Multiplexing in a Cell

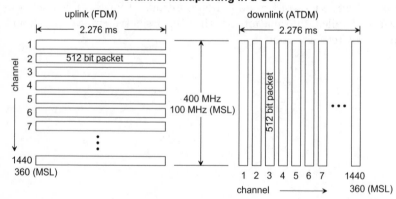

Fig. C.19. Teledesic multiplexing scheme [Stu95]. MSL = mobile satellite link

References

[Abr77]   N. Abramson. The throughput of packet broadcasting channels. *IEEE Transactions on Communications*, vol. **25**, pages 117–128, 1977.

[AGR98]   P. Alexander, A. Grant, and M. Reed. Iterative detection in code-division multiple-access interference suppression algorithms for CDMA systems. *European Transactions on Telecommunications*, vol. **9**, pages 419–426, 1998.

[AN94]   A. S. Acampora and M. Naghshineh. An architecture and methodology for mobile-executed handoff in cellular ATM networks. *IEEE Journal on Selected Areas in Communications*, vol. **12**, pages 1365–1375, 1994.

[Ash97]   G. R. Ash. *Dynamic Routing in Telecommunications Networks*. McGraw-Hill, New York, 1st edition, 1997.

[Ass99]   L. T. Associates. *The Complete Book on Mobile Satellites: Systems, Services and Markets*. Phillips Business Information, 1999.

[ATM96]   Traffic management specification version 4.0. Technical report, ATM Forum, 1996.

[ATM98]   A progress report on the standards development for satellite ATM networks. Technical report, ATM Forum, 1998.

[AV95]   R. Akturan and W. J. Vogel. Optically derived elevation angle dependence of fading for satellite PCS. In *Proceedings 19th NASA Propagation Experimenters Meeting (NAPEX XIX)*, pages 127–132, 1995.

[Bai99]   N. Bains. The ICO system for personal communications by satellite. In *Proceedings 6th International Mobile Satellite Conference (IMSC '99)*, pages 88–93, 1999.

[BB95]   A. Bakre and B. R. Badrinath. I-TCP: Indirect TCP for mobile hosts. In *Proceedings International Conference on Distributed Computing Systems*, pages 136–143, 1995.

[BDM97]   R. Bolla, F. Davoli, and M. Marchese. Bandwidth allocation and admission control in ATM networks with service separation. *IEEE Communications Magazine*, vol. **35**, pages 130–137, May 1997.

[Bel63]   P. A. Bello. Characterization of randomly time-variant linear channels. *IEEE Transactions on Communications Systems*, pages 360–393, December, 1963.

[BER92]   G. Butt, B. G. Evans, and M. Richharia. Narrowband channel statistics from multiband propagation measurements applicable to high elevation angle land-mobile satellite systems. *IEEE Journal on Selected Areas in Communications*, vol. **10**, pages 1219–1226, 1992.

[Bes78]   D. C. Beste. Design of satellite constellations for optimal continuous coverage. *IEEE Transactions on Aerospace and Electronic Systems*, vol. **14**, pages 466–473, 1978.

[BG87]   D. Bertsekas and R. G. Gallager. *Data Networks*. Prentice Hall, Englewood Cliffs, NJ, 1987.

[BJL93]   A. Böttcher, A. Jahn, and M. Lazzari. Performance evaluation of channel-sensitive, stabilized multiple access schemes for land-mobile satellite services, using a hybrid simulation tool. *IEEE Journal on Selected Areas in Communications*, vol. **11**, pages 443–453, 1993.

[BJL96a]   H. Bischl, A. Jahn, and E. Lutz. Wideband channel model for UMTS satellite communications – detailed model. In *ETSI/SMG5 #16, March 18-22*, 1996.

[BJL96b]   H. Bischl, A. Jahn, and E. Lutz. Wideband channel model for UMTS satellite communications – tapped delay model. In *ETSI/SMG5 #16, March 18-22*, 1996.

[BJL96c]   H. Bischl, A. Jahn, and E. Lutz. Wideband channel model for UMTS satellite communications. In *ITU, REVAL, ITU-TG81*, 1996.

[BL95]   H. Bischl and E. Lutz. Packet error rate in the non-interleaved Rayleigh channel. *IEEE Transactions on Communications*, vol. **43**, pages 1375–1382, 1995.

[BL99]   H. Bischl and E. Lutz. ATM cell error rate in non-interleaved fading channels. *Archiv für Elektronik und Übertragungstechnik*, vol. **53**, pages 179–186, 1999.

[Bla83]   R. E. Blahut. *Theory and Practice of Error Control Codes*. Addison–Wesley, Reading, MA, 1983.

[BMW71]   R. Bate, D. Mueller, and J. White. *Fundamentals of Astrodynamics*. Dover Publications, New York, 1971.

[BN96]   B. Bjelajac and A. Nöllchen. Dynamic transceiver, carrier and time-slot allocation strategies for mobile satellite systems. In *Mobile and Personal Satellite Communications 2 – Proceedings 2nd European Workshop on Mobile/Personal Satcoms (EMPS '96)*, pages 270–285. Springer, London, 1996.

[Bos99]   J. Bostic. MAC protocol issues for multimedia satellite systems. In *Proceedings 6th International Mobile Satellite Conference (IMSC '99)*, pages 303–309, 1999.

[BÖW⁺98]   J. Bostic, T. Örs, M. Werner, H. Bischl, and B. Evans. Multiple access protocols for ATM over low earth orbit satellites. In *Proceedings 1st Joint COST 252/259 Workshop*, pages 77–86, 1998.

[Bur97]   K. Burchard. Application of virtual path concepts to the broadband LEO satellite system M-Star. Master's thesis, Technical University Munich, Institute of Communication Networks, Munich, Germany, 1997.

[BW94]   A. Böttcher and M. Werner. Strategies for handover control in low earth orbit satellite systems. In *Proceedings 44th IEEE Vehicular Technology Conference (VTC '94)*, pages 1616–1620, 1994.

[BW97]   H. Bischl and M. Werner. Channel adaptive satellite diversity for non-geostationary mobile satellite systems. In *Proceedings 5th International Mobile Satellite Conference (IMSC '97)*, pages 25–31, 1997.

[BWL96]   H. Bischl, M. Werner, and E. Lutz. Elevation-dependent channel model and satellite diversity for NGSO S-PCNs. In *Proceedings Vehicular Technology Conference (VTC '96)*, pages 1038–1042, 1996.

[CB95]   P. Carter and M. A. Beach. Evaluation of handover mechanisms in shadowed low earth orbit land mobile satellite systems. *International Journal of Satellite Communications*, vol. **13**, pages 177–190, 1995.

[CBTE94]   C. Cullen, X. Benedicto, R. Tafazolli, and B. Evans. Network and common channel signalling aspects of dynamic satellite constellations. *International Journal of Satellite Communications*, vol. **12**, pages 125–134, 1994.

[CC81]   G. C. Clark and J. B. Cain. *Error-Correction Coding for Digital Communications*. Plenum Press, New York, 1981.

[CCF$^+$92]   C. Caini, G. E. Corazza, G. Falciasecca, M. Ruggieri, and F. Vatalaro. A spectrum- and power-efficient EHF mobile satellite system to be integrated with terrestrial cellular systems. *IEEE Journal on Selected Areas in Communications*, vol. **10**, pages 1315–1325, 1992.

[CCI90]   Satellite antenna patterns in the fixed-satellite service. CCIR Report 558-3 (MOD F), Document 4/1045-E, CCIR XVIIth Plenary Assembly, Genf, 1990.

[CGH$^+$92]   W. J. Ciesluk, L. M. Gaffney, N. D. Hulkover, L. Klein, P. A. Oliver, M. S. Pavloff, R. A. Pompoui, and W. A. Welch. An evaluation of selected mobile satellite communications systems and their environment. Esa contract no. 9563/91/NL/RE, MITRE Corp., 1992.

[CI95]   R. Caceres and L. Iftode. Improving the performance of reliable transport protocols in mobile computing environments. *IEEE Journal on Selected Areas in Communications*, vol. **13**, pages 850–857, 1995.

[CI996]   When worlds collide. *Communications International*, pages 25–26, December 1996.

[CJLV95]   G. E. Corazza, A. Jahn, E. Lutz, and F. Vatalaro. Channel characterization for mobile satellite communications. In F. Ananasso and F. Vatalaro, editors, *Mobile and Personal Satellite Communications*, pages 225–250. Springer, London, 1995.

[Cla45]   A. C. Clarke. Extra-terrestrial relays. *Wireless World*, vol. **51**, pages 305–307, 1945.

[CLS96]   T. Chen, S. Liu, and V. Samalam. The available bit rate service for data in ATM networks. *IEEE Communications Magazine*, vol. **34**, pages 56–71, May 1996.

[Col89]   R. E. Collin. *Antennas and Radiowave Propagation*. McGraw–Hill, New York, 1989.

[CTW91]   J. P. Campbell, T. E. Tremain, and V. C. Welch. The federal standard 1016 4800 bps CELP voice coder. *Digital Signal Processing*, vol. **1**, pages 145–155, 1991.

[CV94]   G. E. Corazza and F. Vatalaro. A statistical model for land mobile satellite channels and its application to non-geostationary orbit systems. *IEEE Transactions on Vehicular Technology*, vol. **43**, pages 738–742, 1994.

[Dav85]   M. R. Davidoff. *The Satellite Experimenter's Handbook*. The American Radio Relay League, 225 Main Street, Newington, CT 06111, 1985.

[Dav94]   F. Davarian. Earth-satellite propagation research. *IEEE Communications Magazine*, pages 74–79, April, 1994.

[Del95]   E. Del Re. The GSM procedures in an integrated cellular/satellite system. *IEEE Journal on Selected Areas in Communications*, vol. **13**, pages 421–430, 1995.

[DFG95]   E. Del Re, R. Fantacci, and G. Giambene. Efficient dynamic channel allocation techniques with handover queueing for mobile satellite networks. *IEEE Journal on Selected Areas in Communications*, vol. **13**, pages 397–405, 1995.

[DGL96]   R. De Gaudenzi, F. Giannetti, and M. Luise. Advances in satellite CDMA transmission for mobile and personal communications. *Proceedings of the IEEE*, vol. **84**, pages 18–39, 1996.

[DH95]   A. Duel-Hallen. A family of multi-user decision-feedback detectors for asynchronous code-division multiple access channels. *IEEE Transactions on Communications*, vol. **43**, pages 421–434, 1995.

[DH98]   B. Doshi and P. Harshavardhana. Broadband network infrastructure of the future: Roles of network design tools in technology deployment strategies. *IEEE Communications Magazine*, vol. **36**, pages 60–71, May 1998.

[Die97]   F. J. Dietrich. The Globalstar satellite cellular communication system design and status. In *Proceedings 5th International Mobile Satellite Conference (IMSC '97)*, pages 139–144, 1997.

[Die98]   F. J. Dietrich. The Globalstar satellite cellular communication system: design and status. In *Proceedings 17th AIAA International Communications Satellite Systems Conference and Exhibition*, pages 47–56, 1998.

[Dij59]   E. W. Dijkstra. A note on two poblems in connection with graphs. *Numerische Mathematik*, vol. **1**, pages 269–271, 1959.

[Dos95]   F. Dosière. *Disponibilité des services d'un réseau de télécommunications par satellites en orbites basses*. PhD thesis, Ecole Nationale Supérieure des Télécommunications, Toulouse, France, 1995.

[Dum98]   P. Dumont. The regulatory status of the Ka-band. In *Proceedings 4th Ka Band Utilization Conference*, pages 613–620, 1998.

[EGMG97]   E. Elizondo, R. Gobbi, A. Modelfino, and F. Gargione. Evolution of the Astrolink system. In *Proceedings 3rd Ka Band Utilization Conference*, pages 3–7, 1997.

[ESA98a]   Wideband CDMA option for the satellite component of IMT-2000 "SW-CDMA". Proposal to ITU of a candidate RTT, ESA/ESTEC, 1998.

[ESA98b]   Wideband hybrid CDMA/TDMA option for the satellite component of IMT-2000 "SW-CTDMA". Proposal to ITU of a candidate RTT, ESA/ESTEC, 1998.

[EV98]   J. Eberspächer and H.-J. Vögel. *GSM – Switching, Services and Protocols*. John Wiley, Chichester, 1998.

[EVW96]   M. Eisenschmid, H.-J. Vögel, and M. Werner. Handover signalling in LEO/MEO satellite systems. In *Proceedings 2nd International Conference on Personal, Mobile and Spread Spectrum Communications (ICPMSC '96)*, pages 117–120, 1996.

[FPEH+95] R. J. Finean, D. Polymeros, A. El-Hoiydi, F. da Costa, M. Dinis, A. Saidi, and B. Vazvan. Impact of satellites on UMTS network. In *RACE summit*, pages 388–392, 1995.

[FRM97]   M. Forest, S. Richard, and C. A. McDonald. ACeS antenna feed arrays. In *Proceedings 5th International Mobile Satellite Conference (IMSC '97)*, pages 387–391, 1997.

[GEV92]   R. D. Gaudenzi, C. Elia, and R. Viola. Bandlimited quasi-synchronous CDMA: A novel satellite access technique for mobile and personal communication systems. *IEEE Journal on Selected Areas in Communications*, vol. **10**, pages 328–343, 1992.

[GGV93]   D. J. Goodman, S. A. Grandhi, and R. Vijayan. Distributed dynamic channel asignment schemes. In *Proceedings IEEE 43th Vehicular Technology Conference (VTC '93)*, pages 532–535, 1993.

[Gib96]   J. D. Gibson. *The Mobile Communications Handbook*. CRC Press & IEEE Press, Boca Raton, FL, 1996.

[Gif96]   J. M. Gifford. What makes a little LEO tick? *Satellite Communications*, pages 39–43, June 1996.

[Gir90]   A. Girard. *Routing and Dimensioning in Circuit-Switched Networks*. Addison–Wesley, Reading, MA, 1990.

[GJ90]   I. A. Gerson and M. A. Jasiuk. Vector sum excited linear prediction (VSELP) speech coding at 8 kbps. In *Proceedings IEEE International*

Conference on Acoustics, Speech and Signal Processing (ICASSP '91), pages 461–464, 1990.

[GJPW90]   K. S. Gilhousen, I. M. Jacobs, R. Padovani, and L. A. Weaver. Increased capacity using CDMA for mobile satellite communications. *IEEE Journal on Selected Areas in Communications*, vol. **8**, pages 503–514, 1990.

[GKOH94]   L. M. Gaffney, L. Klein, P. A. Oliver, and N. D. Hulkover. An evaluation of the Spaceway and Teledesic systems. Esa contract no. 9563/91/NL/RE Rider 3, MITRE Corp., 1994.

[Glo95]   Description of the Globalstar system. Technical information note, Globalstar Inc., August 1995.

[GM93]   G. D. Gordon and W. L. Morgan. *Principles of Communications Satellites*. John Wiley, New York, 1993.

[Gol68]   R. Gold. Maximal recursive sequences with 3-valued recursive cross correlation functions. *IEEE Transactions on Information Theory*, vol. **14**, pages 154–156, 1968.

[Gou96]   R. G. Gould. The WRC-95. In *Mobile and Personal Satellite Communications 2 – Proceedings 2nd European Workshop on Mobile/Personal Satcoms (EMPS '96)*, pages 553–563. Springer, London, 1996.

[GST99]   L. Ghedia, K. Smith, and G. Titzer. Satellite PCN – the ICO system. *International Journal of Satellite Communications*, vol. **17**, pages 273–289, 1999.

[GV94]   J. Goldhirsch and W. J. Vogel. Mobile satellite propagation measurements from UHF to K band. In *Proceedings 15th AIAA International Communications Satellite Systems Conference*, pages 913–920, 1994.

[GYG97]   S. A. Grandhi, R. D. Yates, and D. J. Goodman. Resource allocation for cellular radio systems. *IEEE Transactions on Vehicular Technology*, vol. **46**, pages 581–587, 1997.

[Hag80]   J. Hagenauer. Zur Kanalkapazität bei Nachrichtenkanälen mit Fading und gebündelten Fehlern. *Archiv für Elektronik und Übertragungstechnik*, vol. **34**, pages 229–237, 1980.

[Hag82]   J. Hagenauer. Fehlerkorrektur und Diversity-Verfahren bei Fading-Kanälen. *Archiv für Elektronik und Übertragungstechnik*, vol. **36**, pages 337–344, 1982.

[HDL+87]   J. Hagenauer, F. Dolainsky, E. Lutz, W. Papke, and R. Schweikert. The maritime satellite communication channel – channel model, performance of modulation and coding. *IEEE Journal on Selected Areas in Communications*, vol. **5**, pages 701–713, 1987.

[He96]   X. He. *The Signalling System in Satellite Personal Communication Networks*. PhD thesis, University of Surrey, UK, 1996.

[HH91]   R. Händel and M. Huber. *Integrated Broadband Networks: An Introduction to ATM-Based Networks*. Addison–Wesley, London, 1991.

[HJW+97]   P. Höher, A. Jahn, T. Wörz, A. Schmidbauer, and R. Schweikert. A suitability study of satellite emulation by an airborne platform. *International Journal of Satellite Communications*, vol. **15**, pages 51–64, 1997.

[HK97]   T. R. Henderson and R. H. Katz. Satellite transport protocol (STP): An SSCOP-based transport protocol for datagram satellite networks. In *Proceedings 2nd Workshop on Satellite-Based Information Systems*, pages 23–34, 1997.

[HL91]   J. C. Hardwick and J. S. Lim. The application of the IMBE speech coder to mobile communications. In *Proceedings IEEE International Conference on Acoustics, Speech and Signal Processing (ICASSP '91)*, pages 249–252, 1991.

[HL95]   J. Hutcheson and M. Laurin. Network flexibility of the IRIDIUM global mobile satellite system. In *Proceedings 4th International Mobile Satellite Conference (IMSC '95)*, pages 503–507, 1995.

[HM77]   F. Hansen and F. I. Meno. Mobile fading – Rayleigh and lognormal superimposed. *IEEE Transactions on Vehicular Technology*, vol. **26**, pages 332–335, 1977.

[HS99]   Y. F. Hu and R. E. Sheriff. Evaluation of the European market for satellite-UMTS terminals. *International Journal of Satellite Communications*, vol. **17**, pages 305–323, 1999.

[HSD$^+$98]   Y. F. Hu, R. E. Sheriff, E. Del Re, R. Fantacci, and G. Giambene. Satellite-UMTS traffic dimensioning and resource management technique analysis. *IEEE Transactions on Vehicular Technology*, vol. **47**, pages 305–323, 1998.

[Hub96]   J. F. Huber. Integration of satellite PCN's into terrestrial networks and the way towards UMTS. In *Mobile and Personal Satellite Communications 2 – Proceedings 2nd European Workshop on Mobile/Personal Satcoms (EMPS '96)*, pages 564–578, 1996. Springer.

[ICO97]   The ICO system: a technical description. Technical report, ICO, 1997.

[Jah94]   A. Jahn. Measurement programme for generic satellite channels. Final Report, Inmarsat Purchase Order P004001, DLR, Institut für Nachrichtentechnik, 1994.

[JBH96]   A. Jahn, H. Bischl, and G. Heiß. Channel characterisation for spread spectrum satellite communications. In *Proceedings IEEE 4th International Symposium on Spread Spectrum Techniques and Applications (ISSSTA '96)*, pages 1221–1226, 1996.

[JH98]   A. Jahn and M. Holzbock. EHF-band channel characterisation for mobile multimedia satellite services. In *Proceedings 48th IEEE Vehicular Technology Conference (VTC '98)*, pages 209–212, 1998.

[JL94]   A. Jahn and E. Lutz. DLR channel measurement programme for low earth orbit satellite systems. In *Proceedings International Conference on Universal Personal Communications (ICUPC '94)*, pages 423–429, 1994.

[JL96]   A. Jahn and E. Lutz. Channel measurements for EHF-band land mobile satellite systems. In *Mobile and Personal Satellite Communications 2 – Proceedings 2nd European Workshop on Mobile/Personal Satcoms (EMPS '96)*, pages 419–432. Springer, London, 1996.

[JL97]   A. Jahn and E. Lutz. LMS channel measurements at EHF-band. In *Proceedings 5th International Mobile Satellite Conference (IMSC '97)*, pages 183–188, 1997.

[JSBL95a]   A. Jahn, M. Sforza, S. Buonomo, and E. Lutz. Narrow- and wideband channel characterization for land mobile satellite systems: Experimental results at L-band. In *Proceedings 4th International Mobile Satellite Conference (IMSC '95)*, pages 115–121, 1995.

[JSBL95b]   A. Jahn, M. Sforza, S. Buonomo, and E. Lutz. A wideband channel model for land mobile satellite systems. In *Proceedings 4th International Mobile Satellite Conference (IMSC '95)*, pages 122–127, 1995.

[KKB96]   A. Klein, G. K. Kaleh, and P. W. Baier. Zero forcing and minimum mean-square error equalization for multi-user detection in code-division multiple access channels. *International Journal of Satellite Communications*, vol. **14**, pages 276–287, 1996.

[Kle76]   L. Kleinrock. *Queueing Systems, Vol.1 Theory*. John Wiley, New York, 1976.

[KV92]   N. Kleiner and W. J. Vogel. Impact of propagation impairments on optimal personal mobile SATCOM system design. In *Proceedings Mobile/Personal Communications Systems*, 1992.

[LC83]   S. Lin and D. J. Costello. *Error Control Coding: Fundamentals and Applications*. Prentice Hall, Englewood Cliffs, NJ, 1983.

[LCD⁺91]   E. Lutz, D. Cygan, M. Dippold, F. Dolainsky, and W. Papke. The land mobile satellite communication channel — recording, statistics and channel model. *IEEE Transactions on Vehicular Technology*, vol. **40**, pages 375–386, 1991.

[Lee93]   W. C. Y. Lee. *Mobile Communications Design Fundamentals*. John Wiley, New York, 1993.

[LJWB96]   E. Lutz, A. Jahn, M. Werner, and H. Bischl. DLR activities in the field of personal satellite communications systems. *Space Communications*, vol. **14**, pages 111–121, 1996.

[LLV97]   G. Losquadro, M. Luglio, and F. Vatalaro. A geostationary satellite system for mobile multimedia applications using portable, aeronautical and mobile terminals. In *Proceedings 5th International Mobile Satellite Conference (IMSC '97)*, pages 427–432, 1997.

[LMN94]   Y.-B. Lin, S. Mohan, and A. Noerpel. PCS channel assignment strategies for hand-off and initial access. *IEEE Personal Communication*, vol. **1**, pages 47–56, 1994.

[Loo85]   C. Loo. A statistical model for a land mobile satellite link. *IEEE Transactions on Vehicular Technology*, vol. **34**, pages 122–127, 1985.

[Lut92]   E. Lutz. Slotted Aloha multiple access and error control coding for land mobile satellite networks. *International Journal of Satellite Communications*, vol. **10**, pages 275–281, 1992.

[Lut95]   G. Luton. An ATM based concept for handover operation in LEO/MEO satellite systems. Master's thesis, ENST Toulouse/DLR Oberpfaffenhofen, Toulouse, France/Wessling, Germany, 1995.

[Lut96]   E. Lutz. A Markov model for correlated land mobile satellite channels. *International Journal of Satellite Communications*, vol. **14**, pages 333–339, 1996.

[Lut97]   E. Lutz. Other-cell interference in satellite power-controlled CDMA uplink. In *Proceedings 5th International Mobile Satellite Conference (IMSC '97)*, pages 83–87, 1997.

[Lut98]   E. Lutz. Issues in satellite personal communication systems. *Wireless Networks*, vol. **4**, pages 109–124, 1998.

[Mat99]   C. Matarasso. TCP/IP over satellite. Internal report, Deutsches Zentrum für Luft- und Raumfahrt, 1999.

[Maz99]   S. Mazur. A description of current and planned location strategies within the ORBCOMM network. *International Journal of Satellite Communications*, vol. **17**, pages 209–223, 1999.

[MB93]   G. Maral and M. Bousquet. *Satellite Communications Systems*. John Wiley, Chichester, 1993.

[MC94]   P. Monte and S. Carter. The Globalstar air interface: modulation and access. In *Proceedings AIAA International Communications Satellite Systems Conference*, pages 1614–1621, 1994.

[MDER91]   G. Maral, J.-J. De Ridder, B. G. Evans, and M. Richharia. Low earth orbit satellite systems for communications. *International Journal of Satellite Communications*, vol. **9**, pages 209–225, 1991.

[MH81]   K. Murota and K. Hirade. GMSK modulation for digital mobile radio telephony. *IEEE Transactions on Communications*, vol. **29**, pages 1044–1050, 1981.

[ML85]   A. M. Michelson and A. H. Levesque. *Error-Control Techniques for Digital Communication*. John Wiley, New York, 1985.

[MM91]   J. L. Massey and T. Mittelholzer. Technical assistance for the CDMA communication system analysis. Final report, ESTEC contract No. 8696/89/NL/US, 1991.

[Mon95]   P. Monsen. Multiple-access capacity in mobile user satellite systems. *IEEE Journal on Selected Areas in Communications*, vol. **13**, pages 222–231, 1995.

[Mos96]   S. Moshavi. Multi-user detection for DS-CDMA communications. *IEEE Communications Magazine*, pages 124–136, October 1996.

[Mot96]   Motorola Satellite Systems, Inc. Application for authority to construct, launch and operate the M-Star system. FCC filing, Washington, DC, USA, 1996.

[MP92]   M. Mouly and M.-B. Pautet. *The GSM System for Mobile Communications*. Michel Mouly and Marie-Bernadette Pautet, 49, rue Louise Bruneau, F-91120 Palaiseau, 1992.

[MS98]   F. Makita and K. Smith. Design and implementation of ICO system. In *Proceedings 17th AIAA International Communications Satellite Systems Conference and Exhibition*, pages 57–65, 1998.

[MSN96]   *Mobile Satellite News*, July 7, 1996.

[MSN99]   *Mobile Satellite News*, pages 4–5, February 1999.

[MZ96]   D. Mitra and I. Ziedins. Virtual partitioning by dynamic priorities: fair and efficient resource sharing by several services. In *Proceedings International Zurich Seminar on Digital Communications*, pages 173–185, 1996.

[NBA97]   N. P. Nguyen, P. A. Buhion, and A. R. Adiwoso. The Asia cellular satellite system. In *Proceedings 5th International Mobile Satellite Conference (IMSC '97)*, pages 145–152, 1997.

[Nel95]   R. A. Nelson. Satellite constellation geometry. *Via Satellite*, pages 110–122, March 1995.

[Nol97]   P. Noll. MPEG audio coding standards. *IEEE Signal Processing Magazine*, September 1997.

[Pap65]   A. Papoulis. *Probability, Random Variables, and Stochastic Processes*. McGraw-Hill, London, 1965.

[PAP⁺95] S. Paul, E. Ayanoglu, T. F. L. Porta, K.-W. H. Chena, K. K. Sabnani, and R. D. Gitlin. An asymmetric link-layer protocol for digital cellular communications. In *Proceedings IEEE InfoCom '95*, pages 1053–1062, 1995.

[Par91]   J. D. Parsons. Sounding techniques for wideband mobile radio channels: A review. *Proceedings of the IEE*, vol. **138**, pages 437–446, 1991.

[Pat98]   B. Pattan. *Satellite-Based Cellular Communications*. McGraw-Hill, New York, 1998.

[PEBB96] M. A. N. Parks, B. G. Evans, G. Butt, and S. Buonomo. Simultaneous wideband propagation measurements applicable to mobile satellite communication systems at L and S-band. In *AIAA 16th International Communications Satellite Systems Conference (ICSSC '96)*, pages 929–936, 1996.

[PFV⁺99] A. Pandolfi, F. Fedi, S. Vahid, R. Tafazolli, B. G. Evans, N. Blefari-Melazzi, G. Reali, and C. Matarasso. A satellite network for broadband interactive services. In *Proceedings Vehicular Technology Conference (VTC '99)*, pages 2760–2764, 1999.

[PL95]   K. Pahlavan and A. H. Levesque. *Wireless Information Networks*. John Wiley, New York, 1995.

[PR98]   J. N. Pelton and A. U. M. Rae. Global satellite communications tech-
            nology and systems. WTEC panel report, International Technology
            Research Institute, 1998.

[PRFT99]   S. R. Pratt, R. A. Raines, C. E. Fossa, and M. A. Temple. An opera-
            tional and performance overview of the Iridium low earth orbit satellite
            system. *IEEE Communications Surveys*, pages 2–10, 2nd Quarter, 1999.

[Pri84]   W. L. Pritchard. Estimating the mass and power of communication
            satellites. *International Journal of Satellite Communications*, vol. **2**,
            pages 107–112, 1984.

[Pro89]   J. G. Proakis. *Digital Communications*. McGraw-Hill, New York, 1989.

[Pry95]   M. D. Prycker. *Asynchronous Transfer Mode*. Prentice Hall, London,
            1995.

[PS97]   C. Partridge and T. J. Shepard. TCP/IP performance over satellite
            links. *IEEE Network*, vol. **11**, pages 44–49, September/October 1997.

[PSN93]   W. L. Pritchard, H. G. Snyderhoud, and R. A. Nelson. *Satellite Com-
            munication Systems Engineering*. Prentice Hall, Englewood Cliffs, NJ,
            1993.

[Rah93]   M. Rahnema. Overview of the GSM system and protocol architecture.
            IEEE Communications Magazine, pages 92–100, April 1993.

[REE92]   P. P. Robet, B. G. Evans, and A. Ekman. Land mobile satellite com-
            munication channel model for simultaneous transmission from a land
            mobile terminal via two separate satellites. *International Journal of
            Satellite Communications*, vol. **10**, pages 139–154, 1992.

[Ric99]   M. Richharia. *Satellite Communications Systems*. McGraw-Hill, New
            York, 1999.

[RM96a]   J. Restrepo and G. Maral. Analysis and comparison of satellite constel-
            lation configurations for single permanent visibility. In *IEEE Confer-
            ence on Satellite Systems for Mobile Communications and Navigation*,
            pages 102–105, 1996.

[RM96b]   J. Restrepo and G. Maral. Constellation sizing for non-geo 'earth-fixed
            cell' satellite systems. In *Proceedings AIAA International Communica-
            tions Satellite Systems Conference (ICSSC '96)*, pages 768–778, 1996.

[RM96c]   J. Restrepo and G. Maral. Providing appropriate service quality to
            fixed and mobile users in a non-geo satellite-fixed cell system. In *Mobile
            and Personal Satellite Communications 2 – Proceedings 2nd European
            Workshop on Mobile/Personal Satcoms (EMPS '96)*, pages 79–96, 1996.
            Springer.

[RSHP96]   M. Rice, J. Slack, B. Humpherys, and D. Pinck. K-band land-mobile
            satellite channel characterization using ACTS. *International Journal
            of Satellite Communications*, vol. **14**, pages 283–296, 1996.

[RU96]   B. Rimoldi and R. Urbanke. A rate-splitting approach to the Gaussian
            multiple-access channel. *IEEE Transactions on Information Theory*,
            vol. **42**, pages 364–375, 1996.

[SA85]   M. R. Schroeder and B. S. Atal. Code-excited linear prediction (CELP):
            high-quality speech at very low bit rates. In *Proceedings IEEE In-
            ternational Conference on Acoustics, Speech and Signal Processing
            (ICASSP '85)*, pages 937–940, 1985.

[SAI95]   Operations and procedures in integrated UMTS-satellite network. De-
            liverable 15, SAINT, Race project 2117, 1995.

[SBM93]   M. Sforza, S. Buonomo, and A. Martini. ESA research activities in
            the field of channel modelling and simulation for land mobile satellite
            systems. COST 227: Land mobile satellite communications systems,
            COST 227 TD (93) 044, 1993.

[Sch87]   M. Schwartz. *Telecommunication Networks: Protocols, Modeling and Analysis.* Series in Electrical and Computer Engineering. Addison–Wesley, Reading, MA, 1987.

[SIAK95]   H. Shinanoga, H. Ishikawa, N. Araki, and H. Kobayashi. KDD's activities related to Inmarsat P. In *AIAA/ESA Workshop on International Cooperation in Satellite Communications*, pages 325–335, 1995.

[Sie95]   R. Siebenhaar. Multiservice call blocking approximations for virtual path based ATM networks with CBR and VBR traffic. In *Proceedings IEEE InfoCom '95*, pages 321–329, 1995.

[ST99]   Z. Sahinoglu and S. Tekinay. On multimedia networks: Self-similar traffic and network performance. *IEEE Communications Magazine*, vol. **37**, pages 48–52, January 1999.

[Sta98]   W. Stallings. *High-Speed Networks: TCP/IP and ATM Design Principles.* Prentice Hall, Upper Saddle River, NJ, 1998.

[Stu95]   M. A. Sturza. Architecture of the Teledesic satellite system. In *Proceedings 4th International Mobile Satellite Conference (IMSC '95)*, pages 212–218, 1995.

[Tan96]   A. S. Tanenbaum. *Computer Networks.* Prentice Hall, London, 3rd edition, 1996.

[TIA98]   TIA/EIA. Satellite ATM networks: architecture and guidelines. *TIA/EIA Telecommunications Systems Bulletin (TSB)*, 1998.

[Tre82]   T. E. Tremain. The government standard linear predictive coding algorithm: LPC-10. *Speech Technology*, pages 40–49, April 1982.

[Tro97]   B. Troy. The EAST project (Euro African Satellite Communications) an opportunity for the armed forces to use commercial Satcom system. In *Proceedings Milcom*, 1997.

[VA90]   M. K. Varanasi and B. Aazhang. Multistage detection in asynchronous code-division multiple-access communications. *IEEE Transactions on Communications*, vol. **38**, pages 509–519, 1990.

[Ver86]   S. Verdu. Minimum probability of error for asynchronous Gaussian multiple-access channels. *IEEE Transactions on Information Theory*, vol. **32**, pages 85–96, 1986.

[Vit94]   A. J. Viterbi. The orthogonal-random waveform dichotomy for digital mobile personal communication. *IEEE Personal Communications*, vol. **1**, pages 18–24, 1994.

[Vit95]   A. J. Viterbi. *CDMA Principles of Spread Spectrum Communication.* Adison-Wesley, Reading, MA, 1995.

[VTSC+95] J. Ventura-Traveset, I. Stojkovic, F. Coromina, J. Benedicto, and F. Petz. Key payload technologies for future satellite personal communications: a European perspective. *International Journal of Satellite Communications*, vol. **13**, pages 117–135, 1995.

[VV93]   A. M. Viterbi and A. J. Viterbi. Erlang capacity of a power controlled CDMA system. *IEEE Journal on Selected Areas in Communications*, vol. **11**, pages 892–900, 1993.

[VVZ94]   A. J. Viterbi, A. M. Viterbi, and E. Zehavi. Other-cell interference in cellular power-controlled CDMA. *IEEE Transactions on Communications*, vol. **42**, pages 1501–1504, 1994.

[Wal77]   J. G. Walker. Continuous whole-earth coverage by circular-orbit satellite patterns. Technical Report 77044, Royal Aircraft Establishment, 1977.

[Wat99]   T. Watts. Clearing the hurdles: The satcom industry focuses on execution. Research report, Merrill Lynch, 1999.

[WBO+98]   M. Werner, J. Bostič, T. Örs, H. Bischl, and B. G. Evans. Multiple access for ATM-based multiservice satellite networks. In D. W. Faulkner and A. L. Harmer, editors, *Broadband Access and Network Management, Proceedings 3rd European Conference on Networks & Optical Communications (NOC '98), Part I*, pages 290–297, 1998. IOS Press. Invited Paper.

[WDV+97]   M. Werner, C. Delucchi, H.-J. Vögel, G. Maral, and J.-J. De Ridder. ATM-based routing in LEO/MEO satellite networks with intersatellite links. *IEEE Journal on Selected Areas in Communications*, vol. **15**, pages 69–82, 1997.

[Wer95]   M. Werner. Analysis of system connectivity and traffic capacity requirements for LEO/MEO S-PCNs. In E. Del Re, editor, *Mobile and Personal Communications, Proceedings 2nd Joint COST 227/231 Workshop*, pages 183–204, 1995. Elsevier.

[Wer97]   M. Werner. A dynamic routing concept for ATM-based satellite personal communication networks. *IEEE Journal on Selected Areas in Communications*, vol. **15**, pages 1636–1648, 1997.

[WF99]   M. Werner and J. Frings. Network design of the intersatellite link segment in broadband LEO satellite systems. In *Multimedia Satellite Networks, hot topic session at the 16th International Teletraffic Congress (ITC-16)*, page unpaginated, 1999.

[WHS97]   M. Werner, Y. F. Hu, and R. Sheriff. Network dimensioning. In *SEC-OMS – Satellite EHF Communications for Mobile Multimedia Services*, European ACTS Project AC004. Deliverable D22 (Final Report WP 3500), AC004/DLR-NTD/DR-R/022/b1, 1997. EU Commission.

[WJLB95]   M. Werner, A. Jahn, E. Lutz, and A. Böttcher. Analysis of system parameters for LEO/ICO-satellite communications networks. *IEEE Journal on Selected Areas in Communications*, vol. **13**, pages 371–381, 1995.

[WL98a]   M. Werner and E. Lutz. Bandwidth requirements for multiservice satellite systems. In *Proceedings IEEE International Conference on Communications (ICC '98)*, pages 94–99, 1998.

[WL98b]   M. Werner and E. Lutz. Multiservice traffic model and bandwidth demand for broadband satellite systems. In M. Ruggieri, editor, *Mobile and Personal Satellite Communications 3, Proceedings 3rd European Workshop on Mobile/Personal Satcoms (EMPS '98)*, pages 235–253, 1998. Springer.

[WM97]   M. Werner and G. Maral. Traffic flows and dynamic routing in LEO intersatellite link networks. In *Proceedings 5th International Mobile Satellite Conference (IMSC '97)*, pages 283–288, 1997.

[WR99]   M. Werner and P. Révillon. Optimization issues in capacity dimensioning of LEO intersatellite link networks. In *Proceedings 5th European Conference on Satellite Communications (ECSC 5)*, 1999. CD-ROM.

[WWFM99]   M. Werner, F. Wauquiez, J. Frings, and G. Maral. Capacity dimensioning of ISL networks in broadband LEO satellite systems. In *Proceedings 6th International Mobile Satellite Conference (IMSC '99)*, pages 334–341, 1999.

[XSR90]   Z. Xie, R. T. Short, and C. K. Rushforth. A family of suboptimum detectors for coherent multi-user communications. *IEEE Journal on Selected Areas in Communications*, vol. **8**, pages 683–690, 1990.

[ZA99]   B. Zheng and M. Atiquzzaman. Traffic management of multimedia over ATM networks. *IEEE Communications Magazine*, vol. **37**, pages 33–38, Januar 1999.

[ZTE96]   W. Zhao, R. Taffazolli, and B. G. Evans. Combined handover algorithm
            for dynamic satellite constellations. *Electronic Letters*, vol. **32**, pages
            622–624, 1996.

Index

AAL – adaptation layer (ATM), 278, 279, 283
ABR – available bit rate (ATM), 281, 285, 301, 314
ACF – autocorrelation function, 138
ACI – adjacent channel interference, 127
ACK – acknowledgement, 290, 291
ADPCM – adaptive differential PCM, 85
AGCH – access grant channel, 206
altitude, 17, 19, 27, 29, 32, 50
AMBE – advanced multiband excitation, 387, 392
AMPS – advanced mobile phone system, 368
anomaly
– eccentric, 16
– mean, 16
– true, 16, 17
antenna
– aperture, 49
– beamwidth, 56
– characteristic, 49, 61
– diameter, 50
– directional, 48
– diversity, 80
– efficiency, 49, 51
– gain, 48, 50, 59
– isoflux, 249
– multifeed reflector, 246
– parabolic, 50
– pattern, 349, 350
– phased-array, 248
– spot beam, 58
– tapered, 50
AOCS – attitude and orbit control system, 243
apogee, 16, 44, 45
application, 7

argument of perigee, 19
– drift, 21
ARQ – automatic repeat request, 80, 110, 291, 313
ascending node, 19, 22, 30
asynchronous polar constellation, 34
ATM – asynchronous transfer mode, 277, 300
– AAL – adaptation layer, 278, 279, 283
– ABR – available bit rate, 281, 285, 301, 314
– CAC – connection admission control, 284, 285, 287, 304
– CBR – constant bit rate, 281, 285, 301, 314
– CDV – cell delay variation, 283
– CDVT – cell delay variation tolerance, 283
– cell, 280
– cell header, 280
– CLP – cell loss priority, 280
– CLR – cell loss ratio, 283
– congestion control, 305
– CTD – cell transfer delay, 283
– EPD – early packet discard, 313
– error control, 312, 313
– flow control, 306
– GFC – generic flow control, 280
– HEC – header error control, 280, 281, 312
– MBS – maximum burst size, 282
– MCR – minimum cell rate, 283
– mobile, 285
– nrt-VBR – non-real-time variable bit rate, 281, 285, 301, 314
– over satellite, 285
– PCR – peak cell rate, 282
– PHY – physical layer, 278, 279
– protocol architecture, 278, 303

– PT – payload type, 280
– reference model, 279
– resource management, 304
– rt-VBR – real-time variable bit rate, 281, 285, 301, 314
– SCR – sustainable cell rate, 282
– service categories, 281, 282, 302
– service classification, 283, 302
– traffic and congestion control, 284
– traffic descriptors, 282
– traffic shaping, 306
– UBR – unspecified bit rate, 281, 285, 301, 314
– UNI – user-network interface, 280
– UPC – usage parameter control, 284, 285, 305
– VC switching, 280
– VCI – virtual channel identifier, 280
– VP switching, 280
– VPI – virtual path identifier, 280
– wireless, 285

B-ISDN – Broadband Integrated Services Digital Network, 277, 299
backoff factor, 128
band-sharing, 163, 265
bandwidth
– allocation, 312
– cluster, 322
– demand, 315, 327
– requirements, 316
– system, 322
BCCH – broadcast control channel, 205
BCH code – Bose-Chaudhuri-Hocquenghem code, 105
beam-forming, 246
beamwidth, 50, 349
BER – bit error rate, 54, 56, 97
big-LEO, 11, 259, 263
blocking probability, 196, 216
Boltzmann constant, 53
boresight, 49, 50, 56
BPF – bandpass filter, 252
BPSK – binary phase-shift keying, 87
BSC – binary symmetric channel, 107
BSSMAP – base station system management application part, 205
busy hour, 319

C/N, 53, 54
C/N_0, 54
CAC – connection admission control (ATM), 284, 285, 287, 304

call completion probability, 224
call control, 212
call setup, 212
– mobile originating, 212, 213
– mobile terminating, 212, 215
capacity
– OD demand pair, 334
– requirements, 316
– worst-case link (ISL), 332, 334
capture effect, 125
Cartesian coordinate, 21, 22
CAS – carrier assignment strategy, 287, 311
CBR – constant bit rate (ATM), 281, 285, 301, 314
CC – call control, 204
CCCH – common control channel, 205
CCF – cross-correlation function, 139
CCI – co-channel interference, 171
CDM – code-division multiplex, 119
CDMA – code-division multiple access, 81, 136, 309, 310
– bandwidth efficiency, 149
– bit error probability, 147
– channel coding, 148
– chip, 137
– chip rate, 146
– echo, 157
– FH-CDMA – frequency hopping CDMA, 137
– MAI – multiple access interference, 146
– MF-CDMA – multi-frequency CDMA, 150, 190, 309
– other-cell interference factor, 183, 187
– polarization multiplex, 164
– processing gain, 146
– Qualcomm CDMA, 150
– quasi-synchronous CDMA, 169
– SW-C/TDMA – satellite wideband hybrid CDMA/TDMA, 165
– SW-CDMA – satellite wideband CDMA, 165
– synchronous CDMA, 144, 151
– throughput, 149
– voice activation, 164
CDV – cell delay variation (ATM), 283
CDVT – cell delay variation tolerance (ATM), 283
cell, 56
– bandwidth, 127
– cluster, 175

– overlapping, 348
– pattern, 349
CELP – code excited linear prediction, 86
CEPT – Conference Européene des Administrations des Postes et Telecommunications, 264
channel, 61
– allocation, 214
– coherence bandwidth, 70, 71, 80
– correlation, 76
– correlation coefficient, 77
– delay spread, 70
– direct path, 70, 72
– duration of a state, 62, 65
– echo, 70
– echo delay, 73
– echo power decay, 73
– far echo, 70, 72, 73
– impulse response, 70, 72
– line-of-sight, 67
– lognormal fading, 64, 65, 72
– multipath, 68
– narrowband, 62, 67
– narrowband model, 62, 65
– near echo, 70, 72, 73
– number of echoes, 73
– parameter, 65, 72, 74, 353, 356, 357, 360
– power delay profile, 70, 72
– Rayleigh fading, 64, 65, 72, 74
– Ricean fading, 64, 65, 72
– shadowing, 68, 70–72
– state, 62
– tapped delay-line model, 72
– time-share of shadowing, 65, 67, 75, 76
– traffic channels, see GSM signaling channels
– transfer function, 70
– two-state model, 62, 65, 76
– wideband model, 68, 72, 74
channelization, 246
circular orbit, 17
Clarke, Arthur C., 3
closed-loop power control, 164
CLP – cell loss priority (ATM), 280
CLR – cell loss ratio (ATM), 283
cluster
– bandwidth, 322
– size, 175
– worst-case, 322

CM – connection management, 204, 205
coherence bandwidth, 70, 71, 80–81
collision, 34
complementary error function, 95
connection
– end-to-end, 226
– setup, 212
connection-oriented service, 273, 283
connectionless service, 273, 283
constellation, 28
– asynchronous polar, 34, 407
– GEO, 34
– ICO, 42
– inclined, 328
– LEO, 37
– MEO, 41
– polar, 31, 41
– Walker, 30, 328
coordinate systems, 21, 345
– Cartesian, 21, 22, 345
– geocentric, 21, 345
– polar, 345
– spherical, 21, 345
correlation coefficient, 76, 77
country code, 265, 393
coverage, 29, 348, 350
– angle, 26, 27, 350
– area, 26, 27, 50, 56, 58
– double, 29, 74
– global, 28
CPFSK – continuous-phase FSK, 89
CTD – cell transfer delay (ATM), 283

dB-spread, 65
DBF – distributed Bellman–Ford algorithm, 237
DCA – dynamic channel allocation, 216
– C/I-based, 217
– CDMA with user-specific codes, 219
– centralized, 216
– channel borrowing, 219
– channel rearrangement, 218
– channel reuse optimization, 217
– cost function, 218
– hybrid channel allocation, 219
– interference-based, 217
DCAAS – dynamic channel activity assignment system, 397
DCT – discrete cosine transform, 275
decorrelating detector, 161
DECT – Digital Enhanced Cordless Telecommunications, 216

delay spread, 70
DEPSK – differentially encoded
    phase-shift keying, 88
digital speech interpolation, 120
dimensioning
– GEO spot beam capacity, 315
– ISL capacity, 328
– network, 315
direct signal, 61, 64, 70, 72
diversity, 38, 80
– antenna, 80
– frequency, 80
– satellite, 74, 75, 79–81
– time, 81
Doppler shift, 59
downlink, 47, 53, 55
DPCCH – dedicated physical control
    channel, 167
DPDCH – dedicated physical data
    channel, 167
DPSK – differential phase-shift keying,
    88, 95
drift
– argument of perigee, 21
– RAAN, 21
DSPA – Dijkstra shortest-path
    algorithm, 231
dual-mode, 38, 41
duration of a state, 62, 65
DVTR – dynamic virtual topology
    routing, 229, 230, 332
dynamic channel allocation, 195

E_s/N_0, 54
earth central angle, 24, 32, 58, 350
EAST, 36
eccentric anomaly, 16
eccentricity
– numerical, 15
echo, 61
– delay, 70, 73
– far, 70, 72, 73
– near, 70, 72, 73
echo compensation, 58
ecliptic, 19
EHF band – extremely high frequency
    band, 11, 62, 67, 68, 261, 366
– channel parameter, 356
EIC – equipment identification register,
    203
EIRP – equivalent isotropic radiated
    power, 48, 55
elevation, 24, 26, 50, 58, 61, 65, 68, 75,
    77, 79

– minimum, 26, 27, 29, 32, 56
ellipse, 15
elongation, 22
energy, 17, 45
EPD – early packet discard (ATM), 313
equalizer, 81
Erl – Erlang, 122
Erlang-B formula, 196
ET – earth terminal, 203, 205
ETC – earth terminal controller, 203,
    205
ETSI – European Telecommunications
    Standards Institute, 74, 264
exponential distribution, 73

FACCH – fast associated control
    channel, 206
fading, 61, 64, 74
– amplitude, 97
– fast, 108
– frequency non-selective, 68, 80, 95
– frequency-selective, 81
– lognormal, 64, 65, 72
– multiplicative, 96
– Rayleigh, 64, 65, 72, 74
– Ricean, 64, 65, 72
fairing, 44
Faraday rotation, 61
FCA – fixed channel allocation, 216
FCC – Federal Communications
    Commission, 263
FCCH – frequency correction channel,
    205
FDD – frequency-division duplexing,
    117
FDM – frequency-division multiplex,
    119
FDMA – frequency-division multiple
    access, 126
FEC – forward error correction, 98, 313
feeder link, 47
figure of merit, 54
fixed earth station, 202
footprint, 26, 56, 348, 350
forced termination, 216
forward link, 47
free distance, 102
free space loss, 47, 52, 58, 62
frequency diversity, 80
frequency hopping, 80
frequency reuse, 214
frequency reuse distance, 216
frequency reuse factor, 175, 195

FSK – frequency-shift keying, 89
FSS – fixed satellite service, 260
FTP – file transfer protocol, 296

G/T, 54, 55
GAN – global area network, 387
gateway, 38, 79, 202
– block diagram, 203
– home gateway, 202, 214
– link, 47
– PSTN connecting gateway, 212
– responsible gateway, 212
– service area, 207
– visited gateway, 202, 212, 214
Gaussian antenna, 177
GDP – gross domestic product, 317
GEO – geostationary orbit, 3, 20, 34
geocentric coordinate, 21
geosynchronous orbit, 20
GFC – generic flow control (ATM), 280
GII – Global Information Infrastructure, 4, 274
Globalstar, 31, 38, 39, 58, 74, 75, 79, 375
GMDSS – global maritime distress and safety system, 388
GMPCS MoU – global mobile personal communications by satellite memorandum of understanding, 266
GMS – gateway management system, 203
GMSK – Gaussian-filtered MSK, 91, 382
Gold sequence, 140
GPM – gross potential market, 317
GPS – Global Positioning System, 209
graceful degradation, 163
gravitation constant, 17
ground segment, 202
ground track, 21, 22
GSM – Global System for Mobile Communication, 204
– protocol architecture, 205
– protocol layers, 204
GSM signaling channels, 204
– AGCH, 206
– BCCH, 205
– CCCH, 205
– FACCH, 206
– FCCH, 205
– PCH, 206
– RACH, 205
– SACCH, 206

– SCH, 205
– SDCCH, 206
GTO – geostationary transfer orbit, 44, 45
guard band, 127
GW, see gateway

Hadamard sequence, 142
Hamming code, 105
Hamming distance, 105
handheld terminal, 8, 383, 393
handover, 79, 81, 203, 219
– backward handover, 221
– DCA, 223
– delay jitter, 235
– forward handover, 221
– gateway handover, 220
– handover break, 222, 223
– handover decision, 221
– handover-queueing, 223
– in non-GEO S-PCNs, 220
– inter-network, 220
– intercell, 219
– intra-cell, 219
– path handover, 228
– reservation-based, 223
– satellite handover, 219
– satellite handover strategies, 221
– signaling, 223
– spot beam handover, 219
– terminal-initiated procedure, 221
– VPC handover, 228, 232
– with earth-fixed cells, 220
– with satellite-fixed cells, 220
– without priorities, 222
HEC – header error control (ATM), 280, 281, 312
HEO – highly elliptical orbit, 28
hexagon, 173
high-penetration alert, 367, 379, 390
HLR – home location register, 202, 203
HO, see handover
HPA – high-power amplifier, 248

ICO – intermediate circular orbit, 8, 19
ICO system, 31, 42, 58, 74, 75, 379
IF – intermediate frequency, 253
IMBE – improved multiband excitation, 86, 382
impulse response, 70, 72
IMSI – international mobile subscriber identity, 206
IMT-2000 – International Mobile Telecommunication-2000, 9, 259, 363

inclination, 18, 30, 44
Inmarsat – International Maritime
    Satellite Organization, 386
Inmarsat-3, 35, 386
integration
– dual-mode terminal, 240
– network, 240
– system, 240
– terminal, 240
– terrestrial-satellite, 240
interference
– adjacent channel, 54
– cancellation, 161
– co-channel, 54
– intersymbol, 81, 92
interleaving, 81, 97
Internet, 273, 287, 364
intersymbol interference, 81, 92
ionosphere, 19, 61
IP – internet protocol, 288
– congestion control, 290
– datagram, 288
– header, 289
– IPv4 – IP version 4, 289
– IPv6 – IP version 6, 290
– packet, 288
– TOS – type of service, 289
Iridium, 31, 33, 41, 58, 75, 225, 389
ISC – international switching center,
    202
ISDN – Integrated Services Digital
    Network, 201, 300
ISL – intersatellite link, 40, 41, 201,
    225, 261, 299, 328
– capacity, 236
– capacity dimensioning, 328
– inter-orbit, 226, 331, 338, 340
– intra-orbit, 226, 331, 340
– load, 236
– network traffic, 237, 238
– permanent, 328
– pointing, 40, 331
– routing, 228, 332
– segment, 228
– switching, 226
– topology, 225, 329, 331
ISO – International Organization for
    Standardization, 204
isotropic antenna, 48
ITU – International Telecommunication
    Union, 58, 257

JPEG – joint photographic experts
    group, 275

K band, 11, 260
Ka band, 11, 260, 365
Kepler elements, 19
Kepler laws, 15
Kepler, Johannes, 15
KSPA – k-shortest-path algorithm, 332
Ku band, 11, 260, 365

L band, 11, 259, 361–363
LA – location area, 208
LAC – location area code, 208
LAPD$_m$ – link access procedure D
    mobile, 205
launch, 42
launcher, 45
law of gravity, 17
LEO – low earth orbit, 3, 19, 37
line of nodes, 19
linear multiuser detector, 160
link
– availability, 74, 75, 79, 80
– budget, 47, 54, 55, 58
– feeder, 47
– forward, 47
– gateway, 47
– margin, 55, 74, 75, 79
– return, 47
– service, 47
– subscriber, 47
– user, 47
little-LEO, 11, 258, 263, 364, 396
LNA – low-noise amplifier, 53, 248
location area identity, 208
location update, 209
lognormal distribution, 64, 65, 72
LOS – line-of-sight, 67, 70, 72
LP optimization – linear programming
    optimization, 334, 335
– BO – bounded optimization, 336, 337
– ES – equal sharing, 334, 337, 339,
    340
– FO – full optimization, 336, 337, 339
LPC – linear predictive coding, 85
Luneberg lens, 401

m-sequence, 139
M-Star constellation, 329
MAC – multiple access control, 306
– contention-based, 308
– contention-free, 309
– PRMA – packet reservation multiple
    access, 308
– protocol, 306

– reservation Aloha, 308
– reservation-based, 309, 311
main lobe, 50
market, 7, 11
market prediction, 316
Markov chain, 65
mass of earth, 17
matched filter, 144
MBS – maximum burst size (ATM), 282
MCR – minimum cell rate (ATM), 283
MDA – Moore–Dijkstra algorithm, 236
mean anomaly, 16
MEO – medium earth orbit, 4, 19
– constellation, 41
MF-CDMA – multi-frequency CDMA, 150, 190, 309
MF-TDMA – multi-frequency TDMA, 133, 136, 180, 308, 322
minimum elevation, 26, 27, 29, 32, 56
MLS detector – maximum-likelihood sequence detector, 160
MM – mobility management, 203, 205, 206
MMSE detector – minimum mean-square error detector, 161
mobile user link, 47
mobility, 206
mobility management procedures, 206
MOC – message origination controller, 203
Molnija orbit, 21
MOS – mean opinion score, 84
MPEG – moving picture experts group, 275, 301
MS – mobile station, 222
MSC – mobile switching center, 203, 205
MSK – minimum-shift keying, 90
MSS – mobile satellite service, 260
MTP – message transfer part, 205
MTU – maximum transmission unit (TCP), 296
multimedia communications, 10
multimedia satellite system, 299
multipath, 61, 64, 68
multiple access protocol, 120, 306
multiservice
– correlation, 319
– traffic model, 318

nadir, 24
nadir angle, 27, 56, 350

narrowband channel model, 62
NCC – network control center, 202
network
– architecture, 201
– connection-oriented, 273
– connectionless, 273
– control, 203
– design process, 333
– dimensioning, 315
Newton, Isaac, 17
NNI – network-network interface (ATM), 303
noise figure, 53, 54
noise power spectral density, 53
noise temperature, 53–55
nrt-VBR – non-real-time variable bit rate (ATM), 281, 285, 301, 314
NRZ – non-return-to-zero, 87
Nyquist-filtering, 92

O-QPSK – offset-QPSK, 89
OBP – on-board processing, 41
OD – origin–destination
– pair, 226, 231
OD – origin-destination
– capacity, 335
– pair, 332, 335
– traffic, 334
off-boresight angle, 50, 56
on-board switching, 41, 300, 303
open-loop power control, 164
orbit, 15
– altitude, 29
– circular, 17
– co-rotating, 225
– counter-rotating, 225
– drift, 21
– elliptical, 15, 45
– geosynchronous, 20
– Molnija, 21
– orientation, 18
– parameters, 19
– period, 15, 18
– perturbation, 21
– phasing, 34
– phasing factor, 30
– plane, 30
– polar, 20, 31, 41, 225
– spacing, 30, 32, 34
– sun-synchronous, 21
– Tundra, 21
orbital elements, 19
OSI – open systems interconnection, 204

overlapping, 28, 56

paging, 211
paper satellite, 262
PAT – pointing, acquisition, and
    tracking, 40
PCH – paging channel, 206
PCM – pulse code modulation, 84
PCN – personal communications
    network, 8
PCR – peak cell rate (ATM), 282
PDA – private digital assistant, 9
PED – packet error distribution, 106
perigee, 16, 44, 45
– argument of, 19
period, 18, 19
– of earth, 20, 34
– orbit, 15
personal communications, 7
phase modulation, 87
phasing factor, 30
PHY – physical layer (ATM), 278, 279
π/4-QPSK, 89
PL – physical link (ISL)
– traffic distribution, 340
– traffic load, 338, 339
PN sequence – pseudo-noise sequence,
    139
pointing
– ISL, 40, 331
Poisson distribution, 73
Poisson process, 195
polar constellation, 31, 41
polar orbit, 20, 41
polarization, 61
– circular, 61
– linear, 61
– multiplex, 119
position determination, 209, 396
power control, 74, 218, 312
power delay profile, 70, 72
power flux density, 48, 51, 58
preferentially-phased Gold sequence,
    141
PRMA – packet reservation multiple
    access, 308
protocol architecture (ATM), 303
PSTN – public switched telephone
    network, 201, 212
PT – payload type (ATM), 280

QPSK – quadrature phase-shift keying,
    88

quadrifilar helix, 383

RAAN – right ascension of the
    ascending node, 19, 21, 30
– drift, 21
RACH – random access channel, 205
radio resource management, 311
radius of earth, 17, 21
rain attenuation, 52
Rake receiver, 81
Rayleigh distribution, 64, 65, 72, 74
registration procedure, 209
repeater
– bent-pipe, 244
– regenerative, 244
– transparent, 244
resource management, 304, 311
return link, 47
reuse distance, 173
Rice distribution, 64, 65, 72
Rice factor, 64, 67, 68, 96
roaming, 206
roll-off factor, 92
rotation of earth, 23, 59
round-trip delay, 42, 58, 59, 111, 124
routing, 225
– connection-oriented, 225
– dynamic, 228
– ISL routing, 228, 332
– ISL shortest paths example, 233
– off-line routing framework, 332, 333
– on-line, 236
– shortest path search, 231
– sliding-window optimization, 232,
    234
– system period optimization, 232, 234
– traffic adaptive, 228, 236
RR – radio resource management, 203,
    205
RS code – Reed-Solomon code, 106
rt-VBR – real-time variable bit rate
    (ATM), 281, 285, 301, 314
RTT – round-trip time, 292

S band, 11, 259, 363
S-DLC – satellite data link control
    (layer), 301, 303, 313
S-HLR – satellite home location
    register, 241
S-MAC – satellite multiple access
    control (layer), 301, 303, 313
S-MSC – satellite mobile switching
    center, 240, 241

S-PCN – satellite personal communications network, 8, 201,
  204
S-PHY – satellite physical layer, 301,
  303, 313
S-UMTS – satellite UMTS, 165, 316
S-VLR – satellite visitor location
  register, 241
SACCH – slow associated control
  channel, 206
satellite
  – diversity, 38, 74, 75, 79–81
  – orbit, 15
  – regenerative, 56
  – track, 21, 22
  – transparent, 41, 55
  – velocity, 45
  – visibility, 27, 38, 207
SatT – satellite terminal, 323
SCC – satellite control center, 202
SCCP – signaling connection control
  part, 205
SCH – synchronization channel, 205
scheduling, 307–309
SCPC – single channel per carrier, 126
SCR – sustainable cell rate (ATM), 282
SDCCH – stand-alone dedicated control
  channel, 206
SDM – space-division multiplex, 119
SDMA – space-division multiple access,
  119
seam, 31, 41
self-similar traffic, 276
service
  – availability, 75, 76, 79
  – charge, 267
  – quality, 74
service area
  – guaranteed service area, 207, 208
  – Iridium example, 208
  – momentary service area, 207
  – of a gateway, 207
service profile, 318
shadowing, 61, 64, 68, 70–72, 74, 79
  – time-share, 65, 67, 75
signal
  – bandwidth, 80, 81
  – energy, 54
  – processing, 41
signal-to-interference ratio, 177, 191
signal-to-noise ratio, 53–55
  – mean, 98
  – per symbol, 54

signaling channels, 204
SIM – subscriber identity module, 206
SIM card – subscriber identity module
  card, 368, 383, 386
site diversity, 393
slant range, 24, 27, 58
slotted Aloha, 132
slow fading, 108
SOHO – small office, home office, 369,
  399, 407
solid angle, 56
SPA – shortest-path algorithm, 231
space segment, 201
spectrum efficiency, 197
specular reflection, 61, 62
spill-over, 58, 350
spot beam, 56, 58, 348
  – coverage example, 323
  – isolation, 184
  – traffic distribution example, 327
  – worst-case cluster, 322
SPS – service provider station, 323
SSCOP – service-specific connection-
  oriented protocol (TCP), 297
SSMA – spread spectrum multiple
  access, 136
SSP – sub-satellite point, 24
SSPA – solid-state power amplifier, 248
statistical multiplexing, 277, 306, 307,
  311
street of coverage, 219
sun-synchronous orbit, 21
symbol duration, 54
system
  – architecture, 201, 299
  – bandwidth, 195, 315, 322
  – bandwidth demand, 327
  – capacity, 196
  – design process, 316
  – dimensioning, 316
  – license, 262

tapered-aperture antenna, 184
tapped delay-line, 72
TCM – trellis-coded modulation, 113
TCP – transport control protocol, 290
  – congestion control, 292, 296
  – dynamic window sizing, 293, 296
  – error control, 291
  – fast recovery, 294
  – fast retransmit, 293
  – flow control, 292, 294
  – retransmission, 291, 295

– segment, 290
– slow start, 293, 296
TCP/IP – transport control protocol /
   internet protocol, 287, 300
TDD – time-division duplexing, 117
TDM – time-division multiplex, 119
TDMA – time-division multiple access,
   129, 308, 310
– bandwidth efficiency, 132
– burst rate, 131
– frame, 131, 308
– guard time, 131
– header, 131
– MF-TDMA – multi-frequency
   TDMA, 133, 136, 180, 308, 322
– SORF – start of receive frame, 135
– SOTF – start of transmit frame, 135
– throughput, 131
Teledesic, 34, 407
terminal
– dual-mode, 38, 41, 240
– group terminal, 320, 321
– individual terminal, 321
thermal noise, 53
throughput, 123, 127
time diversity, 81
time-share of shadowing, 65, 67, 76
TMSI – temporary mobile subscriber
   identity, 206
TNL – terrestrial network link, 226
topology
– dynamics, 226
– time-varying, 228
– virtual, 228
– VPC, 334
TOS – type of service (IP), 289
track, 21, 22
traffic model
– busy hour, 319, 324
– group terminal, 320, 321
– individual terminal, 321
– multiservice, 318
– multiservice correlation, 319, 324
– service profile, 318, 325
– terminal types, 324
– traffic parameters, 318, 325
transmission equations, 53
transmission rate, 54
transmit power, 48
transparent satellite, 41
trellis, 100
true anomaly, 16, 17, 23

TT&C – telemetry, tracking, and
   command, 243
Tundra orbit, 21
two-state channel model, 62, 65, 76,
   353, 356
TWT – travelling wave tube, 128, 253

UBR – unspecified bit rate (ATM),
   281, 285, 301, 314
UDL – up/downlink, 226
UHF band – ultra high frequency band,
   258
UMTS – Universal Mobile Telecommunication System, 9, 259, 363
UNI – user-network interface (ATM),
   280, 303
UPC – usage parameter control (ATM),
   284, 285, 305
uplink, 47, 53, 55
uplink scheduler, 309
USAT – ultra small aperture terminal,
   10
user capacity, 197
UW – unique word, 134

V band, 11, 261, 366
Van Allen belts, 19
VC – virtual channel, 228, 279
VCC – virtual channel connection, 228,
   279
VCI – virtual channel identifier, 279,
   280
VCT – virtual connection tree, 227
velocity, 17, 45
VHE – virtual home environment, 9
VHF band – very high frequency band,
   258
video traffic, 275
view angle, 345–347, 349
virtual trunking concept, 334
visibility, 27, 38
Viterbi algorithm, 101
VLR – visitor location register, 202,
   203
voice activation, 129, 132
voice activity, 145
VP – virtual path, 228, 279
VPC – virtual path connection, 228,
   279
VPI – virtual path identifier, 228, 279,
   280
VSAT – very small aperture terminal, 4
VSELP – vector sum excited linear
   prediction, 86

Walker, J. G., 30
WAN – wide area network, 4
wavelength, 49, 59
WCL – worst-case link, 237
– capacity, 332, 334
– traffic load, 237, 336, 337
wideband channel model, 68, 72, 74,
    357, 360
WRC – World Radiocommunication
    Conference, 257
WSSUS – wide sense stationary
    uncorrelated scattering, 72

ZF-DF detector – zero-forcing
    decision-feedback detector, 162

Printing: Mercedes-Druck, Berlin
Binding: Buchbinderei Lüderitz & Bauer, Berlin